GREEN CONSUMERISM

Perspectives, Sustainability, and Behavior

GREEN CONSUMERISM

Perspectives, Sustainability, and Behavior

Edited by
Dr. Ruchika Singh Malyan
Punita Duhan

AAP | APPLE ACADEMIC PRESS

Apple Academic Press Inc.
3333 Mistwell Crescent
Oakville, ON L6L 0A2 Canada

Apple Academic Press Inc.
9 Spinnaker Way
Waretown, NJ 08758 USA

© 2019 by Apple Academic Press, Inc.

First issued in paperback 2021

Exclusive worldwide distribution by CRC Press, a member of Taylor & Francis Group
No claim to original U.S. Government works

ISBN-13: 978-1-77463-183-6 (pbk)
ISBN-13: 978-1-77188-694-9 (hbk)

Library and Archives Canada Cataloguing in Publication

Green consumerism : perspectives, sustainability, and behavior / edited by Ruchika Singh Malyan, Punita Duhan.

Includes bibliographical references and index.
Issued in print and electronic formats.
ISBN 978-1-77188-694-9 (hardcover).--ISBN 978-1-351-13804-8 (PDF)

1. Green marketing. 2. Consumption (Economics)--Environmental aspects. 3. Environmental responsibility.
   I. Malyan, Ruchika Singh, editor II. Duhan, Punita, 1973-, editor

HF5413.G74 2018   658.8'02   C2018-904551-5   C2018-904552-3

Library of Congress Cataloging-in-Publication Data

Names: Malyan, Ruchika Singh, editor. | Duhan, Punita, 1973- editor.

Title: Green consumerism : perspectives, sustainability, and behavior / editors, Ruchika Singh Malyan, Punita Duhan.

Description: Toronto ; New Jersey : Apple Academic Press, 2019. | Includes bibliographical references and index.

Identifiers: LCCN 2018036403 (print) | LCCN 2018038652 (ebook) | ISBN 9781351138048 (ebook) | ISBN 9781771886949 (hardcover : alk. paper)

Subjects: LCSH: Green marketing. | Green products. | Consumer behavior--Environmental aspects. | Environmentalism--Economic aspects.

Classification: LCC HF5413 (ebook) | LCC HF5413 .G7254 2019 (print) | DDC 658.8/02--dc23

LC record available at https://lccn.loc.gov/2018036403

Apple Academic Press also publishes its books in a variety of electronic formats. Some content that appears in print may not be available in electronic format. For information about Apple Academic Press products, visit our website at **www.appleacademicpress.com** and the CRC Press website at **www.crcpress.com**

ABOUT THE EDITORS

Dr. Ruchika Singh Malyan, PhD, MBA

Dr. Ruchika Singh Malyan holds a PhD from Mewar University, Rajasthan, India, and is currently working as a faculty member of Business Administration at Meera Bai Institute of Technology, operational under the Department of Training and Technical Education, Government of NCT of Delhi, for the past six years. Prior to this, she has worked as Assistant Manager in Marketing Communications and has a wide array of experience in the field of advertising and marketing communication. She has published research papers with prestigious national and international publishers. She has also attended and presented research papers at national and international conferences and seminars organized by prestigious institutes in India, including the Indian Institute of Management (Lucknow), Bhati Vidyapeeth Institute of Management and Research (New Delhi), and Banaras Hindu University (Varanasi), India. She has been a committee member for curriculum and content development for the National Institute of Open Schooling and has attended various training programs and workshops apart from organizing several national and international seminars. Her research interests include sustainable development, ecological marketing, green consumer behavior, advertising management, and corporate social responsibility.

Punita Duhan, MBA

Punita Duhan is a faculty member in Business Administration at the Meera Bai Institute of Technology, operating under the Department of Training and Technical Education, Government of NCT of Delhi, for the last 18 years. She is currently pursuing research in social media at the Institute of Management Studies, Banaras Hindu University, Varanasi, India. She has published more than 12 research papers with prestigious national and international publishers and has also published an edited book, *Managing Public Relations and Brand Image through Social Media*. Another book, *Radical Reorganization of Existing Work Structures Through Digitalization*, is in press. She is also on editorial boards of prestigious peer-reviewed journals. In addition, she has been actively involved in curriculum development and E-content development for various universities, the National Institute of

Open Schooling, and the Central Board of Secondary Education and Technical Education Departments of Delhi and Haryana state in the management area. She has attended more than 30 training programs and workshops and has organized almost a dozen seminars, conferences, and workshops.

CONTENTS

List of Contributors... *xi*

List of Reviewers... *xiii*

List of Abbreviations...*xv*

Acknowledgment..*xix*

Foreword...*xxi*

Preface.. *xxiii*

Introduction..*xxix*

PART I: Green Marketing Practices: Different Perspectives 1

1. **Green Marketing: The Next Marketing Revolution** 3
   Anitha Acharya and Manish Gupta

2. **Green Pricing: The Journey Until Now and the Road Ahead** 21
   Anjali Karol and C. Mashood

3. **Green Consumerism: Study of Academic Publications in Scientific Journals Indexed in the Web of Science and Scopus** 41
   J. Álvarez-García, C. P. Maldonado-Erazo, and M. C. Del Río-Rama

4. **Green Consumer Behavior** 67
   Kulvinder Kaur Batth

5. **Green Practices for Green(er) Living: The Road Ahead** 83
   George Varghese and Lakshmi Viswanathan

6. **Determinants of Consumer Purchase Intention for Solar Products in Varanasi City** 99
   Om Jee Gupta and Anurag Singh

7. **Environmentally Conscious Consumer Behavior and Green Marketing: An Analytical Study of the Indian Market** 119
   Pradeep Kautish

8. **Empirically Examining Green Brand Associations to Gain Green Competitive Advantage through Green Purchasing Intentions** 143
   Nitika Sharma and Madan Lal

PART II: Sustainability Aspects of Green Marketing 159

9. **Sustainable Green Marketing: A Trend of Consumerism........... 161**
    Harsh Tullani and Richa Dahiya

10. **Analyzing Long-Term Benefits in the Face of Higher
    Upfront Costs for Green Affordable Housing: A Study of
    Ghaziabad, UP (India)... 185**
    Siddharth Jain, Prateek Gupta, and Deepa

11. **Innovation in Green Practices: A Tool for Environment
    Sustainability and Competitive Advantage 209**
    Nomita Sharma

12. **Communicating Sustainability and Green Marketing:
    An Emotional Appeal 229**
    Moturu Venkata Rajasekhar, Krishnaveer Abhishek Challa,
    Dharmavaram Vijaylakshmi, and Nittala Rajyalakshmi

PART III: Ecological Dimensions of Green Consumer Behavior...... 259

13. **Eco-Awareness: Imbibing Environmental Values in
    Consumers.. 261**
    Anjali Karol and C. Mashood

14. **Environmental Marketing and Education 285**
    Kunal Sinha and S. N. Sahdeo

15. **Going Green: Toward Organic Farming and a Plastic-
    Free Eco-Friendly Lifestyle 305**
    Sumit Roy

16. **Effective Utilization of Renewable Biomaterials for the
    Production of Bioethanol as Clean Biofuel: A Concept
    Toward the Development of Sustainable Green Biorefinery 335**
    Geetika Gupta, Pinaki Dey, and Sandeep Kaur Saggi

Index *365*

To my parents, Dr. P. S. Nayyer and Kalpana Kumari Nayyer, for always supporting me in my endeavors.

– Dr. Ruchika Singh Malyan

To my son, Siddharth Bamel, who has always shown maturity beyond his age.

– Punita Duhan

LIST OF CONTRIBUTORS

Anitha Acharya
Department of Human Resources, IBS, a constituent of IFHE, Deemed to be University, Hyderabad, India

José Álvarez-García
Department of Financial Economics and Accounting of the University of Extremadura, Caceres, Spain

Kulvinder Kaur Batth
Department of Commerce, K. C. College, Affiliated to University of Mumbai, Mumbai, Maharashtra, India

Krishnaveer Abhishek Challa
Department of Foreign Languages, Andhra University, Vishakhapatnam, Andhra Pradesh, India

Richa Dahiya
Department of Management Studies, SRM University, Sonepat, Haryana, India

Deepa
KIET School of Management, Ghaziabad, Uttar Pradesh, India

Pinaki Dey
Department of Biotechnology, Karunya University, Coimbatore, Tamil Nadu, India

Geetika Gupta
Department of Biotechnology, Thapar Institute of Engineering and Technology, Bhadson Road, Patiala, Punjab 147004, India

Manish Gupta
Department of Human Resources, IBS, a Constituent of IFHE, Deemed to be University, Hyderabad, India

Om Jee Gupta
Institute of Management Studies, Banaras Hindu University, Varanasi, U. P, India

Prateek Gupta
KIET Group of Institutions, Ghaziabad, U. P, India

Siddharth Jain
Department of Civil Engineering, KIET Group of Institutions, Ghaziabad, U. P, India

Anjali Karol
Institute for Financial Management and Research, Chennai, India

Pradeep Kautish
College of Business Management, Economics and Commerce, Mody University of Science and Technology, Lakshmangarh, Sikar, Rajasthan

Chanchal Kumar
University of Delhi, Delhi, India

Madan Lal
Department of Commerce, Delhi School of Economics, University of Delhi, New Delhi, India

Claudia Patricia Maldonado-Erazo
Universidad Técnica Particular de Loja (UTPL), Loja, Ecuador

C. Mashood
SPI Global, Chennai, India

Moturu Venkata Rajasekhar
Department of Commerce and Management Studies, Andhra University, Vishakhapatnam, Andhra Pradesh, India

Nittala Rajyalakshmi
Department of Commerce and Management Studies, Andhra University, Vishakhapatnam, Andhra Pradesh, India

Sumit Roy
Jadavpur, Kolkata, West Bengal, India

Maria De La Cruz Del Río-Rama
Department of Business Organisation and Marketing, University of Vigo, Ourense, Spain

Sandeep Kaur Saggi
Department of Biotechnology, Thapar Institute of Engineering and Technology, Bhadson Road, Patiala, Punjab147004, India

S. N. Sahdeo
Department of Management, Birla Institute of Technology, Ranchi, Jharkhand, India

Nitika Sharma
Department of Commerce, University of Delhi, New Delhi, India

Nomita Sharma
Department of Management Studies, Keshav Mahavidyalaya, University of Delhi, New Delhi, India

Anurag Singh
Institute of Management Studies, Banaras Hindu University, Varanasi, U. P, India

Kunal Sinha
Department of Management, Birla Institute of Technology, Jharkhand, India

Harsh Tullani
Department of Management Studies, SRM University, Sonepat, Haryana, India

George Varghese
Institute for Financial Management and Research, Chennai, India

Dharmavaram Vijaylakshmi
Gayatri Vidya Parishad, Gayatri Valley, Rushikonda, Visakhapatnam, Andhra Pradesh, India

Lakshmi Viswanathan
Institute for Financial Management and Research, Chennai, India

LIST OF REVIEWERS

Anitha Acharya
Department of Human Resources, IBS, a constituent of IFHE, Deemed to be University, Hyderabad, India

Prachi Aggarwal
Meerabai Institute of Technology, Directorate of Training and Technical Education, New Delhi, India

José Álvarez-García
Department of Financial Economics and Accounting of the University of Extremadura, Caceres, Spain

Kulvinder Kaur Batth
Department of Commerce, K. C. College, Mumbai, India

C. Mashood
SPI Global, Chennai, India

Krishnaveer Abhishek Challa
Department of Foreign Languages, Andhra University, Vishakhapatnam, Andhra Pradesh, India

Geetika Gupta
Department of Biotechnology, Thapar Institute of Engineering and Technology, Bhadson Road, Patiala, Punjab 147004, India

Manish Gupta
Department of Human Resources, IBS, a Constituent of IFHE, Deemed to be University, Hyderabad, India

Om Jee Gupta
Institute of Management Studies, Banaras Hindu University, Varanasi, U. P, India

Siddharth Jain
Department of Civil Engineering, KIET Group of Institutions, Ghaziabad, U. P, India

Anjali Karol
Institute for Financial Management and Research, Chennai, India

Pradeep Kautish
College of Business Management, Economics and Commerce, Mody University of Science and Technology, Lakshmangarh, Sikar, Rajasthan

Chanchal Kumar
University of Delhi, Delhi, India

Sumit Roy
Jadavpur, Kolkata, West Bengal, India

S. N. Sahdeo
Department of Management, Birla Institute of Technology, Ranchi, Jharkhand, India

Nitika Sharma
Department of Commerce, University of Delhi, New Delhi, India

LIST OF ABBREVIATIONS

A and HCI	Arts and Humanities Citation Index
ADI	Acceptable daily intake
AFEX	Ammonia fiber expansion
AMA	American Marketing Association
ASTRA	Application of Science and Technology for Rural Areas
BMI	Business model innovation
BPA	Bisphenol-A
BP	British Petroleum
BSES	Bombay suburban electric supply
CBM	Carbohydrate-binding module
CBO	Community-based organization
CBP	Consolidated bioprocessing
CFA	Confirmatory factor analysis
CFC	Chlorofluorocarbon
CFI	Comparative fit index
CFLs	Compact fluorescent lamps
CII	Confederation of Indian Industry
CLRI	Central Leather Research Institute
CPCB	Central Pollution Control Board
CSR	Corporate social responsibility
DDA	Delhi Development Authority
DDT	Dichlorodiphenyltrichloroethane
DFE	Design for the environment
EAI	Environmental Awareness Index
ECCB	Environmentally conscious consumer behavior
EC	Environmental concern
EE	Environmental education
EK	Environmental knowledge
EMS	Environmental management system
ESCI	Emerging Sources Citation Index
FAO	Food and Agriculture Organization
FMCG	Fast moving consumer good
FSSAI	Food Safety and Standards Authority of India
GBA	Green brand association

GFI	Goodness of fit
GHG	Greenhouse gas
GP	Green practices
GPI	Green purchasing intention
GT	Green trust
HYV	High yielding varieties
IF	Impact factor
IGBC	Indian Green Building Council
IG	Insulated glass
IISc	Indian Institute of Science
IUCN	International Union for the Conservation of Nature and Natural Resources
JCR	Journal citation reports
KMO	Kaiser-Meyer-Olkin
LCA	Life cycle analysis
LED	Light-emitting diode
LEDs	Light-emitting diodes
MI	Meyer's index
MRL	Maximum residue limit/level
NCERT	National Council of Educational Research and Training
NFI	Normed fit index
NGOs	Nongovernmental organizations
NPOP	National Program for Organic Production
OECD	Organization for Economic Co-operation and Development
OLS	Ordinary Least Squares
PCA	Principal component analysis
PCB	Polychlorinated biphenyls
PCE	Perceived consumer effectiveness
PEB	Past environment-friendly behavior
POP	Persistent organic pollutant
PPC	Portland pozzolana cement
PVC	polyvinyl chloride
RCI	Relative quality indices
RF	Radio frequency
RMESA	Root mean square error of approximation
RoHS	Restriction of the use of certain hazardous substances
SBI	State Bank of India
SCI-EXPANDED	Science Citation Index Expanded
SEED	Social Economic Environmental Design

SEM	Structural equation modeling
SHF	Separate hydrolysis and fermentation
SHGs	Self-help groups
SJR	Scimago Journal Rank
SME	Small and medium-sized enterprise
SPM	Suspended particulate matter
SSCI	Social Sciences Citation Index
SSFF	Simultaneous saccharification, filtration, and fermentation
SSF	Simultaneous saccharification and fermentation
SSI	Small-scale industries
ToC	Touch of color
UEC	Ungra Extension Centre
UNEP	United Nation's Environmental Program
UNESCO	United Nations Educational, Scientific and Cultural Organization
VOC	Volatile organic compound
WEEE	Waste electronics and electrical equipment
WEF	Willingness to be environmental friendly
WoS	Web of Science
WWF	World Wildlife Fund

ACKNOWLEDGMENT

We cannot accomplish all that we need to do without working together. And, this academic endeavor in the form of the book is no exception. It was the dedicated, continuous, and combined effort of the publisher, contributors, researchers, reviewers, and editors that ensured that this book sees the light of the day.

The editors, hereby, take this opportunity to acknowledge the effort of all those who have worked vigorously to put this volume together.

First, the editors would like to express their sincere gratitude to the publisher, Apple Academic Press, for providing us the opportunity to publish this work with them and to help us at each step.

Second, the editors thank each one of the authors for their valuable contributions. The authors' priceless research is the backbone of the book and helped the editors to bring forward multiple facets of green marketing. Most of the authors helped with the blind review of the chapters. The editors highly appreciate their double effort to improve the quality of the book.

Next, the editors are grateful to Prof. Bijender K. Punia, Vice Chancellor, Maharshi Dayanand University, for his prefatory note and encouraging words.

Last, but not the least, the editors are grateful to their families and friends for their constant support to help them complete the project in time.

Dr. Ruchika Singh Malyan and Punita Duhan
Meera Bai Institute of Technology, Directorate of Training and
Technical Education, New Delhi, India

FOREWORD

Today, the world is in need of new decisions and innovations, which has led to a green marketing environment and the creation of a new market condition to the potential buyers. Both marketers and consumers are nowadays focusing on green products and services. Companies are adopting green marketing practices and are producing green products that have fewer harmful effects on the environment than conventional products. Consumers are becoming more and more aware of the environmental problems and are actively trying to reduce their impact on the environment by purchasing green products and moving towards a greener lifestyle.

Additionally, strategic marketing actions of companies target their products and services to these kinds of consumer groups to gain market share and minimize their production costs. Green consumerism creates a balance between the expectations of consumer behavior and business profit motives.

This book is a strenuous effort by the editors to understand the area of green consumer behavior in a versatile manner. This book enables readers to comprehend the concept of green consumer behavior, which has led to various green marketing initiatives being taken by companies. This book puts together the work of researchers and consultants working in this field from across the world.

Overall, this book meets the needs of researchers, academicians, professionals, and students working in the domain of sustainable development, green consumer behavior, green products and methods, ecological development, and so forth. This book will help develop the eco-literacy of people working in this area. It will prove to be an excellent reference resource and a handbook for detailed and up-to-date knowledge about green consumer behavior. Editors Dr. Ruchika Singh Malyan and Punita Duhan deserve to be congratulated for this sincere academic endeavor that covers the various facets of green marketing in a comprehensive manner.

Prof. B. K. Punia
Vice Chancellor,
Maharishi Dayanand University, Rohtak – 124001 (Haryana), India

PREFACE

The idea of green consumerism is very important in this decade and is popular among many sections of society in an effort to save the planet Earth. The world over, economic development has come at the cost of the environment. Green consumerism is a movement, expanding rapidly, to encourage people to buy products that are considered environmentally friendly. The needs of the consumers are evolving, and brand loyalty is being continuously redefined.

Nowadays, consumers are taking responsibility and doing the right thing as consumer awareness and motivation continue to drive change in the marketplace. Also, companies are integrating appropriate green strategies into their operational activities, product development processes, and marketing activities to achieve a competitive advantage in saturated markets. This helps companies gain market share and minimize their production costs.

Therefore, green consumerism has been increasingly promoting the practice of purchasing products and services produced in a way that minimizes social and/or environmental damage. It also creates a balance between the expectations of consumer behavior and business profit motives.

Accordingly, this book aims to understand the importance of promoting green products and then tries to explain consumers' buying intentions and decisions in respect to green or ecologically friendly products under the consumer behavior theory. This publication will be beneficial for academicians, researchers, practitioners, and students studying green marketing and for usage by marketers and consumers.

The editors' objective, while proposing this book, was to consolidate the contemporary academic and business research. The editors have made a meticulous effort to present up-to-date research related to green marketing in a comprehensive manner and to provide some ideas for future research avenues. Some of the broad areas this book caters to are green pricing, green consumer behavior, various dimensions of consumer purchase intentions, sustainable marketing, innovation techniques used to go green, eco-awareness, and other ongoing developments in this rapidly expanding area.

The book has been organized in the following three sections:
Section I: Green Marketing Practices: Different Perspectives
Section II: Sustainability Aspects of Green Marketing
Section III: Ecological Dimensions of Green Consumer Behavior

Each section caters to a distinct dimension of green marketing. A total of 16 chapters are spread in the three sections. A brief description of each of the chapter follows.

The first chapter of the book presents a comprehensive evolution journey of green marketing. The authors extensively review the extant literature on green marketing and bring forth the various definitions propounded by the researchers. The term "green marketing" was first discussed in the year 1975 in a seminar on "ecological marketing" organized by the American Marketing Association (AMA) (Henion and Kinnear, 1976). Since then, terms such as environment-friendly, recyclable, ozone-friendly, refillable are often associated with green marketing. The authors lucidly explain the evolution of green marketing along with the opportunities and challenges faced by the marketers in this field. They conclude the chapter with discussing the impact on stakeholders and the best green practices followed by various organizations.

The second chapter focuses on the concept of green pricing. It is a pricing strategy that encompasses environmental responsibility in pricing products and services on the higher side. By charging a premium, consumers feel that superior quality products are offered and they get an emotional contentment from contributing toward environmental protection. The chapter takes the readers through understanding the factors affecting the decisions to adopt green pricing. The authors discuss the reasons for criticism of green pricing despite the much hyped benefits.

In the third chapter, the authors have listed the past research and studies done in the area of green marketing that have been published in scientific journals indexed in the Web of Science and Scopus. This chapter is organized into four sections. It starts with an introduction in which the topic under study is contextualized and the objective is presented; the second section details the methodology used; the third section describes the analysis of the data; and the last section presents the conclusions and limitations of the research. The authors also discuss the limitations of the study as it is the first of its kind of research in this area of study. This also led to identification of areas of future research, including green consumption and green consumer behavior.

In continuation to the research done in the previous chapter, Chapter 4 outlays the objectives and outcomes of green consumer behavior. In this chapter, the author describes green consumer behavior as pro-environment behavior of consumers. It is the environment-friendly behavior of consumers with the objective of supporting the environment. Pro-environment behavior is reflected in the change of attitudes, perception, motives, values, beliefs, and desires which, in turn, are delivered in the changing needs of the consumers. A model of consumer behavior has been proposed by the authors, which consists of four factors. It can be adopted by manufacturers, companies, and marketers of green products. They can make use of the model while framing strategies for green products. The application of the model will help in the appropriate assessment and evaluation process so as to understand the dynamics of consumer behavior.

Chapter 5 talks about green practices that can be followed for a greener living. The authors say that awareness about green practices is progress itself. Therefore, educating oneself and others holds the key to a paradigm shift from business to a one planet living. The authors also suggest the mantra of reuse, reduce, and recycle, which, if followed religiously, would reduce our carbon imprints on the planet. With certain limitations, there are various opportunities discussed for the eco-entrepreneurs. They are those business professionals who keep the concept of going green close to their heart and central to their business, along with the end goal of profit. Eco-entrepreneurs have enormous uncaptured and unexplored markets available to them, and they can certainly do well if they can offer green products that are superior to their nongreen alternatives.

Chapter 6 explains the determinants of consumer purchase intention for solar products in Varanasi city (India). This study is aimed at solar products that are green products and are used for conservation of resources. Even after the gigantic promotion of solar products usage, the growth of the solar product market is very slow. The authors have investigated the key factors that influence the purchase intention of consumers for solar products. The authors suggest that if similar research is conducted in other states as well, marketers and policymakers will be able to serve consumers in a better way and can generate better demand for solar products.

Chapter 7 explores a wide range of literature on environmental concern by synthesizing consumer marketing domain and perspectives of corporations on the issue of environmentally conscious consumer behavior and green marketing. The author has analyzed the issue by conducting in-depth research using various statistical tools. The implications derived from the present study have an impact on a number of marketing-related areas. The results indicate that consumers in India have a positive attitude toward environmental concern, they demand green products, and they attempt to figure out the impact of their consumption on environmental well-being.

Chapter 8 offers an important insight as to how a consumer derives trust from brand associations. The authors have explored the relationship between green brand association and green trust to achieve green competitive advantage. They summarize the literature on three novel constructs, that is, green brand associations, green trust, and green purchasing intentions. A framework of green purchasing intentions in compliance with brand research to explore new environmental trends is proposed, which shall help companies in increasing their green purchase intentions. And they conclude the research by drawing conclusions and mentioning the discussions about the findings, implications, research limitations, and possible directions for future research.

Chapter 9 provides an understanding of the terms sustainability and inclusivity in the context of green marketing. The authors also discuss how green innovation and green promotions are a major tool of green marketing that lead to a firm's performances resulting in a sustainable green market. The outcome of this chapter is a model that shows a mutual interrelation between green innovation, green promotion, and a firm's performance in a sustainable and inclusive green market. The authors conclude that enhancing a company's green innovation and green promotion capacity can provide a new strategic weapon for managers to maintain sustainability and inclusivity in an emerging new level of markets.

Chapter 10 highlights the issues related to green housing and also identifies the reasons as to why affordability of such houses has become unreachable due to higher upfront cost. The authors have analyzed the long-term benefits of green houses against their higher investment costs. This study reveals that India is still lagging behind in effective implementation of green housing concepts due to high costs involved in construction as people have to invest 12.94% more than what they would invest in an ordinary conventional home. The authors have concluded that greenhouses are costlier than conventional houses but only in the short run. As per the need of society and livelihood, green houses are very necessary and will surely be much cheaper than conventional houses in the long run.

In Chapter 11, researchers have highlighted green innovation practices as a tool of environment sustainability and competitive advantage. The chapter also explains their influence on the environment and how they contribute to the competitive advantage of an organization and help in maintaining environmental sustainability. The researchers have proposed that strategies should aim at competitive advantage through combined use of natural resources and cost-saving techniques.

Chapter 12 talks about the challenges faced in the 21st century that have resulted in the extinction of many species and continually threaten human existence. The authors find it a great opportunity to transcend, and hence, this chapter introduces the proposal of strategic sustainable development.

They say that the conceptualization of green consumerism would need to be broadened in order to be inclusive of diverse social, economic, and ecological constraints.

Chapter 13 discusses the concept of eco-awareness. The authors define this as the awareness of ecology. It is explained that eco-literacy and eco-awareness are two sides of the same coin. In the research, it was found that eco-awareness of society depends on geography, demography, socioeconomic culture, and the levels of eco-literacy. To support the concepts, the researchers have quoted various success stories and conclude the chapter with a discussion on the future prospects of eco-awareness.

Chapter 14 talks about the importance of environmental marketing and education in reducing environmental degradation. The authors consider education to play an important role in protecting the environment, and hence, this becomes an important area of study. They conclude that environment education should not only create consciousness, but this consciousness needs to be translated into coherent behavior in which collective action finds a fundamental solution for the problems related to the environment.

Chapter 15 is an effort to inform readers of the way in which the modern technologically advanced "artificial easy system" is jeopardizing the environment in general and all living beings in particular. This chapter illustrates how humans have changed their lives to exploit everything in and around them, expecting to get more out of them, even from nature, and in due course this has started depleting the environment, risking their own lives and that of the future generations. On the basis of international studies, the author talks about the changes starting from something as basic and essential as human food to indiscriminate use of nonbiodegradable products, gadgets, and nonrenewable resources.

The authors discuss the effective utilization of renewable biomaterials for the production of bioethanol as green biofuel using sustainable green technology in Chapter 16. They find that different research challenges are being faced by researchers to finally commercialize bioethanol, and hence, there is a need of critical evaluation.

In a nutshell, this book fulfills the aim to understanding the importance of promoting green products and then tries to explain consumers' buying intentions and decisions of green or ecological-friendly products under the consumer behavior theory. It compiles the latest concepts of green marketing along with the emerging trends of 21st century in a systematic manner. Overall, the editors have made an effort to meet the need of the target readers that will help them to identify areas of future research and channelize their efforts in the right direction.

INTRODUCTION

Business organizations have been using various resources to successfully meet the demand-supply equation at the marketplace. However, today's businesses and consumers have realized the biggest consequence and challenge confronting them is the need to protect and preserve the natural resources and that their production and consumption behavior can directly impact the ecological balance of the environment. The deterioration process of natural resources caused by such activities calls for responsible roles to be assumed by all stakeholders, which includes consumers, governments, institutions, organizations, and the media, so as to meet the environmental crisis. Going by this are the current popular buzz statements: usage of sustainable energy, consuming green products, focus on green marketing practices, and so forth. These social concerns and responsibility toward enhancing sustainability have led to the emergence of marketing strategies that promote green. This phenomenon is rapidly gaining importance in modern marketing. Today, the world is in need of new decisions and innovations, which has led to a green marketing environment and the creation of a new market condition for the potential buyers. Both marketers and consumers are nowadays focusing on green products and services. Companies are adopting green marketing strategies and are producing green products that have fewer harmful effects on the environment than conventional products. Consumers are becoming more and more aware of the environmental problems and are actively trying to reduce their impact on the environment by purchasing green products and by moving toward a greener lifestyle.

It has been observed that consumers have been asking for green products, which means that there has been a clear rise in demand for such products. These products have also increased competition among businesses to generate more environment-friendly products. Some governments have been taking measures to support the organizations that have adopted green practices, such as by generating corporate environmental profiles and by monitoring and evaluating green performance. This ultimately improves the corporate image of the organization. Additionally, strategic marketing actions of companies target their products and services to these kinds of consumer groups to gain market share and minimize their production costs.

Green consumerism creates a balance between the expectations of consumer behavior and business profit motives.

The book is an attempt to present a holistic approach for understanding the contemporary green marketing concepts along with the emerging prototype. The book captures the concepts, innovation, evolution, emergent trends, hindrances, green consumer behaviors, and many other facets of green marketing and emerging applications in this area expansively. Sincere efforts have been made to put together the efforts of researchers across the world that give a complete and comprehensive idea of green marketing to readers. This will help them identify future areas of research and make valuable addition to scarce literature in the realm of "GREEN."

Dr. Ruchika Singh Malyan and Punita Duhan
Meera Bai Institute of Technology
Department of Training and Technical Education
New Delhi, India

PART I

Green Marketing Practices: Different Perspectives

CHAPTER 1

GREEN MARKETING: THE NEXT MARKETING REVOLUTION

ANITHA ACHARYA[1,*] and MANISH GUPTA[2]

[1]*Department of Human Resources, IBS, a constituent of IFHE, Deemed to be University, Hyderabad, India, Tel.: +91 8712290557, E-mail: anitha.acharya@ibsindia.org*

[2]*Department of Human Resources, IBS, a Constituent of IFHE, Deemed to be University, Hyderabad, India, Tel.:+91 8712316252, E-mail: manish.gupta.research@gmail.com*

1.1 INTRODUCTION

Industrialization in England brought mechanization and ease of doing work with it. However, this caused an almost irreparable damage to the environment. The severe consequences of environmental disturbance are visible in the form of increased number of natural disasters and acceleration of global warming. Concern for the environment coupled with government support has led people to look for green or eco-friendly alternatives and limit the exploitation of natural resources. Companies have now noticed the increasing demand for green products and have started directing their marketers to tap this whole new breed of green consumers (Luo and Bhattacharya, 2006).

According to Peattie and Charter (2003), the key challenge for the new millennium is to find more equitable and sustainable ways to produce, consume, and live. One of the visions for the future shared by environmentalists was sustainability. The Brundtland Report in the year 1987 titled "Our Common Future" brought the issue into the mainstream. In the wake of the 1992 Rio Earth Summit, the governments across the world and major corporations increasingly adopted the quest for sustainability as a goal. But the real challenge lies in implementing these goals into significant growth in the face of powerful vested interests, a deeply ingrained, environmentally antagonistic management paradigm, and a global economy. For marketing, the challenge is twofold. In the short term, social and ecological issues have become important external influences on companies and the markets within which they operate. Companies have to react to varying customer needs, new government rules which reflect rising fear about the socio-environmental impacts on business. According to Shrivastava (1995), the quest for sustainability in the long term will require fundamental changes to the management paradigm which underpins marketing and the other business function. This chapter aims to epitomize how going green is influencing current marketing practices and how its implications will require a more profound shift in the marketing paradigm, if the aim of the marketers is to make a profit and provide customer satisfaction.

1.2   NEED FOR GREEN MARKETING

Companies are expected to commit to green marketing strategies because of four main reasons. First, the cost of the natural resources is rising. Second, the consumers have become more aware of the pitfalls of the conventional products not only on the environment but also on them. As a result, the consumers are switching from nongreen product providers to green product providers. Third, the government agencies are taking restrictive measures for the non-green products and supportive measures for green products. Also, the pressure from the nongovernmental organizations (NGOs) has supported the change (Kleindorfer et al., 2005). Moreover, research in this area suggests that the usage of green products has long-term financial benefits (King and Lenox, 2002).

As resources are limited and human needs are infinite, achieving organizational objectives is difficult for marketers. Nevertheless, they can utilize the available resources in an efficient manner by going green. There is a rising interest among the consumers all over the world concerning the protection of the environment. It is evident from prior research that people are concerned about the environment and are changing their behavior (Mishra and Sharma, 2010). As a consequence of this, green marketing has originated, which speaks about rising demand for sustainable and socially accountable products and services. Thus, the growing consciousness among the individuals all over the world regarding protection of the environment in which they live, individuals do want to bestow a clean earth to their offspring. Marketers are becoming more concerned about the environment and would like to introduce environment-friendly products. Green marketing was given prominence in the late 1980s and 1990s after the proceedings of the first workshop on "Ecological Marketing" held in Austin, Texas (US), in 1975.

1.3 WHAT IS IT AND HOW IT WORKS?

The term green marketing was first discussed in the year 1975 in a seminar on "Ecological Marketing" organized by American Marketing Association (AMA) (Henion and Kinnear, 1976). Since then, terms such as environment-friendly, recyclable, ozone friendly, refillable are often associated with green marketing. However, green marketing is a much broader concept which is applicable to services, consumer goods, and even industrial goods. It incorporates a broad range of activities including product modification, changes to the production process, changes in packaging, as well as modifications in the advertisement. Not surprisingly but defining green marketing is a challenging task.

From practitioners' perspective, green marketing refers to the marketing of products and services that are presumed to be environmentally preferable to others whereas, from a researchers perspective, green marketing refers to the "analysis of how marketing activities impact the environment and how the environmental variable can be incorporated into the various decisions of corporate marketing" (Chamorro et al., 2009, p 223). According to Peattie (2001a), green marketing attempts to reduce the negative social and environmental impacts of existing products and production systems. It promotes less damaging products and services. The latest trend in green marketing research assimilates green marketing into the broader framework of corporate social responsibility (Chamorro et al., 2009). This change is the result of a shift in thinking from green marketing as a type of consumerism (the 1970s–1990s) to green marketing as a societal conscientiousness (Porter and Kramer, 2006).

Furthermore, acting in an environment-unfriendly way is unprofitable (Cronin et al., 2011). For example, British Petroleum (BP) lost 55% shareholder value after the deepwater horizon incident which happened on April 20, 2010. Deepwater Horizon was a deepwater, offshore oil drilling rig owned by Transocean (RIG) and operated by BP. While drilling at the Macondo Prospect, there was an explosion on the rig caused by a blowout that killed 11 crew members. On April 22, 2010, Deepwater Horizon sank while the well was still active and caused the largest offshore oil spill in the U.S. history. Market incentives, as this logic goes, support pro-environment corporate decisions (Kassinis and Vafeas, 2006). Also, recent research has found that promoting oneself as a "green company" increases consumers' inclination toward the company and allows companies to charge a premium price (Acharya and Gupta, 2016).

1.4   EVOLUTION OF GREEN MARKETING

The concept of green marketing has evolved over a period of time. According to Peattie (2001a), the green marketing evolution was a three-phase process including ecological, environmental, and sustainable green marketing. Each of these phases has been elaborated in the subsequent paragraphs:

- The first phase termed "ecological green marketing" lasted from the 1960s to 1970s (Delafrooz et al., 2014). In this era, all marketing activities were concerned to find solutions to the environmental problems or the external problems (Grundey and Zaharia, 2008). It was a result of a workshop held by the American Marketing Association (AMA) on "Ecological Marketing" in the year 1975. The main contribution of this phase in establishing the concept by publishing several books on the topics relating to ecological marketing (Nadaf and Nadaf, 2014). It provided a concrete base for analyses and empirical work.
- The second phase was "environmental green marketing" and in this era, the focus shifted to "clean technology" that involved designing of products which are innovative and manage pollution and waste (Peattie, 2011). In this era, for raising customers' green loyalty, the companies were to pay special attention to enhance customers' green satisfaction, green perceived value, and green trust instead of typical satisfaction, perceived value, and trust (Chen, 2013). Eventually, it attracted a lot of investments in the potential resources contributing to the aforesaid factors.
- The third phase was "sustainable green marketing." It came into prominence in the late 1990s and early 2000. This was the result of "sustainable development," which is defined as meeting the needs of the present without compromising the capability of future generations to meet their own needs (Kilbourne, 1998). Research in this phase mainly relates to consumer behavior and is still going on.

In spite of some attention in the 1970s, it was only in the late 1980s that the idea of green marketing emerged. According to Prothero (1990), green marketing consists of the rapid increase in green consumerism at this time as heralding a dramatic and expected shift in consumption toward greener products (Vandermerwe and Oliff, 1990). Prior research bodies were cited as identifying heightened environmental awareness, a growing consumer interest in green products, and a pronounced willingness to pay for green features (Worcester, 1993). Realistic evidence for this came in the form of extremely successful global consumer boycott of aerosols which were chlorofluorocarbon (CFC) driven, and the international success of publications by Elkington and Hailes (1988) titled "The Green Consumer Guide." This resulted in a rupture of corporate activity in the area of green marketing and an upsurge in green business research and writing among academics. Corporate interest in green marketing was indicated by early market research findings suggesting major changes and innovations in products produced by them. The results of the survey conducted by Vandermerwe and Oliff (1990) revealed that 85% of European multinationals claimed to have altered their production systems and 92% claimed to have altered their products in response to green concerns. Regardless of this optimistic picture, by the mid-1990s new market research evidence began to emerge which was less explicit about the growth of green consumerism. A follow-up report by Mintel (1995) on the environment recorded only a very slight increase in green consumers since 1990 and identified a noteworthy gap between concern and actual purchasing. According to Allan (2005), the frequency and prominence of green claims were also found to be in decline and green products seemed to have achieved only limited success (Werther and Chandler, 2005). Even though green product growth continued strongly in certain markets such as financial services, food, and, tourism; there was no longer talk about the impressive growth in launching the green product.

1.5   OPPORTUNITIES AND CHALLENGES INVOLVED

1.5.1   OPPORTUNITIES

The main reason for the emergence of green marketing is the availability of a plethora of opportunities including sustainable competitive advantage, wider consumer base, government subsidies, and corporate social responsibility. These have been described in the points given below:

- Sustainable competitive advantage: Environment-friendly technology can be used to gain competitive advantage (Shrivastava, 1995). If a firm is offering services and products that are able to meet the needs of consumers along with environmental benefits, then it can create a point of difference in the minds of the consumer. This point of difference will force them to use that particular service or product.
- Wider consumer base: Prior studies have highlighted that consumers are aware of green marketing and are concerned about the issues related to the environment (Acharya and Gupta, 2016; Banyte et al., 2015). They are willing to purchase green products. This increasing consumer base can provide firms a large target segment, and can also help the firms in reducing the cost of their product and services because of full utilization of the machine which is used for producing the products. It would be a win-win situation for both the consumers and the firm, the consumers would enjoy the benefit of the reduced price and the firms would benefit in terms of greater profit and market share.
- Government subsidies: Since depletion of natural resources has become a major concern for the world, and it has also become a sensitive part of the decision making for all governments across the globe. The governments are requesting firms to implement environment-friendly practices. For example, in Bangalore, which is the capital city of Karnataka, India, the government has asked plastic bags (Vijaykumar, 2016) manufacturers to stop production of plastic bags and go for production of paper bags; as using environment-friendly practices by an organization would help the organization to meet the legal framework of environmental protection without forceful implementation. The government is also providing subsidies for firms to implement the green and clean technology.
- Corporate social responsibility: Implementing environment-friendly practices by the firms would reflect their concern for the environment as well as for the society. Green marketing is now considered as a new means of corporate social responsibility (Chowdhary and Dasani, 2013). Firms can also use the fact that they are environmentally responsible as a marketing tool. This would enable the firms to attain its environmental as well as profit-related objectives. It would also benefit the employees since they feel happy that their firm is benefiting the society

1.5.2 CHALLENGES

- Homogenous: There is no homogeneity in green products. The results of the study conducted by Yadav and Pathak (2013) revealed that only 5% of the marketing messages from green campaigns are entirely true and there is a lack of standardization to validate these claims. There is no proper standardization currently in place to certify whether the product is organic or not. There should be regulatory bodies in place that can provide certifications.
- Novel concept: Consumers are aware of the benefits of green products. But it is still a novel concept for the public (Ottman et al., 2006). Consumers have to be educated and should be aware of the threats on the environment if they do not use green products or services (Bhalerao and Deshmukh, 2015; Boztepe, 2016; Kumar and Sharma, 2015).
- Endurance and determination: The stakeholders need to view the environment as a major long-term investment opportunity; the firms need to look at the long-term benefits from this new green movement. It requires a lot of endurance and they have to wait for a long time to see the results. Since it is a novel concept, it will take time for consumers to get adjusted to it.

1.6 WHAT GREEN MARKETERS SHOULD AND SHOULD NOT DO?

Green marketing is one of the few areas that are likely to create a lot of growth opportunities for the firms because green marketing plays a key role in making masses aware that in-turn saves the globe from pollution. Green marketing should not be considered as just one more approach to marketing but has to be pursued with much greater force as it has an environmental and social dimension attached to it. With the alarming threat of global warming, it is very important that green marketing becomes the standard rather than an exception. Recycling of plastic, metals, and paper waste in a safe and environmentally harmless manner should become much more systematized and universal. Marketers also have the responsibility to make the consumers understand the need for and benefits of green products as compared to nongreen products. Consumers are willing to pay more for green products in order to maintain a cleaner and greener environment. The government can also have tie-ups with NGOs in order to create awareness among the consumers.

After the first sign of green marketing movement, during the second half of the 1990s marketing scholars began to shift their research focus beyond the original green marketing schedule and its focus on the quest of environment-based competitive advantage. Ideas about what might constitute sustainable marketing began to emerge. This shows that more substantive progress will require a reshaping of a number of marketing elements and practice including strategy formulation, product redefinition, market penetration, highlight on product benefits, campaigns, and emphasis on cost. The details are shared in the points described below:

- Strategy formulation: Sarkar (2012) suggests three ways of formulating a sustainable green marketing strategy. One is by consumer value positioning which involves designing such a product which has distinct green characteristics than that of its counterparts. Second is by calibration of consumer knowledge which requires presenting unique features of the green product and how it is going to be environment-friendly. Another way is by enhancing the credibility of product claim which involves getting green certifications from reputed authorities.
- Product redefinition: Consumers who buy organic food are taking decisions on issues beyond the tangible product. More focus is required on how the product is manufactured in order to gain competitive advantage (Cordano, 1993). The manufactures can give more details on the various stages of production which the product undergoes. This will give more clarity to the consumers and will help them in taking the right decisions as to whether buy the product or not.
- Market penetration: New markets need to be developed in which there is a free flow of raw materials (Kassinis and Vafeas, 2006), and also option of recycling the products and alternative forms of production and consumption (such as *Sante* or farmer's *mandi*) need to be formed, where there are no middlemen. This, in turn, will reduce the cost of the product since the middlemen's share is not included while arriving at the final price for the product.
- Highlight on product benefits: The benefits of using the green products should be highlighted. The manufacturer should focus more on the product benefits, for example, the benefits of using organic beauty products is that it does not harm the skin, and if the customer uses the product then their skin will glow.
- Campaigns: The advertising campaign should be such that it educates the consumers rather than impressing them. The message which is used to communicate the features of the product should be explained in simple terms so that it is easily understood by everyone.

- Emphasis on cost: Products such as houses and jewelry continue to be marketed through competition based on price, not competition in terms of overall costs of ownership and use (Peattie and Crane, 2005). An emphasis on issues like energy efficiency within the context of total costs would mean that our homes will look very different compared to what they are now.

According to Ottman et al. (2006), the only two things that green marketing needs to focus on are satisfying its customers and improving the quality of the environment. The authors also suggest the ways to improve and achieve these two basic green marketing objectives. Consumer value positioning of green products includes five dimensions such as efficiency and cost-effectiveness, health and safety, performance, symbolism and status, and convenience. The details of these key dimensions for a successful green marketing have been explained below:

- Efficiency and cost-effectiveness: The green products need to offer higher efficiency and lower costs. In case of electric bulbs, compared to tungsten bulbs, compact fluorescent lamps (CFLs) are efficient but costly. Similarly, light-emitting diodes (LEDs) are efficient but costly than CFLs. Therefore, even if a product is costly, a price sensitive customer would think twice before purchasing a costly green product. A message by ASKO reflects this value "The only thing our washer will shrink is your water bill."
- Health and safety: Another important criterion is health and safety. Customers would not like to risk their health and compromise with their safety. The marketers need to ask if the green products they offer ensure customers' health and safety. For example, Patanjali Ayurved Limited, an Indian fast moving consumer good (FMCG) company claims that its products are healthy and safe because of the ingredients used in preparation are herbal and natural. Not surprisingly, its share in the Indian FMCG market is increasing exponentially (Rukhaiyar, 2016). A message by Earthbound Farm Organic reflects this value "20 years of refusing to farm with toxic pesticides. Stubborn, perhaps. Healthy, most definitely."
- Performance: The green products are often perceived as a poor-performer compared to that of non-green ones. However, latest green products are more promising and offer a better performing alternative. Ottman et al. (2006) give an example stating that "front-loading washers clean better and are gentler on clothes compared

to conventional top-loading machines because they spin clothes in a motion similar to clothes dryers and use centrifugal force to pull dirt and water away from clothes" (p. 29). A message by Citizen Eco-Drive Sports Watch reflects this value "Fueled by light so it runs forever. It's unstoppable. Just like the people who wear it."

- Symbolism and status: Giving an identity to the product which differentiates it from other products is of paramount importance and a way to position the product in the market. For example, Mahindra e2o, an electric car, from the camp of Mahindra was one of the first e-cars in India. Though it had the first mover advantage, it couldn't meet the stakeholders' expectations. Another example is Cochin International Airport Ltd., an Indian company, which used solar energy to power the entire Cochin Airport and made it the world's first solar-powered airport (Menon, 2015). A message by The Body Shop reflects this value "Make up your mind, not just your face."

- Convenience: The extent to which the product offers ease of using it is what convenience implies here. Energy efficient products generally offer inherent convenience benefits. It helps the companies gain competitive advantage. For example, Toyota highlighting car-pooling benefits on its Prius Website. Another example is companies taking advantage of New Delhi (the capital of India) government's decision of allowing odd and even numbered cars on roads on alternate days. The Economic Times, an Indian national newspaper, reported that "Promto, which provides last-mile connectivity to commuters in and around Connaught Place with its electric bike service, is planning to enhance its fleet to meet the demand during odd-even days" (TNN, 2016). A message by General Electric's CFL Flood Lights reflects this value "Long life for hard-to-reach places."

In addition to the aforesaid five dimensions that are integral to green marketing, Ottman et al. (2006) argued that bundling is critical to the success of a green product. Bundling refers to selling several products in a bundle (Adams and Yellen, 1976). The companies can start with bundling a green product with non-green products and then slowly increase the share of green products in that bundle. Exploratory studies in this area have reaffirmed Ottman et al.'s (2006) claims. For example, Acharya and Gupta's (2016) thematic analysis of semi-structured interviews and focus groups revealed how status plays an important role in the purchase decision of a green consumer even in a developing country such as India.

1.7   IMPACT ON STAKEHOLDERS

Stakeholders play one of the most influencing roles in any firm and market. They persuade all aspects of green strategy including the purchase of a green product, the packaging of the product, the product nature, advertisement, promotion and also in programs related to green awareness. Environmental strategies are associated with a stronger stakeholder orientation (Bansal and Roth, 2000). It is a useful concept to grasp the degree to which a firm understands and addresses environmental stakeholders' demands (Maignan and Ferrell, 2005). Prior research shows that pressure from stakeholder influence managerial choices since such pressures are perceived as limits and opportunities. When pressures are perceived as limits, firms bring about rapid socialization to obtain authenticity from stakeholders, as well as outcomes which are measurable (King and Lenox, 2001). If they are considered as opportunities, pressures create an incentive structure that promotes instrumental corporate green responsiveness as a means to obtaining positive attention from the public and increased support from the stakeholder (Cordano, 1993; McDonagh and Prothero, 2014). Therefore, there are reasons to think that stakeholders influence the formulation and direction of corporate strategies (Polonsky et al., 1994).

1.8   BEST PRACTICES

Due to the surge in the number of green products available in the world markets, services-driven organizations are also looking for green marketing to increase their brand value. Starbucks and Johnson and Johnson have already taken the lead in successfully executing their green marketing strategies. Subsequent paragraphs describe these practices in detail:

1.8.1   STARBUCKS

Starbucks Corporation, a giant American coffeehouse chain, used green marketing to grow its brand. It has long been a promoter of sustainable coffee-growing practices, paying a premium price to encourage farmers to adopt more environment-friendly practices. Its Shared Planet initiative promotes environmental responsibility among its stores, employees, and customers.

They took three main initiatives (1) carrying out physical modifications, (2) harnessing the Earth day opportunity, and (3) launching Facebook campaign (Lesser, n.d.). These initiatives have been elaborated below:

- Carrying out physical modifications: For green marketing, Starbucks started with changing their products to give a "green impression" to the customers. First, it used a recyclable packaging to reduce cup waste. The company has been following this practice since 2009. In 2013, the company gave an offer *"Bring 10 used paper cups to get a coffee tumbler."* Second, it knows that its stores constitute 80% of the carbon footprint of the company. So, it ensures to use as much electricity generated from the renewable sources as possible. Also, it uses efficient lighting for energy conservation (Ripton, 2014). Third, it started another initiative called "green building." Under this initiative, low-flow valves were installed, floor tiles were changed to recyclable ones, among others.
- Harnessing the Earth day opportunity: Starbucks has been making use of the Earth day for years to attract green consumers. In 2011, its scheme was to give a free brewed coffee or tea in exchange for a reusable mug or tumbler. Similarly, in 2012, it launched a campaign *"One person switching can save trees. Together we can save forests"* in which people came with a reusable cup and drew something on a giant poster to convey a green message. In 2013, the offer of giving one coffee tumbler for every set of 10 used paper cups.
- Launching Facebook campaign: Starbucks decided to market its green initiative online using social media, and particularly Facebook. Its official account is country wise. For example, for India, it is https://www.facebook.com/starbucksindia/?fref=ts, whereas for Canada it is https://www.facebook.com/starbuckscanada/?fref=ts. It regularly updates its websites and answers queries and addresses customers' concerns. Starbucks strongly identifies its products with nature as evident from its official Facebook pages.

1.8.2 JOHNSON AND JOHNSON

Another company famous for its green initiatives is Johnson and Johnson, an American multinational medical, pharmaceutical, and packaged goods manufacturer. Al Iannuzzi, company's senior director, shared some green marketing tips with Anna Clark, president, EarthPeople. These tips include researching the relevant, marketing of the green social initiatives, attaching importance to communication, and go beyond eco-labels (Clark, 2011). All of these have been described below:

- Researching the relevant: It is important to brainstorm the significant areas of improvements. Here, the significance is based on the impact a particular area has on the product. For instance, P&G found that the most impactful area of improvement for their detergent was the hot water used for washing clothes. It made them work on detergent that can give expected results in the cold water.
- Marketing of the green social initiatives: A company needs to identify and promote a "social cause." An example of identifying a cause is Häagen-Dazs's support for research to protect the honey bee population. In Häagen-Dazs's case, honey bee population is directly connected to their brand. However, merely working for a sustainable future is not enough. The message needs to be marketed to the existing and potential customers.
- Attaching importance to communication: Johnson and Johnson started a unique process called earthwards that relies on deriving meaning communication regarding sustainability. The company needs to clearly understand the customer requirements and take timely action. Its websites states the following about Earthwards (Johnson and Johnson, 2016):

"At Johnson and Johnson, the Earthwards® approach embodies our commitment to product stewardship and defines how we address our environmental and social impacts. We use it to engage product development teams and drive continuous product innovation by designing more sustainable solutions across a product's lifecycle."

Johnson and Johnson's Earthwards initiative determined their end market and created those scorecards and other tools that helped the company adapt its products as per customer feedback.

- Go beyond eco-labels: Marketers often hang-up on labels. Though eco-label creates distinction and becomes one of the significant means of conveying brand value, the product story also needs to be told. The story is expected to exhibit the green improvements.

1.8.3 OTHER POPULAR INITIATIVES

Other than Starbucks and Johnson and Johnson, there are various other companies such as Ford Motor Company, Disney, and Nike that are also following green practices as detailed in the subsequent paragraphs:

1.8.3.1 FORD MOTOR COMPANY

The Ford Motor Company which was founded by Henry Ford in the year 1903 is an American multinational automaker headquartered in Dearborn, Michigan, a suburb of Detroit. When the company realized that heavy polluters in the world initially were automotive manufacturers, the company decided to revolve its image around with a ten-part environmental policy that has been in force for years. The company for its vehicles uses fabrics which are sustainable; their vehicles are 80% recyclable. Ford is also an innovator in fuel efficiency, particularly in the six-speed transmission category, and offers a clean diesel heavy-duty pickup truck. The paint fumes are recycled as fuel. Its factories also employ geothermal cooling systems.

1.8.3.2 DISNEY

The Walt Disney Company, commonly known as Disney, was founded by Walt Disney and Roy. O Disney in the year 1923 is an American diversified multinational mass media and entertainment conglomerate headquartered at the Walt Disney Studios in Burbank, California. In all its facilities, the company uses a zero net direct greenhouse gas emissions policies, and it is effective in reducing the indirect greenhouse gas emissions by reducing electrical consumption. It also has a zero waste policy, in a sense, none of its refuse ends up in landfills. It also uses water savings technologies and is lowering the footprint of its product manufacturing and distribution.

1.8.3.3 NIKE

Nike, Inc., which was founded in the year 1964 by Bill Bowerman and Phil Knight is an American multinational corporation headquartered in Beaverton, Oregon, is engaged in the design, development, manufacturing and world-wide marketing and sales of apparel, accessories, footwear, equipment, and services. The company highlights the significance of green initiatives in its advertising; it also goes one step further by actually implementing those great initiatives. It makes an entire line of sustainable sporting goods and equipment. Nike also uses renewable energy sources for its manufacturing facilities. In addition to its environmental efforts on U.S. soil, it has pushed its contracted suppliers in different countries to extend and carry out written environmental policies. This helps the environment across the globe.

1.9 CONCLUSION

Given the importance of the environment for human beings, the concept of green marketing and sustainable development is getting attention in India with time, but it is still in an embryonic stage. Even though the government has announced and implemented various policies and regulations for environmental protection, many firms are still not adopting it because of the cost involved. Green marketing is not just another term of marketing since along with the profitability concern; it deals with environmental and social dimensions, too, so it has to be pursued with more anxiety and significance. All the stakeholders of the environment should work together to save the environment, revisiting the ancient history concept "*Vasudev Kutumbakam*," that is, the whole world is one single family and all the entity like trees, animals, and so forth, in the ecosystem have *atma* (soul) and are a part of our *kutumb* (family), and therefore, they require to be preserved (Yadav and Pathak, 2013).

Firms should be transparent in disclosing all the contents of their green products. They should highlight the health benefits for the green consumers. The firms should bring the products' long-lasting and environment-friendly features to consumers' attention. The marketers should make non-green consumers aware of the environmental hazards of using non-green products and suggest green products as an alternative. Marketers need to instill a sense of pride in green consumers. Governments across the globe should promote corporate social responsibility through devising ways and means to encourage private investment in the production and marketing of green products.

Companies producing green products and services should set aside a certain amount of their profits for research and development and they should come out with innovative ideas which will help them in reducing the price of green products. This will help companies to attract more customers. The Indian government can charge less tax for green products and services; this will help the producers of green products to charge lesser prices for their product and services.

Further study can be carried out by the researchers in finding out the factors which affect customers in engaging in green products and services. Research can explore the effect of factors like culture and gender on green consumption. Future research can be carried out to find out the impact of advertisement on green consumption.

KEYWORDS

- **green marketing**
- **renewable products**
- **sales strategy**
- **green consumers**
- **recyclable products**

REFERENCES

Acharya, A; Gupta, M. An Application of Brand Personality to Green Consumers: A Thematic Analysis. *Qual. Rep.* **2016,** *21*(8), 1531–1545.

Adams, W. J.; Yellen, J. L. Commodity Bundling and the Burden of Monopoly. *Q. J. Econ.* **1976,** *90*(3), 475–498.

Allan, B. (2005). Social enterprise: through the eyes of the consumer (prepared for the National Consumer Council). Social Enterprise Journal, 1(1), 57-77.

Bansal, P.; Roth, K. Why Companies Go Green: A Model of Ecological Responsiveness. *Acad. manage. J.* **2000,** *43*(4), 717–736.

Banyte, J.; Brazioniene, L.; Gadeikiene, A. Expression of Green Marketing Developing the Conception of Corporate Social Responsibility. *Eng. Econ.* **2015,** *21*(5), 550–560.

Bhalerao, V. R.; Deshmukh, A. Green Marketing: Greening the 4 Ps of Marketing. *Int. J. Knowl. Res. Manage. E-Commer.* **2015,** *5*(2), 5–8.

Boztepe, A. Green Marketing and its Impact on Consumer Buying Behavior. *Eur. J. Econ. Political Stud.* **2016,** *5*(1), 5–21.

Chamorro, A.; Rubio, S.; Miranda, F. J. Characteristics of Research on Green Marketing. *Bus. Strategy Environ.* **2009,** *18*(4), 223–239.

Chen, Y. S. Towards Green Loyalty: Driving from Green Perceived Value, Green Satisfaction, and Green Trust. *Sustainable Dev.* **2013,** *21*(5), 294–308.

Chowdhary, S.; Dasani, L. Green Marketing—A New Corporate Social Responsibility. *INCON-VIII International Conference on Ongoing Research and IT*, Pune, India, January 2013.

Clark, A. Johnson & Johnson's Strides toward Sustainable Healthcare. 2011. https://www.greenbiz.com/blog/2011/04/15/johnson-johnson-strides-toward-sustainable-healthcare (accessed Sep 20, 2016).

Cordano, M. Making the Natural Connection: Justifying Investment in Environmental Innovation. *Proc. Int. Assoc. Bus. Soc.* **1993,** *4*, 1049–1061.

Cronin, J. J., Jr.; Smith, J. S.; Gleim, M. R.; Ramirez, E.; Martinez, J. D. Green Marketing Strategies: an Examination of Stakeholders and the Opportunities they Present. *J. Acad. Mark. Sci.* **2011,** *39*(1), 158–174.

Delafrooz, N.; Taleghani, M.; Nouri, B. Effect of Green Marketing on Consumer Purchase Behavior. *QScience Connect* **2014,** *5*, 1–9.

Grundey, D.; Zaharia, R. M. Sustainable Incentives in Marketing and Strategic Greening: The Cases of Lithuania and Romania. *Technol. Econ. Dev. Econ.* **2008,** *14*(2), 130–143.

Guitierrez, C. Gulf Oil Crisis: BP Credit Rating Cut by Moody's. 2010. www.forbes.com (accessed Sept 18, 2016).

Henion, K. E.; Kinnear, T. C. *A Guide to Ecological Marketing;* American Marketing Association: Columbus, Ohio, 1976.

Johnson, J. Product Stewardship/Earthwards. 2016. http://www.jnj.com/caring/citizenship-sustainability/strategic-framework/product-stewardship-earthwards (accessed Sep 10, 2016).

Kassinis, G.; Vafeas, N. Stakeholder Pressures and Environmental Performance. *Acad. Manage. J.* **2006,** *49*(1), 145–159.

Kilbourne, W. E. Green Marketing: A Theoretical Perspective. *J. Mark. Manage.* **1998,** *14*(6), 641–655.

King, A. A.; Lenox, M. J. Does it Really Pay to be Green? An Empirical Study of Firm Environmental and Financial Performance. *J. Ind. Ecol.* **2001,** *5*(1), 105–116.

King, A.; Lenox, M. Exploring the Locus of Profitable Pollution Reduction. *Manage. Sci.* **2002,** *48*(2), 289–299.

Kleindorfer, P. R.; Singhal, K.; Wassenhove, L. N. Sustainable Operations Management. *Prod. Oper. Manage.* **2005,** *14*(4), 482–492.

Kumar, D. S.; Sharma, R. A Study of Consumers' Perception and Attitude Towards Green Products. *Int. J. Adv. Res. Manage. Soc. Sci.* **2015,** *4*(8), 285–297.

Lesser, S. In Clean Techies, (n.d.). http://cleantechies.com/2012/02/20/top-ten-sustain-ability-initiatives-of-starbucks-corporation/

Luo, X.; Bhattacharya, C. B. Corporate Social Responsibility, Customer Satisfaction, and Market Value. *J. Mark.* **2006,** *70*(4), 1–18.

Maignan, I.; Ferrell, O. C.; Ferrell, L. A Stakeholder Model for Implementing Social Responsibility in Marketing. *Eur. J. Mark.* **2005,** *39*(9/10), 956–977.

McDonagh, P.; Prothero, A. Sustainability Marketing Research: Past, Present and Future. *J. Mark. Manage.* **2014,** *30*(11–12), 1186–1219.

Menon, S. How Is The World's First Solar Powered Airport Faring? BBC. 2015. http://www.bbc.com/news/world-asia-india-34421419

Mishra, P.; Sharma, P. Green Marketing in India: Emerging Opportunities and Challenges. *J. Eng. Sci. Manag. Educ.* **2010,** *3*(1), 9–14.

Nadaf, Y. B. R.; Nadaf, S. M. Green Marketing: Challenges and Strategies for Indian Companies in 21st Century. *Int. J. Res. Bus. Manage.* **2014,** *2*(5), 91–104.

Ottman, J. A.; Stafford, E. R; Hartman, C. L. Avoiding Green Marketing Myopia: Ways to Improve Consumer Appeal for Environmentally Preferable Products. *Environ. Sci. Policy Sustainable Dev.* **2006,** *48*(5), 22–36.

Peattie, K. Golden Goose or Wild Goose? The Hunt for the Green Consumer. *Bus. Strategy Environ.* **2001a,** *10*(4), 187.

Peattie, K. Towards Sustainability: the Third Age of Green Marketing. *Mark. Rev.* **2001b,** *2*(2), 129–146.

Peattie, K. Towards Sustainability: Achieving Marketing Transformation—A Retrospective Comment. *Soc. Bus.* **2011,** *1*(1), 85–104.

Peattie, K.; Charter, M. Green Marketing. In *The marketing book;* **2003,** *5,* 726–755.

Peattie, K.; Crane, A. Green Marketing: Legend, Myth, Farce or Prophesy? *Qual. Mark. Res. Int. J.* **2005,** *8*(4), 357–370.

Polonsky, M. J. An Introduction to Green Marketing. *Electronic Green J.* **1994,** *1*(2), 1–10.

Porter, M. E.; Kramer, M. R. Strategy and Society: the Link Between Corporate Social Responsibility and Competitive Advantage. *Harv. Bus. Rev.* **2006,** *84*(12), 78–92.

Prothero, A. Green Consumerism and the Societal Marketing Concept: Marketing Strategies for the 1990's. *J. Mark. Manag.* **1990,** *6*(2), 87–103.

Ripton, J. T. 6 Companies that Have Great Environmental Initiatives. Renewable Energy World. 2014. http://www.renewableenergyworld.com/ugc/blogs/2014/03/6-companies-that-have-great-environmental-initiatives.html (accessed Sep 11, 2016).

Rukhaiyar, A. Patanjali Products Find a Growing Market. The Hindu. 2016. http://www.thehindu.com/news/cities/mumbai/business/fastmoving-ayurvedic-goods/article8187124.ece (accessed Sep 18, 2016).

Sarkar, A. N. Green Branding and Eco-Innovations for Evolving a Sustainable Green Marketing Strategy. *Asia-Pac. J. Manage. Res. Innovation* **2012,** *8*(1), 39–58.

Shrivastava, P. Environmental Technologies and Competitive Advantage. *Strategic Manage. J.* **1995,** *16*(S1), 183–200.

TNN. Startups Come Up with APP-Ropriate Response to Delhi's Odd-Even Plan. The Economic Times. 2016. http://articles.economictimes.indiatimes.com/2016–04–09/news/72186438_1_app-surge-50-paise (accessed Sep 20, 2016).

Vandermerwe, S.; Oliff, M. D. Customers Drive Corporations. *Long Range Plan.* **1990,** *23*(6), 10–16.

VijayKumar, N. From 1050 Tonnes of Plastic Garbage a Day to Zero: Bangalore to Introduce Plastic Ban. The Times of India. 2016. http://www.thebetterindia.com/48844/plastic-ban-bangalore-karnataka/ (accessed Sep 25, 2016).

Werther, W. B.; Chandler, D. Strategic Corporate Social Responsibility as Global Brand Insurance. *Bus. Horizons* **2005,** *48*(4), 317–324.

Worcester, R. M. Public and Elite Attitudes to Environmental Issues. *Int. J. Public Opin. Res.* **1993,** *5*(4), 315–334.

Yadav, R.; Pathak, G. S. Green Marketing: Initiatives in the Indian Context. *Indian J. Mark.* **2013,** *43*(10), 25–32.

CHAPTER 2

GREEN PRICING: THE JOURNEY UNTIL NOW AND THE ROAD AHEAD

ANJALI KAROL[1,*] and C. MASHOOD[2]

[1]*Institute for Financial Management and Research, Chennai, India Mob.: 9177170570*

[2]*SPI Global, 6th Floor, Block-9B, DLF-IT Park, Chennai 600089, India, Mob.: 9447436386, E-mail: a.mashood@spi-global.com*

*Corresponding author. E-mail: anjali.k@ifmr.ac.in

2.1 INTRODUCTION TO GREEN PRICING

Green pricing or environmental pricing is a pricing strategy that encompasses environmental responsibility in pricing products and services. It is considered as a tool to price products incorporating the environmental externalities (usually costs and rarely benefits) that arise in their production to consumption supply chain. Companies and state utilities increasingly use it to price products at a premium to pay for the environmental regulations and/or to protect the renewable natural resources.

By setting a higher price, green pricing tries to take advantage of the price-quality effect, which operates when higher prices tend to signal higher quality to consumers. By charging a premium, consumers feel that superior quality products are offered. In addition, they get an emotional contentment of contributing toward environmental protection. With more people expressing their willingness to pay a premium, green pricing options enhance choices available to consumers and helps them become more eco-friendly. This has presumable spillover effects in helping create a society that is eco-friendly, eco-aware and eco-literate. The preference for green pricing could have important policy implications on taxation, infrastructure, budgeting and subsidy decisions. Market research suggests that companies experienced mixed success with green pricing (Lieberman, 2002). Consumer confidence in the quality of such products is also mixed.

As the world gains pace and comfort, the environment is facing its repercussions in the form of environmental degradation and resource depletion. Green pricing is a promising tool to control these negative externalities in a two-fold manner. One way is by penalizing polluting activities and products. The second way is by harnessing revenue to pay for environmental protection and to invest in renewable energies. When green pricing is adopted for eco-friendly products, the premium paid is usually for investments and maintenance of the adopted green technologies and to cover the cost of recycling materials. On the other hand, when green pricing is adopted for products that cause environmental damages, the premium is usually for penalizing polluting activities and products and for building renewable technologies in the future (Kolstad, 2000).

2.1.1   THE NEED FOR GREEN PRICING

The need for green pricing started from the concept of a green economy. Though there exist conflicting views on what a green economy is, often these varying perspectives overlap. Toward reifying the concept of a green economy, a working definition of a green economy was given by United Nations Environment Program (UNEP).

> UNEP defines a green economy as one that results in "improved human well-being and social equity, while significantly reducing environmental risks and ecological scarcities" (UNEP, 2010). In its simplest expression, a green economy is low-carbon, resource efficient, and socially inclusive. In a green economy, growth in income and employment are driven by public and private investments that reduce carbon emissions and pollution, enhance energy and resource efficiency, and prevent the loss of biodiversity and ecosystem services (…). The concept of a green economy does not replace sustainable development, but there is a growing recognition that achieving sustainability rests almost entirely on getting the economy right. Decades of creating new wealth through a brown economy model based on fossil fuels have not substantially addressed social marginalization, environmental degradation and resource depletion…. (UNEP, 2011).

If features of a green economy as visualized by UNEP needs to be materialized, it needs associate and fitting variations in the present brown economic activities viz-a-viz production, consumption, distribution, and exchange. Price is an important strategic component that can affect all the four economic activities and alter the incentives for decision-making. Hence, every effort toward green price claims utmost importance given that price is the most important market indicator and mover. Thus, we have established that to achieve the objectives of a green economy we need green pricing. Green economy, though not a substitute for sustainability, can contribute toward environmental protection and; hence, sustainability. Hence said, the need for green pricing arises primarily from the need for sustainable development. Figure 2.1 depicts this process flow from green pricing to sustainable development.

FIGURE 2.1   Process flow from green pricing to sustainable development.

2.1.2   EVOLUTION OF GREEN PRICING

The concept of green pricing first appeared in a paper by Moskovitz (1993). The paper introduced green pricing as an option given to consumers of power utility companies in the United States to pay a premium of 5–10% of their regular bills, to pay the utility to acquire renewable technology using an agreed formula to generate power. The consumers signal the amount to spend on renewables by expressing their commitment to the program. Studies have shown that consumers showed high enthusiasm for a contract period of up to 2 years beyond which enthusiasm dropped significantly. Both small and big consumers were equally interested in the program but became reluctant to the extent of premium increases.

Fascinatingly, some big consumers started using green pricing as a marketing tool to market their companies. Green pricing was never seen as a substitute for the development of renewable and eco-friendly resources. People believe that right policy and channeling of funds should create eco-friendly products and energy sources. Green pricing in this task should only supplement making our earth go green. Nonetheless, there are now conflicting opinions that since green consumerism is the future, green pricing of products and utilities is the first step toward a shift in the future technologies and ways of life. To add a point of positivism to this, over the world, the demand for green products is soaring confirming that green pricing is the future.

There exists a common belief that green pricing is costly. Recent estimates show that green products are on an average 50–70% more costly than regular products. However, not all green pricing schemes need to collect premium or be costly that only the rich can afford. Surveys suggest that consumers would keep expressing their readiness to pay a premium for green products (even up to an extent of 50% for some products) because they are socially more responsible. However, experience shows that green products, when offered, do not have many takers though the number of green consumers is growing in absolute terms. It is only after the greenness of the product, which has established that the consumers start turning to them in large numbers. Therefore, a premium is not always feasible for a large-scale rollout of green products as sales begin to pick up only gradually for most of the categories of green products. Research shows that the performance and quality of green products when compared to their alternatives, availability, the consumer's concern for the environment and prices of the products; decide their willingness to pay a premium for the green products. This calls for product innovation, enhanced performance, and emphasis on quality coupled with an economical pricing strategy for green products. Green products will sooner have to compete with other products, both on quality and price fronts as consumers are utility maximizers. Changing consumer buying patterns on the onset of global slowdown suggests that sales of green products will not become commonplace until consumers get attractive price discounts on them, both at the stores and at online marketplaces.

2.2   GREEN PRICING: VOLUNTARY VS. IMPOSED

There can be two types of green pricing depending on whether or not consumers are provided with the choice of adopting it. If the government, statutory body, or the company implements green pricing for its products and consumers have to pay the green price to use them, then green pricing is said to be imposed on the consumers. On the other hand, if the consumers are given a choice to choose between green products with green price and nongreen products with market price, we say that green pricing is voluntary. The voluntary green pricing scheme is offered as a choice program to the consumers or alternatively, consumers can negotiate green pricing agreements with the company. There exists a dilemmatic situation where we cannot clearly say whether voluntary green pricing or imposed green pricing is better. However, which model to adopt should ideally be guided by the purpose of introducing green pricing, and the sector and segment of the market for which it is introduced.

The mounting prominence placed on environment coupled with the industry's increasing focus on consumer's desires gives a wide scope for green pricing. If green pricing is offered as an alternative more consumers will pay a premium and join the program. The element of choice available in involuntary green pricing leads to more consumer satisfaction than imposed green pricing. Consumers feel valued and respected when they have the decision-making power as to what products to choose and at what prices. Usually, green pricing is optional. However, sometimes when it is imposed by mandate, the following criticisms have been reported.

Lower marginal utility: Consumers do not have a choice when green pricing is imposed on them. Often, there are instances where they are better off paying a lower price to buy a brown product than going green to buy products at a premium which may not match the expected increase in marginal utility of switching toward the green products.

Affect monthly budgets: Sometimes when consumers are forced to buy the premium-priced green products, their monthly allocations to other areas get affected. Therefore, it puts them at a double disadvantage. The products purchased earlier at lower prices are now priced at a premium and they may have to go without some goods that they earlier could afford with their previous budgets. The cost of living of the lower- and middle-income classes, thus, increases.

May promote unhealthy competition: If companies impose green pricing by mandate, it can lead to unsolicited and stifling competition among companies. One possibility is that the companies in a sector where green pricing is imposed may form a cartel and increase prices substantially higher than what is required to cover the cost of the new green technology or the newly imposed environmental regulation costs. This leads to inflation in that segment of the market. Another possibility is that companies may reduce the quality of the products while increasing the price befitting the increased environmental costs. This can lead to a price war if companies begin to compromise on quality and reduce prices to garner more consumer base. In either case, consumers are in distress.

2.3 FACTORS AFFECTING THE DECISION TO ADOPT GREEN PRICING

All successful green pricing programs have some common enabling factors. These factors ensure that green pricing programs achieve their stated objectives (Swezey and Bird, 2001). Companies and government bodies must ensure that these factors are being taken care of when implementing a green pricing program.

Product design: Often people confuse green pricing programs with charities due to their similar nature. Some researchers believe that since green pricing leads to social benefits and convey awareness messages similar to charity, consumers may be reluctant to pay. But, some researchers feel that there is no scope for this confusion as green products ask for a specific premium whereas charities expect donations what the donors can afford and; hence, they are inherently different even though the causes based on which they are crafted may be similar (Friedman and Miller, 2009). Studies show that whenever green pricing programs are designed as contributions toward society, the participation rates fall drastically. Participation improves if the takers feel that they have perceivable benefits rather than the contentment of providing for a social cause that benefits the environment and society at large (Moskovitz, 1993). This ambiguity persists unless the premium is linked to direct benefits.

The size of the market: Green pricing can only be introduced if a sizable amount of consumers indicate a preference for green pricing programs and have the financial backing to enroll in such programs. Green pricing programs generally incur huge initial outlay to start a new eco-friendly line of production, to acquire less polluting technologies or to convert their storage and packaging designs to more green methods, training staff, and so forth. It also requires a good amount to be spent on marketing and administration expenses to shape up the program and deliver it to the consumers. However, in due course of time as more consumers, join and economies of scale begin to operate; green products may become cheaper than their counterparts. Marketers must ensure an optimum market size for the product before introducing green pricing. This will ensure both success and continuity of the program. However, the optimum size varies from product to product. For organic products, green pricing is easy. It requires a bit of marketing to spread awareness on the health benefits of organic products and have very fewer upfront costs as compared to green energy utilities that require heavy investment in renewable energy technology. Thus, the optimum market size for organic products is much smaller than that of green energy utilities.

Income levels: Participation in green pricing schemes also differ across income levels. High-income consumers sign up for green pricing programs more than low-income consumers primarily because they have the luxury to pay a premium and experiment new programs. The high premiums may affect the household budgets of low- and middle-income consumers and so, they usually wait for green pricing programs to deliver some success before signing up for them.

The level of eco-literacy: Consumers who are aware of the environmental issues and who care for environmental protection as a cause are the ones who take up green pricing programs more readily without much coaxing. Though many consumers are aware of environmental issues, not many understand what green programs mean and how green products differ from conventional products. The cost and availability of green products are major concerns for most consumers. In the case of renewable energy, some consumers feel that since wind and sunlight are free, energy generated from them should also be free. However, when consumers are educated about the installation and maintenance cost of these technologies, participation improved (Moskovitz, 1993).

Gender: Studies show that gender differentials exist in the adoption of green pricing programs. Several studies have shown that men are more aware of environmental issues than women due to their higher access to higher education. However, there is no conclusive evidence that gender differentials exist when it comes to the concern for environmental issues. Studies also indicate that this difference is not visible much in the western world due to a more developed and gender equal nature of the West. In the case of Arabian countries, green awareness is only catching up slowly and women lag behind men in their knowledge about green technologies and green pricing schemes. Besides, women are more positive about green products like organic food and cosmetics while men hold a more positive attitude toward green energy technologies (Tikka et al., 2000; Mostafa, 2007).

Existing green pricing practices and its success: Consumers preferred to take up green pricing programs in sectors that have already established the success of green pricing like organic food, organic clothing, organic cosmetics, renewable energy utilities, and so forth. These sectors also give consumers perceived benefits like better health, less environmental pollution, recyclability and reusability of products and lower purchase and usage costs in the long run than their counterparts.

The provider's commitment to green pricing: Consumers look for the credibility of the provider before adopting for green pricing programs. This depends on facts and figures about the provider's past performance with environmental compliance and how well the consumers perceive the provider's environmental performance. Consumers do not want providers to charge a green premium and use it for their other business channels rather than delivering them real green products. A strong marriage between consumers' environmental expectations and what the company promises is necessary for the long-term success of green pricing programs. Environment savvy consumers are swiftly disillusioned if the company fails to deliver its claims. This is the reason why companies with strong social schemes and trusts that cater to the environment are the ones that have established easy success in cajoling consumers to take up their green pricing programs.

The marketing budget of the program: Consumers are concerned if a large chunk of the premium they pay will be used for marketing and other administrative costs of the program rather than actual acquisition and conversion to green technologies. Usually, consumers are nonchalant if these costs stay within a limit of 10% of their premiums. However, consumers are also careful of programs that spend too little on marketing as they anticipate the companies to fetch more consumers under the scheme to ensure the availability of the same green products at a reduced cost in the future.

Customer retention: Updating customers about the progress, environmental impact, benefits delivered, and future course of action of the program would motivate them to continue with the green pricing program. Once enrolled, customers should not be taken for granted to stay with the program forever without providing them these periodic updates. These updates can be in the form of short reports on the company website, periodic newsletters, local advertisements, SMS or e-mails to the users. These updates could be short feature stories on special events, educational activities, future green action plans, celebrities who have also signed up for the same program, or other environmental issues that the company is interested in or is of current relevance. However, caution should be noticed to not let these contact schemes turn nongreen and consume a lot of resources that may distort the green claims of the green pricing scheme. By giving customers a feeling that they are part of a growing program that contributes to the environment, customers would stay with the program (Holt, 1997).

Product availability: Often the availability of the product is a major concern than its greenness. Consumers want the continued availability of products to prefer using them (Biswas and Roy, 2016). If green products are not easily available, consumers will not be willing to pay a high price for it despite its greenness. This is because they do not trust products that are not visible, common, and available at major stores.

Celebrity endorsement: Green-pricing programs endorsed by celebrities such as community leaders, film stars, and other respected and notable persons of the society can easily earn customer attention and, thereby, increase its visibility. This helps in effective marketing of the program by increasing its credibility, which aids in ensuring the success of the program.

Partner with ECO-NGOs: Green pricing program can also collaborate with NGOs and organizations working toward environmental protection. NGOs, protectionist groups, and environmental protection bodies have stated environmental objectives and a proven record of the accomplishment of backing environmental issues, which can bring in reliability and publicity to the program.

Co-branding: Companies that include local retail chains and online market-places in their green pricing initiatives can increase the visibility of their programs and reduce the program's marketing costs significantly. Free coupons and special introductory offers to retail-chain-club-members can make the green pricing campaign attractive leading to greater signup for the program.

Transparency: The efforts of various NGOs and government bodies along with inexpensive access to the internet have made the consumers increasingly more aware and literate about what they consume. It is critical for consumers to know what exactly is the green pricing program in which they are subscribing to, what constitutes the product they buy, how does it differ from alternative nongreen and competing green pricing programs, what is the formula used to calculate the price, the benefits that accrue to them by signing up for this particular green pricing program, the number of consumers needed for the program to break-even, how do companies spend the premiums collected (Friedman and Miller, 2009), the legal implications and ethical responsibilities as a consumer of this program, etc. Marketing programs for green pricing have to sound credible to warrant support for the transparency of the program.

Accreditation and registration: Green pricing programs have to be certi-fied by the company undertaking the program, an industry regulator, and a third-party accreditor. This multistage accreditation unwinds consumers from the hassles of verifying the claims of the program by oneself. Company certification communicates that the company is committed to the program and is here to stay. Certification from an industry regulator makes not just the green pricing program but also the company as a whole under the scanner of a regulatory body. This compels the company to deliver its promises stipulated in the program. Finally, certification from an independent third-party accreditor conveys that the green pricing program is recognized and meets the requisite safety and legal standards.

Tax deductibility: Some green pricing programs give tax deductions to customers. The solar investment tax credit system in the USA gives 30% tax credit for using solar systems at residential and commercial establish-ments. After implementing this program, the demand for solar power rose by 40% annually and by 2015, the prices touched cost parity with conventional energy sources. The tax deductibility gave the customers a huge incentive to enroll in this green pricing program.

Green board: Green pricing programs that are monitored by a body of reputed environmental rights leaders and scientists are more likely to capture consumer attention as compared to non-monitored pricing programs. If the green board functions independently from the company, more will be the credibility. A consumer generally trusts a board of experts and gets the assurance that their hard-earned money is not misused or wasted.

2.4　METHODS OF GREEN PRICING

Initially, it was power companies that started green pricing, and hence, most green pricing methods available in the literature are suitable only for power companies. Green price is typically calculated in three ways in the case of power companies namely the percentage of use method, levying a green surcharge, and green power block pricing.

Percentage of use: Green power utilities charge consumers green price (a slightly higher price) for the percentage of power that comes from green energy sources and normal price for the rest of the energy generated from conventional sources. Consumers can typically choose the mix of green power and conventional power; say 50% from renewable sources (Bolding, 2003). The power company, in turn, utilizes the green price to build own green energy technologies like solar plant and windmills or signs a contract to source it from a green energy supplier. This method of green pricing has several limitations (Bird et al., 2004). These contracts are long-term in nature but consumers cannot be expected to sign long-term contracts with the power company. Consumers pay only for the green power they use on a monthly basis. However, typically consumers cannot distinguish between green power and brown power as they are usually discharged from a common grid. Often they are charged a higher proportion of green price while a major share of their power consumption may be from brown sources even against the agreed percentage.

Green surcharge: Alternatively, companies also charge consumers a progressive green surcharge if their power consumption levels exceed certain levels. This method covers all consumers and uses the surcharge as a deterrent to reduce power consumption. The surcharge collected will be used to develop renewable energy utilities.

Power block pricing: Consumers pay a fixed green premium per month for a block of power from renewable energy sources. For example, green power will be available in $5 per block of 200 KWh and consumers can buy as many blocks of green power as they choose. Since the price is stable and clear, the cost is predictable. In addition, the non-necessity of metering usage ensures easy accounting (Lieberman, 2002). However, the consumer may end up paying for the portion that he does not use as you pay in blocks and not for actual usage.

Green pricing is not limited to power companies anymore. Today, green products have become near substitutes of brown products both in terms of usage and price competencies while earlier they were only alternatives made available to consumers. In this scenario, most pricing strategies can be applied to price green products as well. In this section, we will discuss few commonly used green pricing strategies that can be applied to a large group of green products.

Percentage markup on prices of brown products: The green premium can be calculated as a percentage markup on the prices of similar brown products that are already in the market. However, this markup can only be seen as a contribution toward establishing green technologies. When green pricing is calculated this way, it suffers from a serious limitation. Either consumer may end up paying less or paying more than the cost markup as the premium is not linked to the additional cost incurred to move toward green technologies. The company will have to constantly monitor the prices of substitute products and suitably adjust the price of their green product applying the percent markup. If the prices of the brown products, which are used as a base to calculate the green premium are volatile in nature, the green pricing scheme will also see volatile prices which will downcast consumers' expectations from the program.

Quantity-based: Green pricing may be applied to a block of green goods. The price charged per unit decreases as the quantity increases. Companies can use this method to lure consumers to the discounts available on bulk purchase and increase the sales of their green products.

Pro rata pricing for different consumer groups/segments: Green pricing programs can employ price discrimination to capture maximum consumer surplus. Different prices can be charged from different segments of consumers based on their ability to pay (high/middle/low income), their position in the retail chain (wholesalers/retailers/consumers), the type of user (government/corporate/household), relationship with the provider (new user/existing user), etc. The government may buy at a higher price due to its commitment to improving social and environmental welfare. On the other hand, a corporate buyer looks for price competitiveness and; hence, may buy at a lower price than the government. However, a household buys a green product if they are convinced about the greenness of the product and depending upon their income levels. Having this kind of stratification within the program helps cater to a wide array of consumers with varying income, preferences, and objectives. Special discounts can be given for less active segments like lower-income households, students, and other such groups which are less likely to sign up for green pricing schemes.

Normal price: If consumers need to embrace green products, they need to be priced normally. Companies need to ensure that green need not be costly and should not see green as a means to reap superior profits. It should be an alternative product priced normally (Bennett and Williams, 2011) but are more eco-friendly. Companies have to devise cost-efficient green technologies and deliver green products at the competitive market price.

Subsidy: Price has always been the number one barrier that prevents people from buying green products. Green products tarnished with the history of inferior quality and high price sends a wave that it is something meant only for the elite. Average consumers have to work within a weekly/monthly budget and; hence, cannot be expected to expand on green products. Considering this, companies can collaborate with governments to subsidize their green products until they get good consumer base and become competent to take on rival products on cost frontiers.

2.5 BENEFITS OF GREEN PRICING

Green pricing provides a multitude of benefits that conventional pricing does not provide. These benefits include tax savings, personal recognition and civic pride, consumer value, free advertising, promotion of sustainability, and some educational and environmental benefits.

Tax deductibility: If green pricing programs are offered by nonprofit organizations or offered in partnership with them, consumers get discounts in the form of tax benefits. Tax authorities in many countries, for example, the Internal Revenue Service in the USA provides tax deductions to consumers of green energy utilities by classifying them as charitable contributions. In the Netherlands and the United Kingdom, increasing energy taxes and tax exemption for green power caused green power to cost lower than the conventional power promoting the use of green power (Bird et al., 2002).

Green investments have high gestation periods, which prompt the government to partially fund them. Companies providing green pricing programs are given tax incentives (Beck and Martinot, 2004) acknowledging their educational purpose and positive externalities. For example, owners, developers, and users of green buildings certified by US Green Building Council are given various subsidies, soft loans, and tax abatement for initial few years. This tax deductibility has been used by companies successfully for marketing and brand building.

Personal recognition: Often green pricing programs give personal recognition to their consumers. Monthly newsletters, pamphlets, and advertisements come with featured user satisfaction experiences of such programs and have proved to be highly efficient in canvassing new consumers. Companies also offer best consumer awards, lucky draws, complementary gift-hampers, special discounts for other green products, dedicated programs, etc. to recognize their consumers.

Choice and consumer value: Green pricing options provide consumers with a choice to select among various competing green and brown product alternatives. This way, both green and brown products will have to be constantly improved to stay in the market. Fearing the loss of consumers with the ever-burgeoning green wave, manufacturers of brown products have to upgrade their products to reduce its detrimental environmental effects. With the increased efficiency of brown products and brown products evolving into green products, green products have to be improved on quality aspects to match the quality of brown rivals. The competition is, hence, a blessing for the consumers as they derive greater value with the existence of green pricing programs. The environment also benefits from improved efficiency of products and greater green compliance, which leads to lesser environmental damages.

Civic pride: Consumers of green pricing programs get the civic pride of contributing toward the environment. Their sense of achievement comes from the knowledge that their actions benefit not just them but the society at large through the positive spillover effects of being in a cleaner and a less polluting environment.

Promoting sustainability: Green energy utilities provide subsidized/free energy efficient lamps and other appliances to combat the high price of green energy. This helps the consumers to stick to their green promises and create a path of sustainable energy development. Likewise, consumers of green food pricing programs like contracts for supplying organic fruits and vegetables are given the assurance that they will be provided their weekly supply from other organic farms and orchards at the same agreed green price in the event of a supply crunch. Another example is that builders of green homes are given the assurance that environment groups or the government would pay rents if it fails to find occupants on completion. These are all ways to ensure sustainable development.

Free advertising: Green pricing initiatives get backing from national and global environmental groups, which provide free advertising to the program. Environment groups, media, and community groups would want to associate with green pricing programs as for fulfilling their stated objectives or merely to build their image. Thus, green pricing gets free marketing. In the Netherlands, a marketing campaign launched by World

Wildlife Fund (WWF) powered the demand and consumer loyalty for green electricity and resulted in a 14% rise in green electricity users from 1990 to 2000.

Educational benefits: Most green pricing programs are usually accompanied with documents stating the objectives, mission, and vision of the program. It also informs stakeholders the reasons for adopting green pricing, the supremacy of green pricing to conventional alternatives, environmental issues if any it aims to tackle and its expected contributions to the environment. Consumers reading these documents would be motivated to learn more about the environment and may educate their friends and family on the environmental issues and aspects that persuaded them to adopt green pricing.

Green pricing campaigns spend heavily on marketing the cause and spreading awareness about environmental dos and don'ts to motivate people to turn green. Environmental groups collaborating with these programs campaign to change people's attitudes to protect and conserve nature. Thus, green pricing helps educate people and spread eco-awareness.

Environmental benefits: Most green pricing programs collect premium above the price of their brown alternatives. This money is used for installing green renewables, developing green technologies to produce more eco-friendly products and to promote eco-awareness by campaigning, marketing, and organizing programs for the cause of the environment. This helps to increase the supply of green products, breed a class of eco-aware consumers, which increase the demand for green products resulting in a larger market-share of green products. This will kick-start a process where more companies would come up with green alternatives to their existing range of products. If this happens we would fasten our steps toward a more clean, green, and sustainable environment.

2.6   CRITICISMS OF GREEN PRICING

Despite the green objectives and the much-hyped benefits, green pricing is not free of criticism. It has always been criticized for its lack of transparency and the ever-looming problem of greenwashing. High marketing expenditures and the lack of clarity on how to maintain the accounts of green pricing programs have put to question the very need for green pricing programs.

Lack of transparency: Environment savvy consumers generally buy products that they perceive to be green. Most companies advertise their products as green and charge premiums on going green. The premium should ideally be charged based on the cost premium to go green. However, in the absence of stringent laws on the extent of markup permissible by law, marketers charge premiums in a nontransparent and excessive manner. Sadly, often consumers end up purchasing products at a premium, which are advertised to be green, or products that come with green labels but in both cases are actually brown. This calls for stringent laws that stipulate what constitutes green and how to charge for green products or else transparency and; thereby, consumer confidence in green pricing programs are in danger.

High marketing expenditure: Consumers are concerned whether a large part of the green premium would be consumed as marketing and administrative costs (Friedman and Miller, 2009). Sometimes green pricing programs levy a surcharge on premium if the premium is inadequate to cover costs. This should be avoided for marketing and administrative fairness (Moskovitz, 1993). In the event, the existing green pricing programs do not break-even, the cost of implementing a new or replacement green pricing program would lead to incremental marketing costs. Existing consumers are averse to pay for it and new consumers do not want to pay for existing green technology acquired. Thus, who pays for marketing costs of green pricing programs is also a matter of discussion.

Green pricing programs do not result in green resource development: It is difficult to verify whether the green premium actually results in the development of green resources. This is because financial and operational additionality are hard to verify in the case of green pricing programs. The said program is financially additional if the premium collected actually leads to the development of eco-friendly resources. It is operationally additional if the program could reduce environmental impacts of brown products by converting brown product users to green alternatives users as opposed to the case where the program was non-existent (Friedman and Miller, 2009). Consumers choose to pay a green premium with an expectation to make an impact but it is difficult to demonstrate the impact of green pricing. Due to lack of transparency and greenwashing, all green pricing programs may not be authentic and; hence, may not result in green technology and eco-friendly resource development skyrocketing (Delmas and Burbano, 2011). In some cases, green pricing is charged only to recompense for the detrimental effects of producing the underlying commodity.

Accounting issues: Presently, there is no legal framework that specifies how green pricing has to be structured and the permissible extent of markup (green premium) on price. Many companies make use of these gaps in the legal system to charge premiums more than double the cost of the products. The ethical and moral aspects of green pricing are; therefore, at stake.

Accounting is a tool, which forces companies to be accountable for their activities by providing periodic reports of its activities to the various stakeholders. This helps consumers to verify the green claims of companies by crosschecking their claims with their actual allocation for purchase and maintenance of green technologies, expenses, and profits on green pricing programs. However, many firms resort to greenwashing and these firms in an attempt to save their reputation are left with no choice but to window dress their accounts. The lack of resilient accounting standards for green pricing, hence, promotes window dressing by firms to make their environmental claims sound legitimate and transparent. Then again many firms resort to the culpable act of fuzzy reporting their green pricing programs by creating a complementary corporate sustainability report or environmental compliance report laterally with their annual reports to deceive the public and to portray an eco-friendly image. These reports are insincere as they are not in synchronization with their annual reports or even with their corporate social responsibility (CSR) reports. Consequently, the actuals of green pricing go unverified by the public if it is not incorporated into the annual reports.

Questions on feasibility: Often green pricing programs do not succeed because they are too ambitious that they are beyond feasibility. Most green pricing programs require a minimum number of consumers to break-even. If the participation is less, the existing pricing structure will have to be altered or even discarded.

Product-adulteration: One sort of malpractice that can increase profits tremendously is product-adulteration. Some green product manufacturers add excess preservatives, chemicals, artificial colors, spurious ingredients to display a green image to their products. Products then become counterfeit and harm consumer health but huge premiums are charged to loot the green-savvy consumers. Green pricing is here deployed as a means to cheat consumers.

Greenwashing: An alarming phenomenon, which is a threat to the authenticity of green pricing, is greenwashing. A term coined by Jay Westerveld in 1986, greenwashing is the practice of using false, dubious and misleading claims about the greenness of the products or the environmental performance of the company as a whole. The prevalence of greenwashing has been skyrocketing (Delmas and Burbano, 2011) exploiting the fact that consumers are not prepared to traverse the eco-bubble. Companies with a bad track-record tend to greenwash more in an attempt to regain their bad reputation, as they are not afraid to be caught as they already carry a bad reputation. To cater to the green demands of consumers in the wake of green consumerism, companies have two options. They can either actually implement green strategies or pretend to do so by greenwashing. The lack of clarity regarding regulatory punishments for greenwashing presents companies with an incentive to greenwash (Dahl, 2010). With no clear measures to identify the extent of the greenness of the products, consumers often end up paying for products that they perceive to be green. Companies make use of this ignorance of consumers to charge exorbitant prices in the name of green pricing while they may still be offering brown products. Hence, not all green pricing is authentic green pricing. If greenwashing is not prevented, prices simply get inflated by false claims of green labels and green pricing. Green pricing, hence, can be used as a strategy to mislead consumers and loot them.

Inability to identify legitimate green pricing programs: The very existence of greenwashing camouflages legitimate green pricing schemes from the illegitimate ones. Companies irrespective of their size, reputation, and environmental compliance image have succumbed to greenwashing. Consumers have minimal beliefs in green claims of green pricing schemes and; therefore, choose the ones that they ascertain to be relatively green (About-greenwashing, 2016). The unearthing of recent green scams by trusted brands like Volkswagen has put the public's faith in a trust dilemma. During 2015–2016, Volkswagen had to recall 11 million Volkswagen, Audi, and Skoda cars with EA-189 family diesel engines that were intentionally preprogrammed and fitted with a cheat device that gets activated during laboratory tests to pass emission standards. Overall, it cost the company above $18 billion to implement the fix.

Psychological issues: Being a green consumer takes time, a disciplined lifestyle, money, efforts, sacrifice, and the mental toughness to move out of one's comfort zones. In addition, one has to be in the right place where green products and green pricing programs are available. Moreover, it needs awareness building on what exactly is green and how a green pricing program can help you become green. In the absence of these enabling factors, consumers prefer to live in their comfort zones. Green pricing programs not only should contribute to the environment but also perform as per people's expectations.

Green pricing is feminine: Men tend to shun green pricing primarily because they do not want to experiment. Men take decisions only after careful research and comparisons. Women on the other hand, shop till they drop and are excited to try out new products. Women (Quinlan, 2003) make the majority of home purchase decisions. Green pricing programs try to emotionally attack the consciousness of consumers for not being eco-friendly. Women are more prone to this emotional marketing and, hence, green pricing programs are usually marketed targeting women. This again makes men think that green pricing is feminine. This kind of gender and role stereotype marketing of green pricing programs have been widely criticized for not catering to the mainstream. Fortunately, these stereotypes are fast diluting.

2.7   GREEN PRICING INDEX

A green price index can be created by modifying the regular cost of living index to include the environment and public goods. This index can better measure the true cost of living and act as a better signal for the national welfare. The divergence of regular price index from green price index has the potential to induce a strategic shift in the inclusion of environment in public policy design and the way environmental policy will be framed. The economic incentives of stakeholders would change if such an index is implemented as a stakeholder internalizes both costs and benefits of an environmental policy (Banzhaf, 2005).

2.8   CONCLUSION

Green pricing programs have the capability to address issues of sustainable development and promote healthy living if providers are committed to the environmental cause and embrace transparency, fairness in pricing and credibility by adhering to laws. Adequate marketing, simple pricing scheme design, consumer education, and tax deductibility are other factors that ensure the success of green pricing. However, a number of issues and challenges act as hurdles in the implementation and adoption of green pricing programs. These issues have the potential to render nugatory the anticipated benefits of green pricing programs.

It is impossible to avoid polluting the environment. Certain activities like driving a car and production activities in factories pollute the environment but are vital for the conduct of everyday activities (Kolstad, 2000). Therefore, it is important to strike a balance between human needs and environmental protection needs.

Green pricing programs have to be transparent in their framework, conduct, and delivery to instill faith in such programs. Transparency can encourage participation. Preventing companies from resorting to green-washing and frauds in the name of green pricing by misusing and diverting green price premiums has always been and continues to be a challenge. We need to put in place a structural framework comprising experts from law, ecology, economics, anthropology, sociology, health, and public policy to advise governments and corporates on matters related to green pricing.

There is a common misconception that chemical products are brown and organic products are green. In fact, many chemicals occur naturally in nature and may be more cost and productivity efficient than the green alternatives. Green pricing when implemented for agro-products, consume a lot of lands, labor, water, and organic manure and require a longer harvest time. The result is huge outlay by farmers and high end-product costs for consumers. This assuredly is not a sustainable model and calls for constant innovation to come up with efficient green technologies.

Green pricing schemes need to ensure that their products on offer should be able to catch up with existing products on both cost and quality frontiers. Increased visibility, continuous supply, and wider availability of green products and services have to be ensured before kick-starting any major green pricing program. Understanding the fact that not all consumers are comfortable with the high green premiums, green pricing should improve its affordability to various income groups. Not all green pricing includes a premium or is costly and affordable only by the elite. Many green products in the market today are cheaper than their substitutes. Marketers have to promulgate the view among consumers that green pricing is affordable pricing.

KEYWORDS

- **green pricing**
- **green surcharge**
- **greenwashing**
- **green pricing index**

REFERENCES

About Greenwashing | Greenwashing Index. Greenwashingindex.com, 2016, http://green washingindex.com/about-greenwashing/ (accessed Nov 5, 2016).

Beck, F.; Martinot, E. Renewable Energy Policies and Barriers. *Encycl. Energy* **2004,** 5(7), 365–383.

Bennett, G.; Williams, F. *Mainstream Green: Moving Sustainability from Niche to Normal.* Ogilvy & Mather: New York, 2011.

Bird, L.; Wüstenhagen, R.; Aabakken, J. A Review of International Green Power Markets: Recent Experience, Trends, and Market Drivers. *Renewable Sustainable Energy Rev.* **2002,** 6(6), 513–536.

Bird, L.; Swezey, B. G.; Aabakken, J. *Utility Green Pricing Programs: Design, Implementation, and Consumer Response.* National Renewable Energy Laboratory: Golden, CO, USA, 2004.

Biswas, A.; Roy, M. A Study of Consumers' Willingness to Pay for Green Products. *J. Adv. Manage. Sci.* **2016,** *4*(3), 211–215.

Bolding, K. *Best Practices in Marketing Green Pricing Programs.* Center for Resource Solutions: San Francisco, California, 2003. http://resource-solutions.org/site/wpcontent/uploads/2015/08/bp.handbook.pdf (accessed Nov 10, 2016).

Dahl, R. Green Washing: Do you Know What you're Buying. *Environ. Health Perspect.* **2010,** *118*(6), A246–A252.

Delmas, M. A.; Burbano, V. C. The Drivers of Greenwashing. *Calif. Manage. Rev.* **2011,** *54*(1), 64–87.

Friedman, B.; Miller, M. *Green Pricing Program Marketing Expenditures: Finding the Right Balance.* National Renewable Energy Laboratory: Golden, CO, USA, 2009.

Holt, E. Green Pricing Resource Guide. American Wind Energy Association: Harpswell, Maine, 1997.

Kolstad, C. *Environmental Economics,* 2nd ed. Oxford University Press: New York, 2000.

Lieberman, D. *Green Pricing at Public Utilities: A How-to Guide Based on Lessons Learned to Date.* Center for Resource Solutions: San Francisco, California, 2002.

Moskovitz, D. 'Green Pricing': Customer Choice Moves beyond IRP. *Electr. J.* **1993,** *6*(8), 42–50.

Mostafa, M. M. Gender Differences in Egyptian Consumers' Green Purchase Behavior: The Effects of Environmental Knowledge, Concern and Attitude. *Int. J. Consum. Stud.* **2007,** *31*(3), 220–229.

Programme des Nations Unies pour l'environnement. *Towards A Green Economy: Pathways to Sustainable Development and Poverty Eradication.* United Nations Environment Programme, 2011.

Quinlan, M. L. *Just Ask A Woman: Cracking the Code of What Women Want and How They Buy.* John Wiley & Sons: New York, 2003.

Swezey, B. G.; Bird, L. *Utility Green Pricing Programs: What Defines Success?* National Renewable Energy Laboratory: Golden, Colorado, 2001.

Tikka, P. M.; Kuitunen, M. T.; Tynys, S. M. Effects of Educational Background on Students' Attitudes, Activity Levels, and Knowledge Concerning the Environment. *J. Environ. Educ.* **2000,** *31*(3), 12–19.

CHAPTER 3

GREEN CONSUMERISM: STUDY OF ACADEMIC PUBLICATIONS IN SCIENTIFIC JOURNALS INDEXED IN THE WEB OF SCIENCE AND SCOPUS

J. ÁLVAREZ-GARCÍA[1], C. P. MALDONADO-ERAZO[2], and M. C. DEL RÍO-RAMA[3,*]

[1]*Department of Financial Economy and Accounting of the University of Extremadura, Caceres, Spain, Ph.: +34 609880141, E-mail: pepealvarez@unex.es*

[2]*Universidad Tŭcnica Particular de Loja (UTPL), Loja, Ecuador, Ph.: +59 3992404740, E-mail: cpmaldonado1@utpl.edu.ec*

[3]*Department of Business Organization and Marketing of the University of Vigo, Vigo, Spain, Ph.: +34 6049667878*

Corresponding author. E-mail: delrio@uvigo.es

3.1 INTRODUCTION

A process that marks the development of the economies is the different processes of buying and selling products/services that demanded by a segment or the whole market. They are processes that at present show a great influence on the current deterioration of the environment in three ways: productive processes, use of resources (overuse of resources, which, in many cases, cause their scarcity), and effects of the products on the environment (use and waste). Thus, in the last decade, the work of different NGOs, governments, and people concerned about this situation and aimed at raising awareness among producers and consumers of the importance of mitigating this environmental impact (Diamantopoulos et al., 2003), which results in a change of values.

The emergence of this new social awareness encourages the beginning of a new lifestyle and, therefore, the origin of a new type of consumer "green consumer" (Young et al., 2010), a new approach to consumerism "green consumerism," and product "green product." There is no single definition of these concepts. In this sense, this new green consumer demands products or services from companies that develop a "sustainable production," and additionally, shows great concern about the effects on the environment and people of the products or services they consume (Barber, 2004; Llopis, 2009).

With regard to green consumerism, according to Elkington and Hailes (1989, p 235), it is one who avoids "products that endanger the health of the consumer or another person; cause significant damage to the environment during their manufacture, use or waste; consume a disproportionate amount of energy; cause unnecessary waste; use materials derived from endangered species or environments; as well as those that involve unnecessary animal abuse or that adversely affect other countries." This concept basically refers to those consumers who have environmental concerns when acquiring their products or services (Dueñas-Ocampo et al., 2014). According to Orozco (2003, p 2), "ecological consumers are those who are willing to change many of their behavior patterns for more environmental-friendly ones." The green product is defined by Pickett-Baker and Ozaki (2008) as products available for purchase and supplied by companies with a reputation for reducing the environmental impact on their manufacturing processes.

In its early stages, this type of consumer was denominated as "green consumer," a concept that has evolved to "ethical consumer" as it incorporates new dimensions such as the ethical and moral aspect (Mintel Research, 1994) and the union of both, together with social aspects, which have an influence on consumers' behavior and decision-making (Dueñas-Ocampo et al., 2014), currently giving rise to the concept of socially responsible consumption and a consumer called "ecologically and socially responsible consumer." This research work focuses solely on the green consumerism approach.

Green consumerism arises from the conjunction and relationship of three factors throughout history (Portilho, 2005), the emergence of environmentalism, the introduction of environmental responsibility in the business sector, and the introduction of concern about the environmental crisis in consumption and lifestyles of elitist social classes. Since the 1970s, the first steps were taken by international organizations, governments, public institutions, private companies, and society, in general, to build this thinking.

Among the various actions and activities that have been developed to introduce this type of consumption the following stand out: the Rio

Declaration, the development of Agenda 21, and the World Summit on sustainable development; in each of these events the main economic powers put on the table the problems that affect environmental balance in order to find mechanisms to solve them, as well as the generation of new and better development models that contribute to reducing environmental degradation. On the other hand, in addition to the actions implemented by governments, institutions, for example, it is important to bear in mind that the role of the consumer is vital for sustainable development because they set new market trends and seek for companies and agencies to make better decisions (Moisander et al., 2010, p 74).

However, the report Greendex 2014: Consumer Choice and the Environment—A Worldwide Tracking Survey (Greendex, 2014) measures consumer behavior in 65 areas relating to housing, transportation, food, and consumer goods and ranks average consumers in 18 countries according to the environmental impact of their consumption patterns and is the only survey of its kind showing that sustainable consumption remains a challenge as progress has been insufficient. "Top-scoring consumers of the 2014 Greendex study are in the developing economies of India and China, in descending order, followed by consumers in South Korea, Brazil, and Argentina. Indian and Chinese consumers also scored highest in 2012. Results show that American consumers' behavior still ranks as the least sustainable of all countries surveyed since the inception of the Greendex study in 2008" (Greendex, 2014, p 5). On the other hand, the report by Greendex (2010, p 2) revealed that a lack of trust in companies' environmental claims, combined with the absence of demonstrated leadership from both companies and governments were major barriers that needed addressing.

Once the topic under study is contextualized, it is important to observe scientific research related to this subject as a starting point for this study. The study of the consumer has become a key point in the literature. Authors such as Jain and Kaur (2006), Bergin-Seers and Mair (2009), Finisterra Do Paço et al. (2010), Mazar and Zhong (2010) have developed some exploratory studies to segment the green consumer. Among the main references they take to develop green consumer segmentation is the delimitation of sociodemographic variables as well as environmental awareness. The analysis of sociodemographic variables as a defining element of human awareness is a very popular research topic among the authors (Jain and Kaur, 2006, p 133; Finisterra Do Paço et al., 2009, p 19). The variables that are most frequently used for segmentation are: gender, age, variables related to personality, education, type of educational centers, monthly family income, and occupation, while among the environmental awareness variables there is a long list, which are used in the studies as dependent variables on sociodemographic variables.

There is no single segmentation universally accepted by researchers on green consumers. In this regard, Bigné (1997) in his work included an extensive review of the different segments of green consumers existing in the literature until that moment. In this sense, due to its relevance, the study carried out by The Roper Organization in 1990 for the consumer goods company SC Johnson and Son, Inc., in the United States, divides consumers, from major to minor environmental concern, into five categories: "true blue green" with very strong environmental concern and leaders of environmental movements; "greenback green," who are willing to pay higher prices for organic products; "Sprouts," who have a moderate concern for the environment, which is reflected moderately in their behavior; "Greens," who justify their lack of environmental behavior and criticize the poor performance of others; "Basic brown," who do not believe that individual behavior patterns can solve environmental problems and, furthermore, they do not want to make the effort.

In a more recent study developed in Portugal by Finisterra do Paco and Barata Raposo (2010), it segments this type of consumers into three groups: the non-committed, the green, and the undefined. The first group is composed of young people who have a high training level and who have very negative positions on environmental awareness, but who comment on having a clear understanding of what environmentalism is all about, are willing to pay a greater amount of money for those products that actually reduce damage to the environment and take into account the advertising campaigns that are made in relation to this type of products.

The second group is composed of young adults and older people, who, unlike the previous group, have a good academic background and middle-level jobs, but who have a preference for consuming organic products, whose origin they can verify, being very difficult for marketing campaigns regarding green products to influence their purchasing decision. In the last group, we find very elderly people characterized by a low education and jobs of better rank, that although they develop a series of simple activities like recycling or classification of waste, they do not consider that their small acts contribute to improving the environment and make it clear that green product campaigns have little or no influence on their purchasing processes.

Another issue that has received great attention is the analysis of the causes that motivate consumers to develop green consumption (Han and Yoon, 2015; Sachdeva et al., 2015; Chekima et al., 2016). In this sense, the motivations that stand out are environmental, use of high technology and marketing

(Jagodič et al., 2016, p 141). According to Vining and Ebreo (1992), the most powerful motivation in green consumption is environmental. The second motivational factor is the use of high technology in the sense that consumers value the introduction of high-quality technologies since they consider that they contribute to a better standard of living (Jagodič et al., 2016, p 141). Finally, the third motivating factor is the influence of marketing or promotional techniques that are used to promote green products to consumers. This motivating element has not been studied much, but it is established that it has greater influence on the consumer depending on its age (Chowdhury and Samuel, 2014, p 562).

In this context, characterized by greater environmental awareness at both consumer and company level, the objective of this research work is to analyze the current interest by the scientific community regarding the research topic "green consumerism" seeking to determine the evolution in the generation of knowledge and delimit the areas of knowledge from which scientific production has been addressed. To fulfill this objective, we examine the scientific production developed to the present day in relation to this thematic area through a bibliometric analysis. In this sense, "bibliometrics is a subdiscipline of scientometrics and provides information on the results of the research process, volume, evolution, visibility, and structure. In this way, scientific activity and the impact of both research and sources can be assessed" (Escorcia-Otálora and Poutou-Piñales, 2008, p 237). The bibliographic material to be analyzed is obtained by searching two databases of great international relevance, Scopus and Web of Science (WoS), which base their positioning on relative quality indices (RCI), which have been constructed to measure content quality and is shown on their portals.

This chapter is organized into four sections. After the introduction in which the topic under study is contextualized and the objective is presented, the second section details the methodology used. The third section describes the analysis of the data, and the last section presents the conclusions and limitations of the research.

3.2 METHODOLOGY

The tool used to achieve the stated objective is the bibliometric analysis of exploratory-quantitative nature that is applied to the scientific literature identified with the prescriber "green consumerism" in the international databases Scopus and WoS. This analysis is based on the application of a set of bibliometric indicators (Spinak, 1996, p 35) that allow to analyze the production by years, authors, institution, co-authorships, thematic areas and to identify the journals used for diffusion together with their dispersion analysis. According to Escorcia-Otálora and Poutou-Piñales (2008, p 237), "bibliometric indicators can be classified into two large groups, activity indicators and impact indicators. Activity indicators show the real state of science and within these are the number and distribution of publications, productivity, dispersion of publications, collaboration in publications, average lifespan or aging of citations, connections between authors, among others. Among the impact indicators are the evaluation of highly cited documents "hot papers" and the impact factor (IF); being the latter the best known (Camps, 2008)."

The first step to begin with this type of studies is the choice of databases, whereby the tracking technique is used to identify the bibliographic material. Among the elements to be taken into account for their selection are: (a) the management of bibliographic references of documents; (b) the coverage of citations available; (c) quality levels and impact of documents and resources; and (d) the international relevance that the data of these have at international level. The number of bases that exists is very broad, but the ability of these to comply with the previously mentioned parameters considerably reduce the list; the bases that comply with these parameters 100% are Scopus of the Elsevier publisher and WoS of Thomson Reuters.

In the case of Scopus, one of the main advantages that make it one of the best options is greater coverage of multidisciplinary content than other databases by including in its catalog more than 53 million references that are published in 21,000 scientific journals (Elsevier, 2017). The Web of Science core collection is: Science Citation Index Expanded (SCI-EXPANDED), Social Sciences Citation Index (SSCI), Arts and Humanities Citation Index (A and HCI) and Emerginf Sources Citation Index (ESCI). In addition, it allows the user to carry out downloadings of a maximum capacity of 2000 document references, authors' data, affiliations, DOI, journal data and more data that are relevant for this type of studies (Costas Comesaña, 2008).

As for WoS, it has greater years of coverage than Scopus, managing to cover articles beyond 1996 (Costas Comesaña, 2008, p 65). It also has an adequate standardization and organization of authors' data, institutions and location of resources and only has indexed journals that have a high impact and exceptional quality (Fernández et al., 1999). Another element that makes them a reference for research is the inclusion of RCI, which in the case of Scopus is known as Scimago Journal Rank (SJR) and for WoS as journal citation reports (JCR).

Once the databases have been selected, the second step is to track documents. In this research, the search is done using the term "Green Consumerism" in the fields "Article Title, Abstract, Keywords" in the case of Scopus and WoS in "Topic and Title" and the articles published in journals are selected, discarding other documents such as: review, conference paper, book chapter, and book. Next, the database built for each database was cleaned up, to finally work with a single data matrix built in Microsoft-Office Excel software containing the 96 articles identified within the two databases.

3.3 RESULTS

3.3.1 DOCUMENTS

Taking into account that we consider the "scientific article" as a unit of analysis, 63 articles have been identified in the Scopus database and 33 in WoS. Once the database, which was object of analysis was cleaned up and the duplications in the two databases were eliminated, a total of 76 articles were created. The period in which we find publications goes from 1990 to 2016, with a very slow growth in Scopus until 2008 and in WoS until 2009. In the Scopus database, there are two upturns in 2009 and 2015 with seven published articles and in WoS in 2010 with six articles and in 2015 with eight (Fig. 3.1). Most of the publications are concentrated in the last 10 years; in the case of Scopus they represent 47.61% of the total publications and in WoS 72.72% (Fig. 3.1).

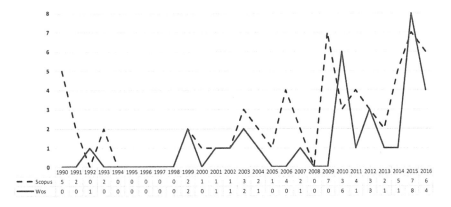

FIGURE 3.1   Evolution of publications in Scopus and Web of Science WoS.

Source: Own elaboration.

If we analyze the publications jointly, we observed two upturns in 2009 with seven articles and in 2015 with nine published. In the last 11 years, 60.52% of the total production is concentrated, which shows an emerging interest of the scientific community in the last decade in this subject. The polynomial trend curve shows a good fit with $R^2 = 0.606$ and shows a significant growth trend over the next 5 years (Fig. 3.2).

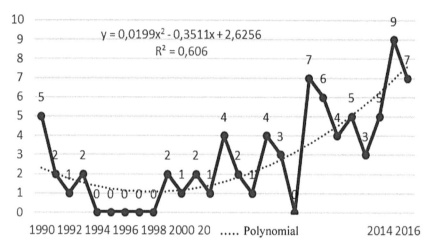

FIGURE 3.2   Trend of publications (Scopus and WoS).

Source: Own elaboration.

3.3.2 AUTHORIAL PRODUCTIVITY AND CO-AUTHORSHIP

A total of 122 authors were identified in Scopus and 74 in WoS, so the productivity index (authors' productivity) is 1.11 articles per author in Scopus and 1.03 in WoS. Table 3.1 shows the top most productive authors. If we analyze both databases together, a total of 143 authors and a productivity index of 1.13 articles per author are identified.

The Lotka Productivity Index (Lotka, 1926) was also used to obtain the decimal logarithm of each author's publications ($IP = \log N$, IP is the personal productivity indicator and N is the number of articles). This value allows the authors to be grouped into three identified productivity levels: (a) small producers characterized by having a single publication and an index equal to zero; (b) medium producers with between two and nine publications and an index greater than zero but less than one; and (c) large producers that have 10 papers and an index equal to or greater than one. Among all the authors shown in Table 3.1 there is an absolute predominance of medium producers, a group composed of 11 authors; the remaining 132 authors are grouped as small producers.

TABLE 3.1 Most Productive Authors.

	Scopus			WoS			
Author	**No. of articles**	**H Index**	**Lotka index**	**Author**	**No. of articles**	**H Index**	**Lotka index**
Agrawal, R.	3	2	0.4771	Mazar, N.	2	2	0.3010
Eden, S. E.	3	1	0.4771	Ozanne, L. K.	2	2	0.3010
Gautam, A.	3	2	0.4771				
Nath, V.	3	2	0.4771				
Sharma, V.	3	2	0.4771				
Adams, R.	2	2	0.3010				
Autio, M.	2	1	0.3010				
Kumar, R.	2	2	0.3010				
Ozanne, L. K.	2	2	0.3010				

Source: Own elaboration.

Among the most productive authors are Agrawal, R., Eden, S. E., Gautam, A., Nath, V. and Sharma, V. with three publications each within Scopus; whereas in WoS only Mazar, N. and Ozanne, L. K. are identified with two publications each. The "h-index" created by Hirsch[55] is also collected in Table 3.1 for each of the authors, which numerically shapes the relevance of each author's production in this thematic area and is obtained by dividing the number of articles produced by the number of citations received so far. This calculation results in a value h=equal to or greater than zero.

Next, an analysis of the authors was developed, establishing 136 authors in Scopus and 76 in WoS, among which 50 belong to duplicate articles, leaving 162 authors in general within the 76 articles of study. Continuing with the analysis, we identified 27 articles with a single author, 30 articles with two authors, eight articles with three authors and 11 articles with four or more authors. In this way, when applying the co-authorship index, we obtain that on average 2.13 authors work per publication, a value that supports the 39% of articles subscribed by two authors. For the calculation of the collaboration index, number of authors, we used the formula $CI = \sum_{i=1}^{n} ji\ ni\ /\ N$ where, N is the total number of documents, j_1 is the total number of documents with multiple authors, n_1 is the number of documents with j authors.

3.3.3 *PRODUCTIVITY BY TYPE OF INSTITUTION AND COUNTRY*

81 affiliations were identified in Scopus with the majority (78%) being university centers. Regarding WoS, the number of affiliations identified was 49 (86% university centers). The top affiliations were the University of Utara Malaysia (Scopus) and University of Malaysia Sabah (WoS), both located in the south-east Asian country, Malaysia.

TABLE 3.2 Most Productive Institutions.

Scopus			WoS		
Institutions	Authors	Author-ship	Institutions	Authors	Author-ship
University of Utara Malaysia	5	5	University of Malaysia Sabah	4	4
University of Malaysia Sabah	4	4	University of Manchester	4	4
University of Manchester	4	3	University of Manitoba	3	3
Pennsylvania State University	4	4	University of Helsinki	3	3
Salem State University	4	4	Lincoln University	3	4
Lincoln University	3	4	North Carolina State University	3	3
Royal Holloway University of London	3	3			
North Carolina State University	3	3			

Source: Own elaboration.

On the other hand, in relation to the affiliation per country to which the authors belong to, Table 3.3 shows that Anglo-Saxon countries have the highest number of affiliations of authors, authorships, and research centers. Regarding the affiliation by country, it is observed that both for Scopus (31 authors, 31 authorships, and 21 research centers) and Wos (17 authors, 17 authorships, and 12 research centers) the United States leads the ranking of affiliations; but if one compares the affiliations of the most productive authors, India is the country with the highest number of affiliations. It has four out of the five affiliations of the most prolific authors in Scopus.

TABLE 3.3 Number of Centers, Authorship, and Authors by Their Country of Affiliation.

	Scopus				WoS		
Country	Authors	Author-ship	No. Centers	Country	Authors	Author-ship	No. Centers
United States	31	31	21	United States	17	17	12
United Kingdom	17	20	10	Canada	9	10	4
Malaysia	12	12	5	United Kingdom	8	8	4
India	10	19	6	Finland	6	6	3
Canada	5	5	3	Malaysia	4	4	1
New Zealand	5	6	2	New Zealand	3	4	1
Slovenia	5	5	3	Sweden	3	3	2
Sweden	5	5	4	Australia	2	2	2
Australia	4	4	3	Austria	2	2	1
Denmark	4	4	3	China	2	2	2
Finland	4	5	3	Germany	2	2	2
Taiwan	4	4	4	Portugal	2	2	1
China	2	2	2	South Korea	2	2	2
Netherlands	2	2	1	Taiwan	2	2	2
Portugal	2	2	1	Colombia	1	1	1
South Korea	2	2	2	Denmark	1	1	1
Colombia	1	1	1	Hong Kong	1	1	1
Germany	1	1	1	India	1	1	1
Hong Kong	1	1	1	Ireland	1	1	1
Ireland	1	1	1	Japan	1	1	1
Japan	1	1	1	Netherlands	1	1	1
Singapore	1	1	1	Singapore	1	1	1
United Arab Emirates	1	1	1	Slovenia	1	1	1
País no definido	1	1	1	Switzerland	1	1	1
Total	122	136	81		74	76	49

Source: Own elaboration.

3.3.4 JOURNALS

Another of the analysis that is developed to examine the resources through which the articles of this thematic area are being published, in such a way that it was identified that the 63 articles of Scopus were published in

50 journals, and in the case of WoS the 33 articles were published in 24 journals. Among all the identified resources, 84% (42) of Scopus and 83% (20) of WoS have published only one article, while only 2% (1) of Scopus and 4% (1) of WoS include journals containing four or more publications on the topic.

We also analyzed the areas and categories in which the journals are classified within each of the databases. Both Scopus and WoS have a wide variety of contents, so the structure for the classification of resources varies depending on the branching they give to each of their general areas, but despite this the two bases have strong similarities. As for the areas, it is possible to observe that within both bases business, management, and accounting (Scopus) lead 42% of the journals and business and economics (WoS) with 36% as can be seen in Table 3.4 followed by environmental science (Scopus) with 28 and 27% in environmental sciences and ecology (WoS). Other areas that stand out in common are psychology, energy, engineering, social sciences among others (Table 3.4).

TABLE 3.4   Number of Journals and Articles by Area of Knowledge.

Scopus				WoS			
Area of knowledge	No. of journals	No. of articles	%	Area of knowledge	No. of journals	No. of Articles	%
Business, Management, and Accounting	21	24	42.0	Business and economics	16	8	36.4
Environmental science	14	20	28.0	Environ. sciences and ecology	8	6	27.3
Economics, econometrics and finance	5	5	10.0	Without area	2	2	9.1
Social sciences	4	4	8.0	Science and technology–OT	2	2	9.1
Agricultural and biological Sc.	2	4	4.0	Engineering	2	1	4.5
Psychology	2	4	4.0	Psychology	1	1	4.5
Energy	1	1	2.0	Social sciences–OT	1	1	4.5
Multidisciplinary	1	1	2.0	Sociology	1	1	4.5
Totals	50	63	100%		22	33	100%

Source: Own elaboration.

On the other hand, with regard to the categories, they are totally different between one base and another. In the case of Scopus, 14% of journals are classified within marketing and 12% in management, monitoring, policy and law; but with regard to WoS, 23% of the resources are classified in Economics and 18% in environmental studies.

The journals with the highest number of articles published on this topic are business strategy and the environment (four articles) in Scopus and International Journal of Consumer Studies (six articles) in WoS.

Another element that was taken into account was the internationally recognized RQI (SJR in the Scopus database and JCR InCites, in the case of WoS). By means of integrating some of their variables, they calculate the quality and impact of each of the journals (Table 3.5).

In this way, it is observed that in the case of Scopus, the journal that leads the ranking is located in the first quartile with an SJR of 1.87 and its country of edition is the United Kingdom; while for WoS the leading journal is in the third quartile with a JCR of 1.08 and its country of publication is the United States. As for resources, in general, 46% of the journals in Scopus are in the Q1 quartile, while in WoS only 36% are within this same quartile. Only 9.1% (two journals) in the WoS database do not have this type of quality index calculated (Table 3.5).

Additionally, Bradford's law analysis was performed, which represents numerically, the concentration phenomenon of a great number of articles in a reduced number of journals (Bradford, 1934). After calculating the Lorenz curve, it can be seen that 52% of the journals have published 78% of the articles in the case of Scopus; while in WoS, 98% of resources have been published in 98% of the journals. The dispersion indicator of the articles was applied to these data, which establishes the number of articles that on average are published per journal, whose result is 1.2 articles per journal for Scopus and 1.5 for WoS. In this way, and supported by the application of three different statistical processes, it can be determined that there is no nucleus of concentration of articles.

3.3.5 KEYWORDS

For the development of all types of research, we establish reference terms to carry out advanced searches and, thereby, locate the articles of greater reference on the research topic. This research was not the exception and after locating the articles for which the different analyses were carried out, the terms used by these publications to index the information they contain were analyzed. The most commonly used keywords within the 76 articles are green consumerism 22, followed by green marketing 10 and consumer

TABLE 3.5 Most Productive Journals.

	Scopus			WoS			
Journals	No. of articles.	%	Quartile	Journal	No. of articles	%	Quartile
Business Strategy and the Environment	4	6.35	1	International Journal of Consumer Studies	6	18.18	3
Working Paper–University of Leeds, School of Geography	3	4.76	4	Business Strategy and the Environment	3	9.09	1
British Food Journal	3	4.76	3				
Journal of Consumer Behaviour	3	4.76	2	Journal of Consumer Behaviour	2	6.06	3
International Journal of Innovation and Sustainable Development	2	3.17	3	Journal of Cleaner Production	2	6.06	1
Journal of Promotion Management	2	3.17	3				
International Journal of Production Economics	2	3.17	1				
Journal of Cleaner Production	2	3.17	1				

Source: Own elaboration.

behavior six, repetitions respectively. It should be noted that 22 of the articles identified did not have any type of keywords within their content nor in the databases.

FIGURE 3.3   Keywords.

Source: Own elaboration.

3.3.6   OVERLAP DATABASES

In order to quantify the intensity of the linear relationship between two variables, in this case, two databases, the parameter that gives us such quantification is the linear correlation coefficient of Pearson's r, whose value ranges between -1 and $+1$. In those cases, where the linear regression coefficient is "close" to $+1$ or -1, it makes sense to consider the equation of a line that "best fits" the point cloud (least squares line). One of the main uses of this line will be to predict or estimate the values of Y that we would obtain for different X values.

Then, a correlation analysis is performed between the databases studied. This process has the purpose of establishing their relationship level or in other words, the intensity of the linear relationship. The parameter that allows to perform the quantification is the linear correlation coefficient of Pearson's r, whose value is between -1 and $+1$. For its interpretation, it is necessary to know that if r <0 there is negative correlation, the two variables are correlated in the opposite direction and a high value of one of them usually corresponds to a low value of the other one and vice versa. The closer the correlation coefficient is to -1, the more evident this extreme covariation will be. If $r > 0$ there is positive correlation, the two variables are correlated in the direct sense. High values of one variable correspond to high values of the other one and the same is for low values. The closer to $+1$ the correlation coefficient is, the more evident is this covariation. And finally, when $r=0$ it is said that the variables are uncorrelated and no sense of covariation can be established.

On the other hand, in the cases where the linear regression coefficient is "close" to $+1$ or -1, it makes sense to consider the equation of a line that "best fits" the point cloud (least squares line). One of the main uses of this line will be to predict or estimate the values of Y that we would obtain for different values of X.

A linear correlation coefficient $=0.51$ was obtained, which establishes a positive but moderate correlation (values between 0 and 0.5 ranging from weak to moderate) between Scopus and WoS, as can be seen in Figure 3.4, where a weak straight-line fit is observed with $R^2=0.2603$.

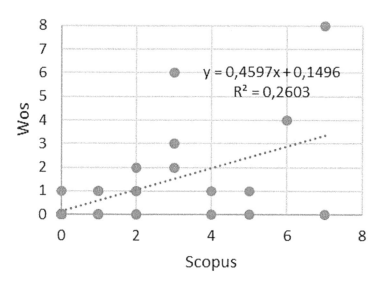

FIGURE 3.4   Point cloud and least squares line.

Source: Own elaboration.

According to Pulgarín Guerrero and Escalona Fernández (2008, p 338), the use of two databases in the same bibliometric study "can generate an overlapping effect between them due to the possibility of superimposing or interweaving the articles that have been collected from each one of them." In order to reduce the effect that can be generated by the use of several bases, the Meyer's index (MI) and the traditional and relative overlapping percentage are applied.

1. MI establishes the number of articles that are repeated in the bases being studied.

$$Meyer's\ index = \frac{\sum Sources*Weight}{Total\ sources}$$

In order to perform the calculations, the articles are first classified by the number of times they are repeated and given a specific weight according to the bases that have been used (Meyer et al., 1983). Single or non-repeated articles have a weight=1, duplicates weight=0.5, triplicates weight=0.3, showing that the more repetitions the articles have, the weight will be lower (Costas et al., 2008, p 332).

The result obtained determines the singularity of each database; establishing that the higher this value the greater the ability of the databases to have single articles indexed among their data (Meyer et al., 1983, p 36).

2. Traditional overlapping (TO) and relative overlapping.

$$\%\,Overlapping\ in\ A = 100*\left(\frac{|A \cap B|}{|A|}\right), \%\,Overlapping\ in\ B = 100*\left(\frac{|A \cap B|}{|B|}\right)$$

The first one consists of an algorithm that integrates all the possible unions that can be developed between the selected databases, resulting in the equality percentage of base X on base Y (Abad et al., 1995, Gluck, 1990). In contrast, the second percentage, relative coverage, takes into account the weight of the overlapping or repeated articles with regard to the single ones, with which the percentage of overlapping that base X has on the base Y can be established (Bearman and Kunberger, 1977).

The results show that out of the 96 articles (63 Scopus and 33 Wos) 56 are single articles in one base or another and that 20 of them are repeated in the two bases. With these data, the MI was applied to each base, which determined that Scopus has a greater singularity than WoS, with MI=0.84 versus 0.70 for WoS.

Continuing with the traditional overlapping percentage (% TO), it was determined that there is a 26.3% equality between Scopus and WoS regarding the articles; while for resources this is 29.8%. Finally, the percentage of relative coverage (% C) is determined, with 34% of Scopus being overlapped by WoS, while 71% of WoS is overlapped by Scopus.

3.3.7   RESEARCH APPROACHES

We proceeded to develop a brief bibliographical analysis and review of the literature with the purpose of identifying and grouping the approaches followed up to the moment. Finally, three clearly defined approaches were identified (Table 3.6).

3.4   CONCLUSION

After the detailed review of the scientific production that is related to green consumerism, it can be established that there is a short path within this thematic area regarding the number of articles (76 articles in 26 years), so the bases have not yet been laid and, therefore, there is a great ignorance of what this issue actually covers, creating stereotypes about this type of consumption and a wide variety of terms to refer to it.

As for the analysis of the results of the bibliography collected, consulted, and cataloged, as per the reference points for the development of future studies, the following conclusions are drawn:

- When analyzing the data base, it is observed that the first article of the study was collected in 1990 which means that it has a research track of only 26 years, being a very recent subject. The article, as documentary typology, was selected for this study due to the relevance that this one has for the scientific community by allowing faster transfer of knowledge among members of the scientific community.
- Based on the development that this topic has, its youth is observed as a factor of influence, leading to highlight the presence of only one article with 168 citations, titled "Do green products make us better people?"
- As for the authors, only one coincidence of the most prolific authors between the bases was found, being part of the low production that each author has, since 90% of the authors only have one article related to the topic.
- In addition, a consolidated working group is identified, which is made up of authors of Indian affiliation, who present three articles

TABLE 3.6 Approaches Followed by Research.

Approaches	Description of the research approach	Authors
		Articles of Scopus and WoS
General survey	It provides an introduction to what addresses this topic, through the conceptualization of the topic and development of the definitions of the main terms	Autio et al., 2009; Koh and Lee, 2012; Akenji, 2014; Zhu and Sarkis, 2016
Motivations and variables of analysis	They establish the types of variables that mark or influence within the development of environmental awareness and green consumption in the population	Cunningham,1990; Chan, 2002; Buttel, 2003; Haanpaa, 2007; Bergin-Seers and Mair, 2009; Isenhour and Ardenfors, 2009; Carlsson et al., 2010; Bartels and Hoogendam, 2011; Chatzidakis et al., 2012; Chang and Hung, 2013; Hudson et al., 2013; Chowdhury and Samuel, 2014; Han and Yoon, 2015; Lin and Hsu, 2015; Nath et al., 2015; Sachdeva et al. 2015; Chekima et al., 2016; Jagodič et al., 2016; Mohr and Schlich, 2016; Schuitema and De Groot, 2015
	They describe different types of motivations that influence people to develop green purchases. This is analyzed within different segments of the world population and types of trade	Hynes, 1991; Springett and Kearins, 2001; Wong, 2001; Antweiler and Harrison, 2003; Elias and Carney, 2004;Autio and Wilska, 2005; Jain and Kaur, 2006; Alsmadi, 2007; Finisterra Do Paço et al., 2009; Finisterra do Paco and Barata Raposo, 2010; Moisander et al., 2010; Coleman et al., 2011; Hassan, 2014
Factors driving green consumerism	Different studies that include a series of actions driven by key factors such as public, private, or NGO bodies to develop an environmental awareness culture within the population. In addition, specific actions are mentioned that promote green consumption by these actors	Adams, 1990; Eden,1990a; Eden,1990b; Eden, 1990c; Jones,1991; Pettit and Sheppard, 1992; Adams,1993; Grunert,1993; Lockie, 1999; Ozanne et al.,1999; Hardner and Rice, 2002; Buttel, 2003; Seyfang, 2003; Eriksson, 2004; Harrison, 2006; Pedersen and Neergaard, 2006; Muldoon, 2007; Sloan, 2007; Bieak Kreidler and Joseph-mathews, 2009; Klintman, 2009; Gupta and Ogden, 2009; Huttunen and Autio, 2010; Mazar and Zhong, 2010; Sandhu et al., 2010; Diaz-Rainey and Ashton, 2011; Doran, 2011; Noor et al., 2012; Willis and Schor, 2012; Thøgersen and Noblet, 2012; Nath et al. 2013; Nath et al., 2014; Chander and Muthukrishnan, 2015; Dobernig and Stagl, 2015; Geels et al., 2015; Gu et al., 2015; Chen et al., 2016; Muralidharan and Xue, 2016; Tseng, 2016

Source: Own elaboration.

jointly on the topic. They are Agrawal, R., Gautam, A., Nath, V., and Sharma, V.

• Regarding country affiliations, the leadership in both bases of authors, authorships, and research centers is of Anglo–Saxon origin (United States, United Kingdom, and Canada), but it stands out that they contain the great majority of contributing authors with only one article to the developed production. Additionally, the study confirms that the majority of authors within both bases are affiliated to universities.

• When analyzing two databases, it can be supported much more strongly that within this thematic area the Bradford Law is not fulfilled, supported firstly by not having a group of articles that are listed as references in terms of their number of citations and, on the other hand, because the journals that the group identified articles have only published on average 1.2 articles for Scopus and 1.5 for WoS, supporting that there is no publication group that is a reference within green consumption.

• As for the overlap study between both bases, it is shown that most of the scientific production is published in Scopus, establishing that 71% of the articles found in WoS are also published in Scopus.

The originality of the study focuses on being the first bibliometric and overlap study on green consumerism. In addition, this research provides a series of data of great relevance for the authors who are starting in this topic, shaping the current state of this thematic area and detailing the main approach to which the studies developed in relation to green consumerism.

However, like any study it has limitations throughout its development. It could be stated that the main limitation was that it did not include other search terms in its analysis, which allow collecting more information on this thematic area. This term was chosen as it was considered that it would be the most used term, as it was used to baptize this new trend on consumption and respect for the environment. A second limitation refers to the use of other bibliometric indicators that have not been used, such as the price index, institutional index, transient index, obsolescence index or Burton-Kebler index, insularity index, self-citation index. These limitations lead to future research, including new search terms, which are much more specific and refer to more specific topics such as green consumption, green consumer behavior, and so forth, and to expand the number of bibliometric indicators to deepen the state of the art on this topic, being this study the first approach to the subject.

KEYWORDS

- **green consumerism**
- **green consumer**
- **green product**
- **bibliometric study**
- **database Scopus**
- **database Web of Science (WoS)**

REFERENCES

Abad, M. F.; Benavent, R.; Bonet, R. Estrategias de búsqueda de artículos de revistas españolas. Estudio de un caso: evaluación de la calidad de los sistemas de información. *Gaceta Sanitaria* **1995,** *9*(51), 363–370.

Adams, R. Green Consumerism and the Food Industry: Early Signs of Big Changes to Come. *Br. Food J.* **1990,** *92*(9), 11–14.

Adams, R. Green Consumerism and the Food Industry: Further Developments. *Br. Food J.* **1993,** *95*(4), 9–11.

Akenji, L. Consumer Scapegoatism and Limits to Green Consumerism. *J. Cleaner Prod.* **2014,** *63*, 13–23.

Alsmadi, S. Green Marketing and the Concern over the Environment: Measuring Environmental Consciousness of Jordanian Consumers. *J. Promot. Manage.* **2007,** *13*(3–4), 339–361.

Antweiler, W.; Harrison, K. Toxic Release Inventories and Green Consumerism: Empirical Evidence from Canada. *Can. J. Econ.* **2003,** *36*(2), 495–520.

Autio, M.; Wilska, T. A. Young People in Knowledge Society: Possibilities to Fulfil Ecological Goals. *Progr. Ind. Ecol. Int. J.* **2005,** *2*(3–4), 403–426.

Autio, M.; Heiskanen, E.; Heinonen, V. Narratives of 'green' consumers—The Antihero, the Environmental Hero and the Anarchist. *J. Consum. Behav.* **2009,** *8*(1), 40–53.

Barber, A. Seven Case Study Farms: Total Energy and Carbon Indicators for New Zealand Arable and Outdoor Vegetable Production. AgriLINK New Zealand Ltd: New Zealand, 2004; p 288.

Bartels, J.; Hoogendam, K. The Role of Social Identity and Attitudes Toward Sustainability Brands in Buying Behaviors for Organic Products. *J. Brand Manage.* **2011,** *18*(9), 697–708.

Bearman, T. C.; Kunberger, W. A. A Study of Coverage Overlap Among Fourteen Major Science and Technology Abstracting and Indexing Services. National Federation of Abstracting and Indexing Services: Philadelphia, 1977.

Bergin-Seers, S.; Mair, J. Emerging Green Tourists in Australia: Their Behaviours and Attitudes. *Tourism Hospitality Res.* **2009,** *9*(2), 109–119.

Bieak Kreidler, N.; Joseph-Mathews, S. How Green Should You Go? Understanding the Role of Green Atmospherics in Service Environment Evaluations. *Int. J. Cul. Tourism Hospitality Res.* **2009,** *3*(3), 228–245.

Bigné, J. E. El Consumidor Verde: Bases de un modelo de comportamiento. Esic-Market **1997**, (96), 29–43.

Bradford, S. C. Sources of Information on Specific Subjects. *Engineering* **1934**, *137*, 85–86.

Buttel, F. H. Environmental Sociology and the Explanation of Environmental Reform. *Organ. Environ.* **2003**, *16*(3), 306–344.

Camps, D. Limitaciones de los indicadores bibliométricos en la evaluación de la actividad científica biomédica. *Colombia Médica* **2008**, *39*(1), 74–79.

Carlsson, F.; García, J. H.; Löfgren, Å. Conformity and the Demand for Environmental Goods. *Environ. Resour. Econ.* **2010**, *47*(3), 407–421.

Chan, K. Environmental Consideration in Purchase Decisions of Hong Kong Consumers. *Environ. Pract.* **2000**, *2*(1), 15–22.

Chander, P.; Muthukrishnan, S. Green Consumerism and Pollution Control. *J. Econ. Behav. Organ.* **2015**, *114*, 27–35.

Chang, A.; Hung, H.-F. A Consumer-Cognition-Based Measurement Model of Corporate Green Brand Image. *NTU Manage. Rev.* **2013**, *24*(1), 129–154.

Chatzidakis, A.; Maclaran, P.; Bradshaw, A. Heterotopian Space and the Utopics of Ethical and Green Consumption. *J. Mark. Manage.* **2012**, *28*(3–4), 494–515.

Chekima, B.; Wafa, S.; Igau, O. A.; Chekima, S.; Sondoh, S. L. Examining Green Consumerism Motivational Drivers: Does Premium Price and Demographics Matter to Green Purchasing? *J. Cleaner Prod.* **2016**, *112*, 3436–3450.

Chen, X.; De la Rosa, J.; Peterson, M. N.; Zhong, Y.; Lu, C. Sympathy for the Environment Predicts Green Consumerism but not more Important Environmental Behaviours Related to Domestic Energy Use. *Environ. Conserv.* **2016**, *43*(02), 140–147.

Chowdhury, P.; Samuel, M. S. Artificial Neural Networks: A Tool for Understanding Green Consumer Behavior. *Mark. Intell. Plann.* **2014**, *32*(5), 552–566.

Coleman, L. J.; Bahnan, N.; Kelkar, M.; Curry, N. Walking the Walk: How the Theory of Reasoned Action Explains Adult and Student Intentions to go Green. *J. Appl. Bus. Res.* **2011**, *27*(3), 107.

Costas, R.; Moreno, L.; Bordons, M. Solapamiento y singularidad de MEDLINE, WoS e IME para el análisis de la actividad científica de una región en Ciencias de la Salud. *Revista española de documentación científica* **2008**, *31*(3), 327–343.

Costas Comesaña, R. Análisis bibliométrico de la actividad científica de los investigadores del CSIC en tres áreas, Biología y Biomedicina, Ciencia de los Materiales y Recursos Naturales: una aproximación metodológica a nivel micro (Web of Science). Universidad Carlos III: Madrid, 2008.

Cunningham, S. L. From Blue Angels to Variable Can Rates. *Biocycle* **1990**, *31*(6), 44–45.

Diamantopoulos, A.; Schlegelmilch, B. B.; Sinkovics, R. R.; Bohlen, G. M. Can Socio-Demographics Still Play a Role in Profiling Green Consumers? A Review of the Evidence and an Empirical Investigation. *J. Bus. Res.* **2003**, *56*(6), 465–448.

Diaz-Rainey, I.; Ashton, J. K. Profiling Potential Green Electricity Tariff Adopters: Green Consumerism as an Environmental Policy Tool? *Bus. Strategy Environ.* **2011**, *20*(7), 456–470.

Dobernig, K.; Stagl, S. Growing a Lifestyle Movement? Exploring Identity-Work and Lifestyle Politics in Urban Food Cultivation. *Int. J. Consum. Studies* **2015**, *39*(5), 452–458.

Doran, P. Is there a Role for Contemporary Practices of Askēsis in Supporting a Transition to Sustainable Consumption? *Int. J. Green Econ.* **2011**, *5*(1), 15–40.

Dueñas- Ocampo, S. D.; Perdomo-Ortiz, J.; Castaño, L. E. V. El concepto de consumo socialmente responsable y su medición. Una revisión de la literatura. *Estudios Gerenciales* **2014**, *30*(132), 287–300.

Eden, S. E. Green Consumerism and the Response from Business and Government. Working Paper–University of Leeds. *Sch. Geogr.* **1990a,** *542,* 1–44.

Eden, S. E. Environmental Issues in the Green Consumer Debate: A Contempary Guide. Working Paper—University of Leeds. School of Geography, 1990b.

Eden, S. E. Voluntary Organisations and the Environment. School of Geography, University of Leeds, 1990c.

Elias, M.; Carney, J. The Female Commodity Chain of Shea Butter: Burkinabe Producers, Western Green Consumers and Fair Trade. *Cahiers de Geographie du Quebec* **2004,** *48*(133), 71–88.

Elkington, J.; Hailes, J. *The Green Consumer Guide: From Shampoo to Champagne: High-Street Shopping for a Better Environment.* V. Gollancz: London, **1989.**

Elsevier, B.V. *Scopus: Content Coverage Guide.* Elsevier, August 2017, pp 2–25. https://www.elsevier.com/__data/assets/pdf_file/0007/69451/0597-Scopus-Content-Coverage-Guide-US-LETTER-v4-HI-singles-no-ticks.pdf.

Eriksson, C. Can Green Consumerism Replace Environmental Regulation? A Differentiated-Products Example. *Resour. Energy Econ.* **2004,** *26*(3), 281–293.

Escorcia-Otálora, T. A.; Poutou-Piñales, R. A. Análisis bibliométrico de los artículos originales publicados en la revista Universitas Scientiarum (1987–2007). *Univ. Sci.* **2008,** *13*(3), 236–244.

Fernández, M. T.; Bordons, M.; Sancho, I.; Gómez, I. El sistema de incentivos y recompensas en la ciencia pública española. In *Radiografía de la investigación pública en España;* En Sebastián, J., Muñoz, E. Ed.; Biblioteca Nueva: Madrid, 1999.

Finisterra do Paco, A. M.; Barata Raposo, M. L. Green Consumer Market Segmentation: Empirical Findings from Portugal. *Int. J. Consum. Studies* **2010,** *34*(4), 429–436.

Finisterra Do Paço, A.; Barata Raposo, M.; Filho, W. Identifying the Green Consumer: A Segmentation Study. *J. Targeting Meas. Anal. Mark.* **2009,** *17*(1), 17–25.

Geels, F. W.; McMeekin, A.; Mylan, J.; Southerton, D. A Critical Appraisal of Sustainable Consumption and Production Research: The Reformist, Revolutionary and Reconfiguration Positions. *Global Environ. Change* **2015,** *34,* 1–12.

Gluck, M. A Review of Journal Coverage Overlap with an Extension to the Definition of Overlap. *J. Am. Soc. Inf. Sci.* **1990,** *41*(1), 43–60.

Grunert, S. C. Green Consumerism in Denmark: Some Evidence from the ØKO Foods-Project. *der Markt* **1993,** *32*(3), 140–151.

Gu, W.; Chhajed, D.; Petruzzi, N. C.; Yalabik, B. Quality Design and Environmental Implications of Green Consumerism in Remanufacturing. *Int. J. Product. Econ.* **2015,** *162,* 55–69.

Gupta, S.; Ogden, D. T. To Buy or Not to Buy? A Social Dilemma Perspective on Green Buying. *J. Consum. Mark.* **2009,** *26*(6), 376–391.

Haanpää, L. Consumers' Green Commitment: Indication of a Postmodern Lifestyle? *Int. J. Consum. Stud.* **2007,** *31*(5), 478–486.

Han, H.; Yoon, H. J. Hotel Customers' Environmentally Responsible Behavioral Intention: Impact of Key Constructs on Decision in Green Consumerism. *Int. J. Hospitality Manage.* **2015,** *45,* 22–33.

Hardner, J.; Rice, R. Rethinking Green Consumerism. *Sci. Am.* **2002,** *286*(5), 88–95.

Harrison, B. Shopping to Save: Green Consumerism and the Struggle for Northern Maine. *Cultural Geogr.* **2006,** *13*(3), 395–420.

Hassan, S.. The Role of Islamic Values on Green Purchase Intention. *J. Islamic Mark.* **2014,** *5*(3), 379–395.

Hirsch, J. E. An Index to Quantify and Individuals Scientific Research Output. *Proc. Natl. Acad. Sci.* **2005,** *102,* 16569–16572.

Hudson, M.; Hudson, I.; Edgerton, J. D. Political Consumerism in Context: An Experiment on Status and Information in Ethical Consumption Decisions. *Am. J.* **2013,** *72*(4), 1009–1037.

Huttunen, K.; Autio, M. Consumer Ethoses in Finnish Consumer Life Stories–Agrarianism, Economism and Green Consumerism. *Int. J. Consum. Studies* **2010,** *34*(2), 146–152.

Hynes, H. P. The Race to Save the Planet: Will Women Lose. *Women's Studies Int. Forum* **1991,** *14*(5), 473–478.

Isenhour, C.; Ardenfors, M. Gender and Sustainable Consumption: Policy Implications. *Int. J. Innovation Sustainable Dev.* **2009,** *4*(2–3), 135–149.

Jagodič, G.; Dermol, V.; Breznik, K.; Vaupot, S. R. Factors of Green Purchasing Behaviour. *Int. J. Innovation Learn.* **2016,** *20*(2), 138–153.

Jain, S.; Kaur, G. Role of Socio-Demographics in Segmenting and Profiling Green Consumers: An Exploratory Study of Consumers in India. *J. Int. Consum. Mark.* **2006,** *18*(3), 107–146.

Jones, B. Green Consumerism and the Supermarkets. *Br. Food J.* **1991,** *93*(3), 8–11.

Klintman, M. Participation in Green Consumer Policies: Deliberative Democracy under Wrong Conditions? *J. Consum. Policy* **2009,** *32*(1), 43–57.

Koh, L. P.; Lee, T. M. Sensible Consumerism for Environmental Sustainability. *Biol. Conserv.* **2012,** *151*(1), 3–6.

Lin, H. Y.; Hsu, M. H. Using social cognitive theory to investigate green consumer behavior. *Bus. Strategy Environ.* **2015,** *24*(5), 326–343.

Llopis, R. Consumo responsable y globalización reflexiva: Un estudio referido al comercio justo en España. *Revista Española del Tercer Sector* **2009,** *11,* 145–165.

Lockie, S. Community Movements and Corporate Images: "Landcare" in Australia. *Rural Sociol.* **1999,** *64*(2), 219–233.

Lotka, A. J. The Frequency Distribution of Scientific Productivity. *J. Wash Acad. Sci.* **1926,** *16*(12), 317–323.

Mazar, N.; Zhong, C. B. Do Green Products Make us Better People? *Psychol. Sci.* **2010,** *21*(4), 494–498.

Meyer, D. E.; Mehlman, D. W.; Reeves, E. S.; Origoni, R. B.; Evans, D.; Sellers, D. W. Comparison Study of Overlap Among 21 Scientific Databases in Searching Pesticide Information. *Online Rev.* **1983,** *7*(1), 33–43.

Mintel Research. Shopping Patterns and Store Loyalty. *Retail Intell.* **1994,** *3,* 33–39.

Mohr, M.; Schlich, M. Socio-Demographic Basic Factors of German Customers as Predictors for Sustainable Consumerism Regarding Foodstuffs and Meat Products. *Int. J. Consum. Stud.* **2016,** *40*(2), 158–167.

Moisander, J.; Markkula, A.; Eraranta, K. Construction of Consumer Choice in the Market: Challenges for Environmental Policy. *Int. J. Consum. Stud.* **2010,** *34*(1), 74–79.

Muldoon, A. Where the Green is: Examining the Paradox of Environmentally Conscious Consumption. *Global Environ. Probl. Policies* **2007,** *1,* 26.

Muralidharan, S.; Muralidharan, S.; Xue, F.; Xue, F. Personal Networks as a Precursor to a Green Future: A Study of "green" Consumer Socialization among Young Millennials from India and China. *Young Consum.* **2016,** *17*(3), 226–242.

Nath, V.; Kumar, R.; Agrawal, R.; Gautam, A.; Sharma, V. Consumer Adoption of Green Products: Modeling the Enablers. *Global Bus. Rev.* **2013,** *14*(3), 453–470.

Nath, V.; Kumar, R.; Agrawal, R.; Gautam, A.; Sharma, V. Impediments to Adoption of Green Products: An ISM Analysis. *J. Promot. Manage.* **2014,** *20*(5), 501–520.

Nath, V.; Agrawal, R.; Gautam, A.; Sharma, V. Socio-Demographics as Antecedents of Green Purchase Intentions: A Review of Literature and Testing of Hypothesis on Indian Consumers. *Int. J. Innovation Sustainable Dev.* **2015,** *9*(2), 168–187.

National Geography and Globescan. Greendex: 2010 Consumer Choice and the Environment—A Worldwide Tracking Survey. **2010.** http://images.nationalgeographic. com/wpf/media-live/file/GS_NGS_2010GreendexHighlights-cb1275487974.pdf.

National Geography and Globescan. Greendex: 2014 Consumer Choice and the Environment—A Worldwide Tracking Survey. **2014.** http://images.nationalgeographic. com/wpf/media-content/file/NGS_2014_Greendex_Highlights_FINAL-cb1411689730. pdf.

Noor, N. A. M.; Jamil, C. Z. M.; Mat, N.; Mat, N.; Kasim, A.; Muhammad, A.; Salleh, H. S. The Relationships between Environmental Knowledge, Environmental Attitude and Subjective Norm on Malaysian Consumers Green Purchase Behaviour. *Malays. J. Consum. Family Econ.* **2012,** *15*(1), 1–20

Orozco, A.; et al. *Mercadotecnia ecológica: actitud del consumidor ante los productos ecológicos. Universidad Politécnica de Cataluña. Departamento de proyectos de ingeniería.* España: Barcelona, 2003.

Ozanne, L. K.; Humphrey, C. R.; Smith, P. M. Gender, Environmentalism, and Interest in Forest Certification: Mohai's Paradox Revisited. *Soc. Nat. Resour.* **1999,** *12*(6), 613–622.

Pedersen, E. R.; Neergaard, P. Caveat Emptor—Let the Buyer Beware! Environmental Labelling and the Limitations of 'green' Consumerism. *Bus. Strategy Environ.* **2006,** *15*(1), 15–29.

Pettit, D.; Sheppard, J. P. It's Not Easy Being Green: The Limits of Green Comsumerism in Light of the Logic of Collective Action. *Queen's Q.* **1992,** *99*(2), 328.

Pickett-Baker, J.; Ozaki, R. Pro-Environmental Products: Marketing Influence on Consumer Purchase Decision. *J. Consum. Mark.* **2008,** *25*(5), 281–293.

Portilho, F. Consumo verde, consumo sustentável e a ambientalização dos consumidores. *Cuadernos de EBAPE.BR* **2005,** *3*(3), 1–12.

Pulgarín Guerrero, A.; Escalona Fernández, M. I. Medidas del solapamiento en tres bases de datos con información sobre ingeniería. *Anales de Documentación* **2008,** *10*, 335–344.

Sachdeva, S., Jordan, J., Mazar, N. Green Consumerism: Moral Motivations to a Sustainable Future. *Curr. Opin. Psychol.* **2015,** *6*, 60–65.

Sandhu, S.; Ozanne, L. K.; Smallman, C.; Cullen, R. Consumer Driven Corporate Environmentalism: Fact or Fiction? *Bus. Strategy Environ.* **2010,** *19*(6), 356–366.

Schuitema, G.; Groot, J. I. Green Consumerism: The Influence of Product Attributes and Values on Purchasing Intentions. *J. Consum. Behav.* **2015,** *14*(1), 57–69.

Seyfang, G. From Frankenstein Foods to Veggie Box Schemes: Sustainable Consumption in Cultural Perspective. Centre for Social and Economic Research on the Global Environment, Working Paper EDM 03-13 (CSERGE, Norwich), **2003.**

Sloan, A. E. Consumer Trends-New Shades of Green. *Food Technol.* **2007,** *61*(12), 16.

Spinak, E. Diccionario enciclopédico de bibliometría, cienciometría e informetría. UNESCO CII/II: Caracas, 1996.

Springett, D.; Kearins, K. Gaining Legitimacy? Sustainable Development in Business School Curricula. *Sustainable Dev.* **2001,** *9*(4), 213–221.

The Roper Organization. *The Environment: Public Attitudes and Individual Behaviour,* Roper Organization and SC Johnson and Son: New York, NY, **1990.**

Thøgersen, J.; Noblet, C. Does Green Consumerism Increase the Acceptance of Wind Power? *Energy Policy* **2012,** *51*, 854–862.

Tseng, C. H. The Effect of Price Discounts on Green Consumerism Behavioral Intentions. *J. Consum. Behav.* **2016,** *15*(4), 325–333.

Vining, J.; Ebreo, A. Predicting Recycling Behaviour from Global and Specific Environmental Attitudes and Changes in Recycling Opportunities. *J. Appl. Soc. Psychol.* **1992,** *20*, 1580–1607.

Willis, M. M.; Schor, J. B. Does Changing a Light Bulb Lead to Changing the World? Political Action and the Conscious Consumer. *ANNALS Am. Acad. Political Soc. Sci.* **2012,** *644*(1), 160–190.

Wong, K. K. Taiwan's Environment, Resource Sustainability and Green Consumerism: Perceptions of University Students. *Sustainable Dev.* **2001,** *9*(4), 222–233.

Young, W.; Hwang, K.; McDonald, S.; Oates, C. J. Sustainable Consumption: Green Consumer Behaviour when Purchasing Products. *Sustainable Dev.* **2010,** *18*(1), 20–31.

Zhu, Q.; Sarkis, J. Green Marketing and Consumerism as Social Change in China: Analyzing the Literature. *Int. J. Prod. Econ.* **2016,** *181*, 289–302.

CHAPTER 4

GREEN CONSUMER BEHAVIOR

KULVINDER KAUR BATTH*

Department of Commerce, K. C. College, Affiliated to University of Mumbai, Churchgate, Mumbai, Maharashtra 400020, India, Mob.: 9833060902

Corresponding author. E-mail: kulprofessional@gmail.com

4.1 INTRODUCTION

The changing environment has been a cause for concern. It is a burning issue which is melting many debates and discussions all over the world. Economics has taught us that we have limited means to satisfy our unlimited wants. The reclamation of green areas, natural lakes, ponds, and alteration in the natural flow of water bodies has caused turbulent consequences on the environmental orientation. A large amount of global discussion platforms are being used to address environment-related causes and concerns. Awareness programs are being held to sensitize people about the future environmental impact on life. Large-scale campaigns and integrated initiatives could only bring fruitful results to achieve the desired objectives. Grassroot level efforts will link the chain of a sustainable partnership between all the stakeholders of the society. All the stakeholders will have to work together to fulfill the goals of environmentally sustainable development.

4.1.1 PRO-ENVIRONMENT BEHAVIOR

Pro-environment behavior is the environment-friendly behavior of the consumers with the objective of supporting the environment. Pro-environment behavior is reflected in the change of attitudes, perception, motives, values, beliefs and desires which in turn are delivered in the changing needs of the consumers. It is involved with the internal variables and external variables influencing the green behavior. The dependent variables are the ones which are creating a more direct impact on the changing consumer behavior. The research chapter discusses the objectives and outcomes of the green consumer behavior.

Today's well-informed consumers are aware of the harmful effects of various products, processes, and their packaging. The fast and quick access to the availability of information is leading to a tremendous change in the behavior. The inherent consciousness of the consumers toward their health and overall physical well-being is leading to the construction of new phases in the innovative era of marketing. Consumers play a major role in the environment orientation. Their attitudes, perception and needs lead to decisions which directly or indirectly affect the environment. The choices of the consumer, categorically such as food, packaging, clothing, transportation, holiday trips, hotels, telecommunication plans, billing deliveries, investments and housing, contribute toward the climatic changes happening all over the world. The planet evolves and revolves around the temperament and sentiments of the Green consumer behavior.

Their behavior stands as a path for the marketer's all over the world while framing the product lines, product mixes, packaging, and so on. Therefore, the core lies at the heart of the consumer behavior.

4.2   GREEN CONSUMER BEHAVIOR

Pro-environment consumer behavior, also termed as green consumer behavior, favors the issues and concerns of the environment. It will lead the companies to design and structure their policies which will sail safely with the environment. Green consumer behavior can be defined as a behavior where the consumers understand and acknowledge the issues, causes and concerns associated with the depletion and degradation of the ecosystem, consequently leading to harmful effects on the planet. A large number of human actions have deteriorated the environmental structuring leading to the creation of environmental disparities. Artificial alterations in the natural setup can be laid as a precedent to avoid such discourse in the future.

Green consumer behavior could be inherent in some instances and in some other places; focused efforts could be taken by implementing the machinery to imbibe such values among people. On the one hand, there are environmental laws to punish, fine or make people liable for their anti-environmental actions. On the other hand, a positive approach could also bear fruitful results, such as awarding, appreciating or motivating people for their environmental concerns. This will be a precedent for many others to join the movement and get recognized.

Active participation in the environmental revival process will create consumers, concerned and actively involved in taking initiatives to save mother earth and therefore termed as green consumers. When the consumers will start caring for the environment, the transformation will follow, creating a movement by the people, for the people, and with the people.

The need to touch the right chord in the hearts will transform the consumers into agents of change. Green consumer behavior will be reflected in the buying. The green consumers will indirectly force the companies to change the products, packages and processes to fit within the parameters of green consumerism. Green consumers will have their personality and lifestyles in tune with sustainable development goals. The change in the consumer psychology will get displayed in their decisions in the form of products bought by them. The concerns of the consumers will apparently build a push strategy for the corporate to inculcate the value system in their charter and thus, lay the foundation for ethics in vision and mission statements.

4.2.1   IMPACT OF PSYCHOLOGICAL FACTORS

Previous research has proved that the influences of psychological and social factors are more powerful than the demographic factors on consumer's green behavior. Psychological factors include consumer belief, perception, images, attitudes, values, lifestyle, and so forth. Green consumer behavior beliefs are in turn influenced by the level of knowledge and information about the environmental norms and relevant issues. The consumption habits of the consumers largely influence the environment. Consumption habits are changing with the consumers choosing the food available locally over the food at the national or international markets. This will reduce demand and thus, the amount of transportation involved in the food will decrease. The more distant the food travels, the more will be the usage of preservatives for the survival of the food leading to an increase in the cost factor. A large amount of shipping of food is leading to water pollution and a great degree of harmful effects on the enumerable species under water. Consumers should prefer locally available food rather than vying for the imported food products. Further, maximum efforts should be undertaken so as to consume the food produced locally, so as lessen the burden on import of food supplies, consequently reducing the impact of air pollution.

4.2.2 ENVIRONMENTAL KNOWLEDGE AND ATTITUDES

To understand the green consumer behavior, it is important to understand the manner in which the green products are perceived and evaluated by the consumers. The right way of analyzing the green consumer behavior is to understand the approach of consumers toward green products. The first step toward this approach would be awakening and realization, which is possible by inculcating the required background for pro-environment products.

Awareness and knowledge is the foundation for a pro-environment behavior. The right step toward a positive environment would be the enlightenment initiated among the consumers at the right stage. Awakening could be engendered at a tender age among the children to imbibe the right values of nurturing and caring toward their surroundings.

Early models of pro-environmental behavior (Kollmuss and Agyman, 2002) emphasized on the two major aspects of knowledge and the favorable attitude leading to environment-friendly behavior. The model discusses the influence of environmental knowledge and attitudes in converting the behavior into pro-environment behavior.

The participation and involvement of the stakeholders will create a sense of belongingness and thus, will lead to an ownership for all the causes and concerns associated with the environment. The knowledge and awareness will create a positive attitude among the participants and increased sensitization will, thus, lead to adjoining of hands which will foster protective initiatives.

Barr and Gilg (2007) discuss the involvement and participation by the stakeholders as an important segment for the protection of the environment by the government. The research emphasizes the different initiatives which have been taken, such as waste management, energy efficiency, water conservation, and so on. The impact of these actions on the conservation of the environment and especially the effect of the participation from all the stakeholders, including the citizens and the government bodies, will create a society on the road to well-forecasted sustainable development goals. The emphasis on the participation of each and every citizen is being discussed at various organizational platforms, (UNDC, 1993). United Nations has also discussed the need for a united approach from the stakeholders as a more efficient approach toward curbing the problem of environment degradation. Barr and Gilg (2006) cover the study conducted in the United Kingdom and the United States of America. Campaigning in the two nations is channelized to produce more sustained efforts. The initial hesitations later become a part and process of the system, such as waste segregation, whereas the concept has just started evolving in the developing nations. The British government has been keenly working toward environmental conservation by continuous efforts, comprising various segments. It has been almost two decades since the concentrated efforts have played a positive role in the policy designing and framing process.

Shrum et al., (1995) present the need to understand the characteristics of the consumers and their impact on advertising strategies. There exists a vast difference between the general attitude and the specific environment behavior. The clubbing of both is a big challenge which the policy makers have to adhere to. The willingness of the consumers depends on a range of categories which reflect the intensity of consumers while accepting the environment-based innovative products (Jansson et al., 2010). The findings of the study showcase that the personal norms have a bigger impact than the power exercised by the habits.

Behavioral attitudes are reflected in the sense of environmental owner-ship and accountability. Attitudes are governed by the behavioral patterns such as values, ethics, motives, and so forth.

4.3   GREEN CONSUMER BEHAVIOR MODEL

A model has been devised to understand, evaluate and an attempt has been undertaken to articulate green consumer behavior. Green consumer behavior model could be adopted by the manufacturers, companies, and marketers of green products. They can make use of the model while framing strategies for green products. The application of the model will help in the appropriate assessment and evaluation process so as to understand the dynamics of consumer behavior. Previous research results were taken into consideration to derive the model. The model presents the four factors which influence the green consumer behavior. Information and the level of knowledge about the environmental causes and concerns change the perception of the consumers toward the environment. Psychological factors such as attitudes, values, and beliefs are both influenced by as well as influence the consumer behavior. Social factors, also termed as group behavior, to some extent, define the behavioral tendencies. For example, family as a unit, friends or peer group, plays an important role. The buzz created by opinion leaders creates a mark on the decision-making aspects of choosing among the alternatives.

Green consumer behavior should be adopted as a process, the benefits of which can be derived over a long period of time. Secondly, it has to be a collective process. The initiative has to be inculcated by all the participants together. Right from a young age the values, morals, and ethics have to be imbibed to create sensitization toward this preventive cause. Thus, each one becomes an agent of change campaigning for it. The participatory model will contribute toward the success of this war against environmental degradation. Global awakening and arising will create the active involvement of all the stakeholders in the journey of saving our very own planet earth. The sense of belongingness has to be from within and not forceful. Carrington et al., (2010) argue that intentions of the consumers are more so motivated by the ethics, values, and beliefs. They want to be protectors of the environment as it sounds morally true and ethically significant.

The marketers are not leaving a single stone unturned to reach out to the customers. The marketing strategies are meticulously designed to ensure and convince the customers and lure their ethical mind-set. The products are articulated to have green technology, green innovation, green processes, green production techniques, green policies, and so forth. The green technology is equipped with organic procedures and processes which build a sense of confidence among the consumers. The sense of belongingness develops and leads to a positive brand image and ethical behavior toward the product.

The world of media has knowingly or unknowingly a large responsibility attached to it. The responsiveness of diligently dealing with the issues and concerns is due to the commitment held by the media. The media should take care of the role it plays. The representation of environmental aspects to the target audience should involve a message well-crafted and efficiently delivered. Well programmed and truly laid animated videos could be used while addressing the environmental issues to the children. Use of opinion leaders should be carefully undertaken as they play a major role in leading the campaign.

Behavioral change is a function of various characteristics. It is an area which is influenced by a variety of determinants and consumer desires and interests. Demographic characteristics of age, education, and income have a strong correlation with pro-environment behavior. Previous researches have also proved the fact that younger to middle age, educated and good income respondents have shown positive tendencies while reflecting their pro-environment behavior. Further, Figure 4.1 also represents the facets of green consumer behavior. Green consumer behavior is a function of demographic, sociological and psychological determinants leaving a profound impact and influence on the green consumer behavior.

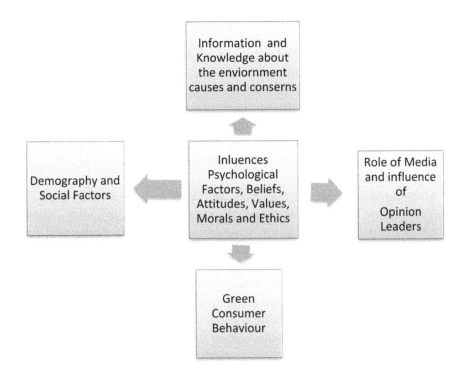

FIGURE 4.1 Research results.

4.4 ENVIRONMENTAL INCLUSION

Financial inclusion gave birth to financial literacy with the objective of including each one in the ambit by offering financial plans and policies befitting all. Environmental inclusion should also cover each stakeholder under its ambit. Environmental awareness, information, and knowledge will lead to incorporating each one right from a very tender age. The sensitization could be done from kindergarten level to postgraduation level by organizing relevant events such as fancy dress, drawing competitions, nature walk, debates, discussions, forums, environmental associations, clubs, competitions, conferences, seminars, and so forth. The human resources departments in the corporate world should take initiatives on their behalf by imbibing the environmental values in the workforce. Green human resource policies could be targeted toward less use of papers and more electronic communication by employing digitization. They could also initiate and encourage eco-friendly policies and plans to promote digitization.

4.4.1 BEHAVIORAL CHANGE AND GREEN EXAMPLES

A large amount of behavioral change is also reflected in the consumer's attitudes while purchasing products. Companies invented the marketing idea of charging the customers for each package used. Brands like Shopper's Stop, Fabindia, Pantaloons, and Lifestyle have started charging extra for packing. It has proved to be a large relief to the environment, thereby shielding the hoarding habit of the consumers. Some of the brands like Fabindia, Starbucks, Metro, and McDonalds have gone a step further by providing paper bags and newspaper bags.

Malaysia is a wonderful example. They want to be known as a green country and have actually made an attempt to engage in green technology. They have been promoting green businesses and green consumerism. Apparently, the underlying focus is also to improve their ranking in environment ratings.

Here begins an era where the consumers carry bags while planning for shopping. Apparently, a number of brands were promoting cloth bags and paper bags for a long time. Cloth bags, apart from being eco-friendly, are also reusable as well as recyclable without causing any harm to the environment. The realization and sensitization in society will lead to a strong partnership, which will contribute toward changing the attitudes of the people. Marketing efforts will then be adjudged from the point of view of pro-environmental behavior.

The corporate world is taking genuine efforts to publicize the value they place upon environmental initiatives and these efforts will see the bigger light of the day. The companies which have these programs equipped and imbibed in their corporate culture should be supported and appreciated by the consumers, government, and suppliers. Goodwill generated by these companies will lay a flowery path for a strong brand loyalty foundation.

Anything which initially comes as a surprise slowly takes over as a habit. The same theory applies to packing as well. Initial resistance has now become the part and parcel of shopping. Consumers are acceptable to the idea of buying cloth bags and paper bags. They preserve and further use them for gifting purposes. Sooner, the eco-friendlier bags will become a new trend and a style statement.

4.4.2 LAW AND ORDER

Stringent laws, strict actions and strong authority need to be in place so as to ensure law enforcement. A proper law enforcement machinery to check the implementation of following environmental laws during all the stages of product processing, as well as product life cycle, is necessary. Government authorities should necessitate an Enforcement Directorate, especially for the environment, exercising their authority by penalizing and holding the lawbreakers responsible for anti-environmental actions.

Plastic packaging, which is perceived to be the most environmentally harmful, should stand banned and any kind of procurement, production, marketing, distribution or usage could be encountered as breaking of the law, attracting relative penalties. The penalties could be more so corrective by allowing the necessary compensation to the environment in the form of tree plantation, cleaning the environment, planting of trees, and so forth.

The environmental laws have broad classification from general, forest laws, air, and water conservation policies. There are three key policies for the protection of the environment in India. They are National Forest Policy, pollution policy, and a conservation policy. Apart from these three major policies, there are 22 legislations or statues for environment protection in the order. The Indian Penal Code also has sections for environmental protection.

The laws have not been effective as the punishment and the penalties associated are not too severe. In fact, the financial penalties are too meager, ranging from Rs. 200 to 500. The environmental policies should incorporate implementing machineries which are more effective and with higher financial penalties.

4.4.3 NUTRITIONAL CONSCIOUSNESS AND INFLUENCING FACTORS

The awareness and enlightenment about the dietary facts are leading the consumers to be more nutrition-conscious, consequently giving birth to the rise of organic products and brands. Deteriorating health conditions and increase in the number of diseases are turning the consumers toward organic products.

Mazar et al. (2010) relate the green purchase behavior to social and moral behavior. Consumption has a large association with the consumer's ethical behavior, which is reflected in the green consumer behavior. Compared to the normal conventional products, society lays high regards for the consumption of green products. Shepherd et al. (2005) discuss a large contradiction in the attitudes and behavior of the people. The research reflects that often people are very positive about organic products, but they do not consider purchasing these products. The organic foods are considered healthy and of high nutritional value, but consumers do not consider organic foods to be good in taste. The fact is that even a strong positive attitude cannot lead to a positive conviction among the consumers. The survey results carried among Swedish consumers showed that a vast majority of consumers still continue buying conventional products. Thus, previous literature also assures the fact that consumers are willing to change to organic products but are not yet comfortable with the idea of paying a higher price for them. Thus, nutritional consciousness is, in turn, getting diverted toward green products.

4.5   NEW TRENDS OF GREEN CONSUMERISM

Larger numbers of citizens these days are turning toward healthier lifestyles. They are incorporating green lifestyles by being more close to nature. There is an increase in participation in marathons, cycling expeditions, adventure clubs, trekking wanderers, jogging circles, gym workouts, walking partners and so on. The trend of allotting a specific time in the busy schedules toward a healthy body and mind is no more a fashion but a necessity. The demographics does not matter when it comes to workouts, whether yoga, Zumba or any other physical training. People from all age groups, gender, occupations and educational backgrounds are imbibing fitness as a lifestyle. The green behavior has become an attitude adjoining the nature to self-wellbeing.

Moorman and Matulich (1993) present the model of consumer's preventive health behavior without any use of dangerous chemicals. There are different sources which consumers have been using to gather health-related information. Large efforts apparently involve varied factors such as diet, exercise, checkups, consumption habits, and so on.

Various factors play a positive role in enhancing the motives of consumers to be more health conscious though Tobler (2011) points out that overall health remains a powerful criterion for a sustainable consumption habit. Health consciousness has become a remarkable factor playing a significant role in consumers' purchase decision-making process.

4.5.1   ROLE OF SOCIAL MEDIA

Increase in the number of diseases as well as new emerging terminologies are not only scaring the society but also training them to be more aware, informed and tread the road toward prevention. Strong public opinions against the harmful products, preservatives, processes, technologies, and hybrids are creating loud people's movement on various social media platforms.

The messages going viral with the touch screen technology in a matter of few seconds are leading to innovative revolutions. Regular uploads, updates, and ongoing videography all over, is, in turn, forcing companies to be on their toes all the time. The world is becoming smaller, united and integrated in the true sense. Mergers and acquisitions are all over-diversified and encompassing in our lives. The access and ease of the innovations on the planet are laying the initial and the aftermath theories. Deadly diseases catering to almost everyone from pre-birth to every other stage of life of human beings have created horror in the minds of consumers. Social media is playing a supplementary role in all of this. Fear plays a great role in the purchase decision-making process and the companies selling green products and services are leaving no chance to capitalize on it.

4.5.2 INTENTIONS VERSUS CONVICTION

There are large numbers of consumers with green purchase intentions. In-depth interview results reflect that around 40% of consumers are keen toward a green environment, and proclaim themselves as pro-environment consumers. But in reality, the detailed analysis presented astonishing facts that these consumers have actually just taken a few baby steps toward a green environmental approach. Though the attitudes and behavior reflect that if the marketers provide them with attractive incentives and luring offers then the market will widen much more in the near future.

Tobler (2011) argues that lack of environmental knowledge remains a criterion for consumer's resistance toward a caring concern for the environment. Therefore, for a positive focus toward the environment, education and the right information remains a pertinent factor to promote environmental concern. Results of the previous research are endearing to the fact that environment knowledge has a positive correlation with the pro-environmental behavior. Carrington et al., (2010) evaluate the gap between the intentions and actual buying behavior of the consumers which is named as ethics. Researchers have discussed this under a different term known as ethical consumer behavior.

There is a need to tap this untapped market with serious strategic efforts to capture the nerve of the consumers. The inner instincts of the consumers could be capitalized on and retained powerfully in the business performance to reach out to this niche market segment. The future which lies ahead could be a turnaround for a large number of companies if they intend to introduce the right policies and programs matching their needs and desires. Consumers should develop the taste buds to consume a large amount of food grown locally and available seasonally. The desire to have everything, everywhere all the time forces the marketers to find all the ways and means to capture the untapped market and capitalize on it. Environment consciousness is not only required but also desirable. The need to emancipate collective efforts of all the stakeholders will definitely fetch fruitful results. The increase in the demand to store, refrigerate and preserve food for longer times, is leading to an increase in the usage of preservatives and various other harmful chemicals. Environmental consciousness includes a change in the consumption habits and lifestyles.

4.6   BEHAVIOR AND PRICE

The idea which creates a little restrain in the minds of the consumers and thus makes them hesitant is also the price. A large number of consumers also think that the organic products come with a price. The price barriers act as a resistance toward green consumer behavior, leading to suppression of their green buying intentions. The other products are cheaper, easily accessible, have a larger variety and altogether come with an attractive combo, which makes it easier for the consumers to convince themselves and their behavioristic self about compromising their green attitudes.

Manaktola and Jauhari (2007) conclude that consumers are keen on promoting and fostering the usage of the hotels in India which are pro-environment. The lodges and hotels which have been continuously implementing eco-friendly practices are favored more by the consumers. On the contrary, a vast majority of consumers are not willing to pay anything extra for the eco-friendly packages that they are offered.

According to Tobler (2011), consumers influence the environment in a prominent manner. The influence might be in the form of their decisions, actions, attitudes or more broadly, their behavior. In some cases, there is a lack of information among the consumers and in some other cases, in spite of the information, the consumers are not keen to follow pro-environment behavior due to the cost involved in it. A large amount of research has been conducted to examine, assess and evaluate the level of knowledge of the consumers with regard to the environment and various factors surrounding it. Apparently, part of the answer to the question lies in the willingness of the consumers to act in favor of the environment.

4.7   CONCLUDING REMARKS

The need to instigate a larger sense of belongingness and ownership will create a much greater partnership among the different stakeholders. Marketers need to address, accommodate and adopt green consumer behavior and the prominent factors playing a contributory role. The emphasis on demographic factors could also be shifted to sociological and psychological factors. Group behavior emancipates among the individuals in the form of decision making and the choices over a range of product categories.

Research could be a foundation for a large number of marketers concerned about green products and the corporate world focusing on green policy framing. The trend toward green finance, green human resource, green marketing, and so on could be revived to create a win-win situation for companies, consumers, and the overall society. The research could be suggestive for the policy makers, authorities and government representatives to involve and include a strategically equipped framework for a larger green perspective.

Green consumer behavior is a process, attitude, perception and a lifestyle, which if catered from the early stage of life, will reap a larger benefit than being imposed on by authorities. Self-regulation is the best regulation. And if the sensitization and self-realization builds up among the citizens, it will touch the right chord. The value system will imbibe the commitment of caring and nurture toward the environment leading to the fulfillment of goals of sustainable development. The viability of the concept will materialize if it is delivered in the right form and the stakeholders will adopt it with the right spirit.

Price remains an important determinant for consumers while selecting green products. Positive behavior and attitude are suppressed when it comes to considering the price of the product or service. Green attitude and green behavior are much behind the price conviction. Therefore, the consumers are keen to adopt green policies in their buying process without compromising on the pricing aspects.

Green consumerism should be inculcated in the consumers by creating awareness with respect to sustainable development and protection of the environment. Awareness among consumers will lead to sustainable development and these values in their buying behavior will create a pro-environment scenario.

Green consumer behavior is for the innovative, adventurous and adoptive consumer, willing to adapt to changing patterns of life, health, and physical well-being. The new age consumers belong to the touch screen world, having excess to every minute detail of information. They are living in the world where every other thing is viral and traveling faster than the speed of light, where information is just one touch away and spreads among millions on the social media platforms, Wherein anything happening anywhere gets clicked, recorded and shared across the world with comments pouring in. The world is more receptive and expressive in the form of likes and dislikes. Walls are now not limited to the interiors but extend to social platforms being read, posted and commented by all. Shortening of distances has decreased the barriers and created a new era of marketing. The current digital scenario is welcoming the technical knowhow, wherein change is the only constant. The world has seen a 360° revolution, innovation, and transformation. Consumers are smarter, informed and technologically updated. Any information is compared, assessed, evaluated and most importantly, analyzed for further processing.

Green consumer behavior will survive the tough battles within cutthroat competition if it overcomes few weaknesses while developing strong persuasion among the consumers. The war with conventional products could be won if the field is leveled and tested first, before batting of successful innings. The right chord needs to be touched to succeed and then, even the sky would not be a limit. The bridge between the intention and conviction has to be built strong enough and well connected for a great win.

KEYWORDS

- pro-environment behavior
- green consumer behavior
- green processes
- eco-friendly packaging
- green products
- organic products

REFERENCES

Barr, S.; Gilg, A. A Conceptual Framework for Understanding and Analyzing Attitudes Towards Environmental Behaviour. *Geografiska Ann. Series B Hum. Geogr.* **2007,** *89*(4), 361–379. http://www.jstor.org/stable/4621594

Carrington, M. J.; Neville, B. A.; Whitwell, G. J. Why Ethical Consumers Don't Walk their Talk: Towards a Framework for Understanding the Gap between the Ethical Purchase Intentions and Actual Buying Behaviour of Ethically Minded Consumers. *J. Bus. Ethics* **2010,** *97*(1),139–158.

Gilg, A.; Barr,S. Behavioural Attitudes Towards Water Saving? Evidence from a Study of Environmental Actions. *Ecol. Econ.* **2006,** *57*(3), 400–414.

Jansson, J.; Marell, A.; Nordlund, A. Green Consumer Behaviour: Determinants of Curtailment and Eco-Innovation Adoption. *J. Consum. Mark.* **2010,** *27*(4), 358–370.

Kollmuss, A.; Agyeman, J. Mind the Gap: Why do People Act Environmentally and What are the Barriers to Pro-Environmental Behaviour? *Environ. Educ. Res.* **2002,** *8*(3), 239–260.

Manaktola, K.; Jauhari, V. Exploring Consumer Attitude and Behaviour Towards Green Practices in the Lodging Industry in India. *Int. J. Contemp. Hospitality Manage.* **2007,** *19*(5), 364–377.

Mazar, N.; Zhong, C.-B. Do Green Products Make us Better People? *Psychol. Sci.* (Sage Publications, Inc.) **2010,** *21*(4), 494–498.

Moorman, C.; Matulich, E. A Model of Consumers' Preventive Health Behaviors: The Role of Health Motivation and Health Ability. *J. Consumer Res.* (Oxford University Press) **1993,** *20*(2), 208–228.

Shepherd, R.; Magnusson, M.; Sjödén, P.-O. Determinants of Consumer Behaviour Related to Organic Foods. *Ambio* (Springer) **2005,** *34,* 352–359.

Shrum, L. J.; McCarty, J. A.; Lowrey, T. M. Buyer Characteristics of the Green Consumer and their Implications for Advert. *J. Advertising* **1995,** *24,* 71.

Tobler, C. Green Consumer Behaviour: Consumers' Knowledge and Willingness to Act Pro-Environmentally a dissertation, Eth Zurich, University of Zurich, Wetzikon, ZH, 2011.

United Nations Development Commission. *Agenda 21: Action Plan for the Next Century.* 1993.

FURTHER READING

Allen, F.; et al. The Foundations of Financial Inclusion: Understanding Ownership and Use of Formal Accounts. *J. Financ. Intermediation* **2016.** http://dx.doi.org/10.1016/j. jfi.2015.12.003

Gatersleben, B.; Steg, L.; Vlek, C. Measurement and Determinants of Environmental Significant Consumer Behaviour. *Environ. Behav.* (Sage Publications) **2002,** *34,* 335–362.

Grunert, S. C.; Juhl, H. J. Values, Environmental Attitudes, and Buying of Organic Foods. *J. Econ. Psychol.* **1995,** *16*(1), 39–62.

Hines, J. M.; Hungerford, H. R.; Tomera, A. N. Analysis and Synthesis of Research on Responsible Environmental Behaviour: A Meta-Analysis. *J. Environ. Educ.* **1986/87,** *18*(2), 1–8.

Huddart, K. E.; Beckley, T. M.; McFarlane, B. L; Nadeau, S. Why we don't "Walk the Talk": Understanding the Environmental Values/Behaviour Gap in Canada. *Hum. Ecol. Rev.* **2009,** *16*(2), 151–160.

IPCC. *Contribution of Working Groups I, II and III to the Fourth Assessment Report of the Intergovernmental Panel on Climate Change;* Climate Change 2007: Synthesis Report; Geneva, 2007.

Känzig, J.; Jolliet, O. *Umweltbewusster Konsum: Schlusselentscheide, Akteure und Konsummodelle [Ecological Consumption: Key Decisions, Actors and Consumption Models];* Bundesamt fur Umwelt: Bern, 2006.

Kothari, H. *Individual Investors Behaviour Towards Financial Decision Making: A Conceptual Model,* Altius Institute of Universal Studies, Indore, 2005.

Lea, E.; Worsley, A. Influences on Meat Consumption in Australia. *Appetite* **2001,** *36*(2), 127–136.

Lea, E.; Worsley, A. Australian Consumers' Food-Related Environmental Beliefs and Behaviours. *Appetite* **2008,** *50*(2–3), 207–214.

CHAPTER 5

GREEN PRACTICES FOR GREEN(ER) LIVING: THE ROAD AHEAD

GEORGE VARGHESE[1,*] and LAKSHMI VISWANATHAN[1,2]

[1]*Institute for Financial Management and Research (IFMR), 24, Kothari Road, Nungambakkam, Chennai 600034, India, Mob.: 8179314798*

[2]*Mob.: 9486443137, E-mail: lakshmi.viswanathan@ifmr.ac.in*

*Corresponding author. E-mail: george.v@ifmr.ac.in

"Only when I saw the Earth from space, in all its ineffable beauty and fragility, did I realize that humankind's most urgent task is to cherish and preserve it for future generations."

Sigmund Jahn (German Cosmonaut)

5.1 INTRODUCTION

How to live well and within the means of one planet is the question of the 21st century. Awareness of the ecological limits, use of human ingenuity to design ways to live within the bounds, adoption of practices that reduce human imprint and keeping "going green" central to decision making holds the key to a sustainable future. Unless we shift from "business as usual" to a "transition to a sustainable society," sustainability will remain elusive and would render humanity as well as the economies at stake. In this backdrop, the paper reflects the stark reality of ecological scarcity and discusses strategies and ways of reducing individual and business footprints as well as of improving one's quality of life, leading to a rewarding and sustainable future.

We begin with what ecological footprint is and what the footprint measurements over the years have taught us. This would substantiate the subsequent sections of our study that deal with the need for green practices and the ways to go about in a more efficient modus and optimal fashion possible. Ecological footprint measures the amount of biologically productive land and sea area a person, community, town, country or humanity as a whole uses to produce the resources that it consumes as well as to absorb the waste human economy generates and compares it against what is available for the same (Wackernagel and Rees, 1998). In other words, ecological footprint attempts to measure the magnitude of the human impression on earth's ecosystems. Thus, ecological footprint represents the critical natural capital requirements of a defined economy or population in terms of the corresponding biologically productive areas (Wackernagel et al., 1999).

These measurements try to answer unambiguously the research question as to what extent the planet's biological capacity is being demanded by a certain population. The ecological footprint thus reflects the stark reality of ecological scarcity, which can be rather distressing and startling (Wackernagel and Rees, 1998). According to the Global Footprint Network, an international think tank that provides ecological footprint accounting tools, our present consumption is much more than what the planet can replenish in the same time frame. Their calculations show that, on average, it takes the planet about 1 year and 6 months to regenerate everything we consume in a single year. Further, it is predicted that we would require an equivalent of nearly two whole planets by 2030 to support the human activity if the present state of consumption and lifestyle continues.

Exhausting the very resources that sustain human life and biodiversity at a rate higher than what is required for them to replenish themselves leads to an ecological overshoot. Thus, overshoot occurs when the resource regenerating capacity of the ecosystem and its ability to absorb carbon dioxide emissions is exceeded by the human imprint on the ecosystem. The day on the calendar when the human impression exceeds the replenishing capacity of the earth in that given year is known as the Earth Overshoot Day (Wackernagel et al., 2002). According to the Global Footprint Network, the Earth Overshoot Day has moved up from early October in 2000 to August 13 in 2015.

The very existence of this ecological overshoot indicates that significant changes from "business as usual" to a "transition to a sustainable society" have to be made by the human society if it wants to build a sustainable future. At this juncture, individuals and institutions the world over must realize the ecological limits and start living within the replenishing capacity of our planet. It goes without saying that saving is better than paying afterwards. Several studies have revealed that even though we cannot inverse the damaging effects of an ecological footprint, we can positively embark on its future impacts by bringing changes to our lifestyle and attitudes. By using human ingenuity in finding ways to live within our bounds, adopting practices that reduce ecological footprint and by keeping sustainability central to our decision making, we can assure that our future generations continue to get the same opportunities to enjoy rich and rewarding lives.

5.2   AWARENESS—THE BEGINNING OF CHANGE

The awareness in itself is progress. The realization of the impact one has on the planet and his ability to reduce it, though minute, holds the key to a paradigm shift from consumerism to sustainable living. We are often discouraged by our own conviction that no significant difference can be made by a single person on account of saving resources while millions are ruthlessly contributing to its depletion. But little do we realize that if that one person lowers his footprint, it is in itself an improvement, even if it is only a minute difference.

While being self-aware of the ecological bounds and limits within which we live, we should make our own strides towards a more self-reliant and sustainable lifestyle. Educating oneself of our own impression on the planet thus forms the base for sustainable thinking and when done collectively, it will lead to a change in lifestyle that would eventually reflect positively on the Earth's ecosystem. Leading by example is often thought to be the best method of educating and so will it be when it comes to teaching our kids about the environmental glitches that are looming upon them and the need and ways of delaying (if not reversing) the detrimental effects of "environment taken for granted" lifestyle. Leading by example could range from saying no to superfluous purchases to sticking to one family car to making gifts instead of buying gifts to even growing your own vegetables. A few good practices, to begin with, will give more ideas and a stronger desire for coming up with more and more ways of living a greener life.

Inviting and encouraging the rest of the family, friends, and community to factor in the environment and its sustainability into their everyday practices and decision making would be the next best thing one could do in their stride towards a better future. Thus, the more we do to educate more people and encourage them to educate a lot more, in the course of time, we will already have made a significant difference that we desire.

In the age of electronics and mass communication, ample use of social media can be used to network, educate and spread the good word and testimonials. Sharing your own experiences in reducing the ecological footprint as well as passing on the techniques adopted or learned is easier than ever before. A green thinking dynamics can be created altogether by connecting and networking with people with common interests toward sustainable development and discussing as well as initiating collective actions thereof. Thus, educating oneself and others holds the key to a paradigm shift from business as usual to a one planet living.

5.3   THE CONSUMER POWER—REDUCE, REUSE, AND RECYCLE

"The more, the better" is a common belief. It would not be wrong to say that we are living in a world that places a lot of importance on consumerism and immediacy—we want to acquire all that we "want" (not need) from wherever it is, at the earliest. This doggedness on having all that we want at any time we want has put a huge human impression on our ecological systems and has generated a disproportionate amount of waste. Shopping has now evolved to be a mindless activity to elude boredom or in some cases, to attain a buyers' high while acquiring the season's latest (Morgan and Birtwistle, 2009). Consumption is ever increasing and is at an all-time high with consumers buying more things than they actually need. These excessive spending habits call forth the imminent need for a shopping diet if sustainability is what we are looking forward to.

As an environmentally responsible consumer, this awareness brings along with it a trade-off where we have to make a choice between the environment and needless spending. This internal trade-off battle leads to a moral and puritanical perspective that consumption is a bad thing one should do without (Slater, 2003). However, it should be made clear that awareness does not mean making consumers regretful of the choices that they make. Instead, we should be looking forward to an attitudinal change more than a behavioral change. That is to say, people should be enthusiastic to make lifestyle deviations in a green direction without conceding on their consumption.

Reducing the amount of consumption would be the simplest way to reduce our impact on our environment. However, it is not always possible. In situations where consumption cannot be compromised, consumers should look for better alternatives that are eco-friendly. To this end, we need to become increasingly aware of the power of a consumer and the indirect impact that our purchases have on the ecology. Such initiatives from the consumers would encourage firms to come up with more green products as well as to resort to green business practices such that they could leverage their business from the green image that they thereby create. For example, if we reject goods produced in an unsustainable manner, and alternatively choose eco-friendly products, it would send strong signals to the manufacturers who would then change their practices and come up with more eco-friendly products, replacing the existing ones (Zsidisin and Siferd, 2001). Green consumerism, as it is commonly known, is the act of making purchasing as well as non-purchasing decisions by consumers, based at least partially on environmental and social criteria (Peattie and Ratnayaka, 1992). Thus, any decision about purchases has to be thought about at least twice. By forgoing superfluous consumption, you not only save resources but also evade the waste generated out of it. That is to say, we can effectively reduce our footprint just by being an informed and selective consumer.

Every day we generate more waste than what our planet could absorb, which eventually gets piled up in landfills and in oceans (Chawla and Rajaram, 2016). Paper and paperboard create a bulk of our trash mainly because of its indispensable role in our day-to-day and business activities. However, the wastage can be significantly reduced at greater ease by bringing small changes to our lifestyle and practices. As often thought and taken for granted, recycling of papers is not a good and viable option for the paper since it often uses chemicals for this process which might create more economic cost than the benefits. Further, against the common notion, paper is not substantially recyclable as lower quality paper cannot be recycled.

Inarguably, plastics are one of the biggest threats to human life as well as the ecology. Deliberate and conscious efforts in avoiding plastics as a consumer are possibly the best way to keep plastic wastes from damaging the environment we live in. Choose nonplastic alternatives whenever possible or in unavoidable circumstances, reuse them to the fullest to avoid creation of additional waste. Shifting to stainless steel bottles from plastic bottles, keeping an eco-friendly bag always at disposal, avoiding plastic containers are some of the simple changes that we can make at ease. Just like paper, plastics too are not infinitely recyclable. In general, plastics can only be recycled once and the resultant plastics are downgraded and used in items like plastic lumber, fleece, and so forth, which are not further recyclable (Hopewell et al., 2009).

Though we have wholeheartedly incorporated the throwaway culture into our lifestyle, of late there is a rise of another set of environmentalists who are finding innovative and implausible ways of using the things that hit the trash to build a better world. It might appear difficult at first to live a waste-free or waste-reduced life, but it is incredibly rewarding and definitely possible. If we are not able to find best ways to use (or handle) the waste we generate, we should at least keep the possibility of minimizing waste generation in account at every stage of our decision making. The waste that we generate will eventually come to haunt us. The rubbish that piles up in the middle of the ocean, the landfills that leak toxins into the groundwater and the chemicals that get released into the environment from plastics are to name a few (Ross and Evans, 2003). In fact, most of the chemicals released by the waste we generate, such as methane, mercury, and other hazardous chemicals, are more potent and unsafe for human life as well (King et al., 2006).

Reduce, reuse, and recycle: The three R's mantra in this very sequence should govern our decisions as part of our daily lives (Barr et al., 2001). If observed religiously, this in itself will decrease our imprint on the planet. People are often misguided by their conviction that as long as they recycle the waste they generate, they are doing the right thing. However, reducing the quantum of resources consumed in the first place rather than recycling them afterward should be the norm. Forgoing all the superfluous and unnecessary purchases would be much better than recycling all the wastes generated thereof. Next would be to reuse. Look for ways of reusing the waste that we are about to throw away. Many a time the waste generated out of replacements can be sold off to someone who might find good use of it. There is a whole new industry that creates works of art out of trash and many a time, converts them into useful things. Lastly, before you try to hit the trash, see the possibility of recycling. This would elude the extraction of virgin resources and would also certainly reduce the waste generated as well as the area needed to absorb it (King et al., 2006). For example, if we recycle 1 t of paper, it can reduce the total amount of timber cut down to meet its global demand by 3.3 t and save up to 7000 gal of water as well as 68% of the energy needed to make it from virgin timber.

5.4 PRACTICAL TIPS FOR GREEN LIVING

The very concept of ecological footprint can be effectively used to show how mankind is working towards lessening its imprint on the planet and this concept of developing in a sustainable manner requires us to take deliberate actions individually as well as collectively to make our footprint as small as possible. Thus, the concept of sustainable development will necessitate us to progress at every step of life without causing additional cost to our future generations and manage resources within our means. The best way to make it happen is to keep track of all the resources we use and make well-informed choices at all possible times.

5.4.1 FOOD

An easy starting point for waste reduction is food waste and yard waste. One would require only a minimal effort to divert from food and yard waste both at a personal and municipal level. As a result of our overindulgent consumption, supply is higher than ever, and our farmers are left with excessive unsalable farm products that ultimately rot in fields. Further, our requirement for perfect looking products would render tons of perfectly edible but imperfect looking food to go waste even before it reaches our homes. Food wasted at home is only an addition (Lundqvist et al., 2008).

There is a lot that we can do to reduce this wastage. One way to reduce the waste that is generally created while getting foods from fields to our plates, such as refrigeration, packing, and transportation, is to grow our own vegetables and fruits whenever and wherever possible. Once we have a garden, it is easy to manure our produce using our own compost too. Home gardening is not only an interesting solution to reducing wastage but also beneficial to our health and nutrition as the organic foods thus cultivated are least contaminated with preservatives and chemical additives. Further, organic farming involves less energy wastage because of its low input and low/minimal pesticides cultivation methods while consumption of imported and processed food necessitates mass production, processing, transportation, packing as well as an advertisement before it becomes available to the end consumer. Thus, if it is not possible to keep a garden, the next best thing one can do is to buy locally produced, organic, in-season foods from the local farmers market that would then encourage community producers.

Planning meals ahead of time not only prevents food wastage but also saves money on provisions. While shopping, look for food with least packing, processing, and transportation to minimize other food-associated wastages. Going on a meat diet is another way to reduce food wastage to a large extent. Going on a meat diet for 2–3 days a week can bring down greenhouse gas emissions considerably, as livestock industry (inclusive of cattle food, fertilizer, land, transport, growth hormones, butchering, etc.) remains a major contributor to greenhouse emissions. It is estimated that meat eaters have an ecological footprint that is double that of the vegetarians. Thus, we can all put a dent in our food and associated wastages by consuming organic and in-season foods, encouraging local production, reducing meat and dairy products consumption, reducing consumption of processed foods such as soft drinks, packed snacks, and other junk foods.

5.4.2 WATER

Clean water scarcity is a global problem affecting life in many ways. With only less than 1% of fresh and usable water on earth, the other 99% being either salt water or permanently frozen, usable water is a precious limited resource indispensable for the sustenance of life on earth and needs to be conserved by all means. With increasing demand for this limited resource on account of an ever increasing population, urbanization, and industrialization, it is of prime importance that we minimize water wastage and use it wisely.

Water, the universal solvent, is a key element that warrants the quality of lives. People's demand for fresh water has increased six-fold over the last decade while there are several regions suffering from acute water scarcity. The difficulties of people in such areas are testimonies for us to avoid misuse of water and conserving it through proper water management practices. The pressure on this precious resource is expected to increase drastically with growing population, industrialization, urbanization, agriculture, and so forth. (Dietz et al., 2007) There is a lot that we could do individually as well as collectively to save water for our future generations as well as to ensure sustainability.

A small effort by everyone ensures big results. Thus, a huge change can be brought forth by small changes in the lifestyle of millions of people around the world. Some positive changes that are easy to implement are closing taps tightly after use, avoiding showerheads and instead using buckets for taking bath, using washing machines and dishwashers to its full capacity each time, using a pan or mug instead of running water to wash fruits and vegetables, and so forth. Further, we can check on water wastage by installing low-flow showerheads and faucets, using water-efficient appliances, checking on water leakages and fixing them, taking shorter showers, using commercial car washes that use much less water than what is required, if washing at home, and so on.

A proper water management system needs to be implemented at every home, office, apartment, and institution to conserve water and to increase usage efficiency. Finding alternative solutions for water requirements like rainwater harvesting, a water-wise design of landscape, water filtering solutions are some interesting ways of water management. The efficiency of water usage can be ensured by using dual-flush converters, installing water monitoring systems, innovative garden watering systems, using rainwater for all purposes other than drinking and cooking, and so forth. We should plant trees during the rainy season when they are water-fed naturally. Further, during other seasons, we can ensure saving water by avoiding watering the garden between 10 and 5 pm when the evaporation rate is high and even better by planting drought-tolerant plants and trees. Watering plants in the mornings and evenings helps them in better absorption too (Lundqvist et al., 2008).

We need to be increasingly aware of the need to conserve water and also about the difficulties and struggles of a vast multitude of people around the globe who are deprived of access to and availability of clean drinking water. The fight for survival is going to be more intense and dreadful if we do not act now and bring positive changes to the way we use water. More water could then be saved by inspiring other family members, friends, neighbors, and colleagues to think on the same line. We should take it as our individual responsibility to save water in our own capacity and also to promote and encourage activities associated with water conservation and spreading awareness.

5.4.3   ENERGY AND TRANSPORT

Energy is an indispensable part of our daily activities. We require energy to light our cities, power our machines and vehicles, warm our houses, cook food, use our favorite gadgets and electronics (Herring, 2006). While some form of energy is renewable, we are heavily dependent on nonrenewable sources of energy for most of our daily activities. These fossil fuels are finite, limited, and require millions of years to replenish themselves. Once they are exhausted, all of our activities will be at stake and hence, we must all carry out our part in conserving them. Saving on energy would necessarily mean that you help the environment by preserving the finite resources, making them last longer and sustaining them for the future generations. There are several ways in which we can responsibly consume and conserve energy and go green.

Simple practices such as turning off lights, fans, electronics, and other gadgets when not in use, installing energy-saving appliances at home, drying clothes outside whenever possible, defrosting refrigerators and freezers regularly, and tuning down air-conditioning during summers, keeping thermostats at low temperatures during winter, and so forth, can save energy to a great extent. It is advisable to unplug electronics using a switchable power strip when not in use in order to avoid power leaks as electronics generally sip power in small amounts even when they are switched off. While running a washing machine, a disproportionately large amount of energy is used up in heating the water alone which can be avoided by using cold water instead, along with better detergents.

However, the best way to go green is by resorting to green power sources (Dincer, 2000). Though wind turbines, hydroturbines, solar panels, biogas plants, and so forth, require an upfront financial outlay and are out of reach for most people, they are getting closer and closer. Solar power has emerged, over the recent past, as a strong contender in the overall energy sector by efficiently powering up homes and offices. If we can think of these green power sources as long-term investments, they will surely get paid back as savings on energy bills in few years' time. With greater pressure on ecology and the imminent need for going green in all walks of life, people along with the governments have been largely turning to and giving more importance to green power over the recent past. Thus, the installation costs of green power sources are expected to come down drastically in the near future.

Transportation is another category that has crucial ramifications on our ecology. The transportation industry has direct detrimental effects on the environment through greenhouse gas emissions and consumption of fossil fuels. However, it is relatively easy to go green in this sector through small changes in the lifestyle and practices. Carpooling is the most efficient way to reduce individual footprints. Not only is it eco-friendly but also saves money on gasoline, parking, maintenance, and so forth, along with providing a platform to socialize with people. In most situations, avoid driving whenever there is an alternative. Walking to your nearest store, using a bike or even better, a public transport, will reduce traffic as well as pollute the air less.

While using cars, maintain them in optimal conditions all the time. Service them regularly to keep a check on emission control system and to ensure peak efficiency. The tires should be sufficiently inflated to maximize mileage. Turn off the engine while the car is idle and while parking, look for shades as they lower the temperature on gas tanks and thus minimize fuel evaporations. Few such conscious efforts by millions can make a significant difference in conserving energy as well as in reducing our eco-footprints. It is up to each one of us to ensure a continual supply of energy in the future and to find more innovative and creative ways of putting an efficient use of the limited resources that we have.

5.4.4 HOUSING AND LIFESTYLE

Housing and lifestyles are key areas that everyone could focus on in their stride towards a sustainable, healthier, and happier environment. There is a lot that one could do in order to go green such as designing rainwater catchment areas, nurturing a garden, maintaining compost, installing solar panels and solar water heaters, purchasing sustainably produced goods and dressing appropriately for the climate (White, 2013).

Dressing appropriately for the season is one creative way of saving on energy. Putting on more layers of clothes in winter to keep warm and dressing light during summer can cut down the thermostat and air condition bills. Energy can be saved on lighting by preferring natural light over electric lighting during the day. Installing water purifiers can avoid the purchase of bottled water and can not only save money in the long run but also is environment-friendly as it can minimize plastic waste. Aluminum or stainless steel containers should be preferred over plastic containers as they are healthier for the environment as well as for human health. Further, always choose non-toxic and biodegradable products that are healthier and environment-friendly.

Sharing is another common way of reducing the use of almost any scarce resources. For example, by sharing books, we can save on paper, ink, other chemicals, and the energy required for its production. While purchasing, make sure that the product is absolutely required and buy those that are of high quality so as to avoid frequent purchases. Use electronics such as mobile phones and computers as long as possible and donate or recycle them once the time comes. By purchasing items that are recyclable and by reusing the existing ones, we can save resources in large quantities. Replace the existing products only when they cannot be repaired and reused. As far as possible, one should always prefer secondhand and recycled materials as long as it serves the purpose.

5.5   GREEN BUSINESS OPPORTUNITIES

Though it may sound very direct and efficient, the idea of making changes in lifestyle for an environmental cause does not always assure the desired results if they are not adequately supported. Ishak and Zabil (2012) point out that the growing consumer awareness may not essentially convert to effective consumer behavior. For that, it requires a collaborative effort from the governments, corporations, and the public. If the firms could incorporate the concern for ecological balance into their business strategies, they can not only take advantage of the increasing green consumer base but also ensure sustainability and profitability to their business. This would then substantiate the awareness among consumers and help them in transforming knowledge to practice.

The rising wave of people the world over seeking ways of going green, both at their homes as well as their workplaces, is a huge market opportunity for entrepreneurs who wish to carry out eco-friendly and sustainable business. Such businessmen who keep the concept of going green close to their heart and central to their business, along with the end goal of profit are called eco-entrepreneurs. Eco-entrepreneurs harness their profits while ensuring the changes they desire in their communities. They operate their enterprise in a way that reinstates or reconciles the planet and nurtures more equitable and fair interactions among everything touched by their business (Ninlawanet al., 2010). These triple bottom line (people, planet, and profit) enterprises foster both a return on investment and a return on the environment while carrying out business with a sense of a fair trade.

We see a rising trend among the people to seek ways to be more self-sufficient, to save money and other resources as well as a desire to lower their ecological footprint. Slowly, people have started preferring healthy, green and sustainable products over their conventional counterparts (Eriksson, 2004). A lot more eco-entrepreneurs are going to be needed to meet the demands of the emergent green economy—developing products and services that provide and facilitate sustainability on a large scale as well as on an individual scale (Min and Galle, 1997). In the future, opportunities are even brighter for eco-entrepreneurs as more and more people move closer to a sustainable mode of living.

Eco-entrepreneurs have enormous uncaptured and unexplored markets available to them and they can certainly do well if they can offer green products that are superior to their non-green alternatives (Menon and Menon, 1997). For example, nowadays, new brands of cleaning products that are green, healthier and more effective are stealing the market from the conventional petroleum-derived, toxic cleaning products. Similarly, there are a multitude of business ideas surrounding each and every aspect of life, such as air and water monitoring systems, automated watering system for yard and garden, harvesting seeds of drought-resistant plants and trees, eco-friendly landscaping and backyard gardening systems, developing indoor growing systems, farm to table restaurant systems, to name a few, which surround food and yard management.

Opportunities are even higher for energy-saving appliances and electronics such as intelligent insulation solutions, smart thermostats, energy-efficient technology, and so forth. Designing and developing products to retrofit homes and offices to make them greener and function optimally will be a profitable as well as a promising endeavor in the coming years. Rooftop solar systems are of increasing demand and the demand is expected to escalate if more such products hit the market with a lower cost. Necessitating a heavier investment, a residential wind power service is a viable business in cities and towns where people can afford such services. As people are progressing from smartphones and smart gadgets to smart homes, energy-auditing services and other energy storage and management systems will have a higher demand in the near future.

Along with the already trending items like dual-flush converters, water monitoring systems, low-flow shower and faucet heads and other energy and water efficient appliances, some other ideas that can be explored are water filtering solutions, water-wise landscape design, landscape mulch and compost, rainwater harvesting systems, and so forth. Such green business ideas can be both a remedy for the ailing environment as well as a huge market opportunity waiting to be optimally explored. If we open our eyes to the world around with the notion of going green central to our business plan, then the opportunities are abundant. The very idea of "profits with a purpose" has the power to make the world we live in, a better place, ensuring both fair trade as well as a fair return on the environment. Once the green business flourishes, it would be more evident in the environment as to how much we can gain control over our imprints on the planet we live in.

5.6 CONCLUSION

Ecological footprint invigorates innovation and inspires a one planet living. It aids us in our decision making towards the common goal of sustainability and quantitatively reflects the positive bearing that people's decision making has on our stride towards living within the means of one planet. The existence of ecological overshoot and the overshoot day scaling up the calendar are stark reminders that humanity is surpassing the sustainable ecological footprint on a global scale. The repercussions of these outcomes are even more demanding as we comprehend that at most times the ecological footprint measurements are clear underestimates of the impression humanity places on our earth's ecosystem. Hence, managing our ecological assets is crucial for the survival and sustenance of humanity as a whole in this resource-constrained world.

Reducing our ecological footprint is one step closer to a more sustainable and happier future. If this act of sustainable living and the increased awareness could become a global phenomenon, our descendants and we would be relishing a healthier life. Our stride towards a more self-reliant and sustainable lifestyle begins with the awareness of ecological bounds and limits within which we live. The realization of the impact one has on the planet and his ability to reduce it, though minute, holds the key to a paradigm shift from consumerism to sustainable living. There are several creative and simple ways of reducing our footprint, but the major challenge is to incorporate them into our daily life.

This doggedness on having all that we want at any time we want has put a huge human impression on our ecological systems and has generated a disproportionate amount of waste. To this end, we need to become increasingly aware of the power of a consumer and the indirect impact that our purchases have on the ecology. Thus, any decision about purchases has to be thought about at least twice. By forgoing superfluous consumption you not only save resources but also evade the waste generated out of it. Hence, you can effectively reduce your footprint just by being an informed and selective consumer. Reduce, reuse, and recycle; the three R's mantra in this very sequence should govern our decisions as part of our daily lives. If observed religiously, this in itself will decrease our imprint on the planet.

More often than not, the burden of "saving the planet" gets transmitted to the consumers and the governments, whereas the corporate companies limit their participation in this regard. That is to say, we shift the burden of environmental conservation to a few fragments of the society, mainly to the consumers. In this regard, a more holistic approach comprising of participation of the private, public and the corporate elements is required to enhance the effectiveness of consumerist movement in the green direction and also to boost green consumerism in an all-inclusive format rather than shifting the burden to a few segments of the society. The change in lifestyle choices of individuals should thus be complemented not only with public policies which prioritize preservation of ecological balance but also socially and environmentally through responsible corporate practices without fail.

Additionally, this rising wave of people world over seeking ways of going green, both at their homes as well as their workplaces, opens up a huge market opportunity for entrepreneurs who wish to carry out eco-friendly and sustainable business. With a largely unexplored market, a lot more eco-entrepreneurs are going to be needed to meet the demands of the emergent green economy—developing products and services that provide and facilitate sustainability on a large scale as well as on the individual scale. Small steps in the right direction are always better than doing nothing at all and if the small actions taken individually and collectively become a global norm, then the results would be more welcoming and evident.

KEYWORDS

- **ecological footprint**
- **sustainability**
- **green practices**
- **ecological overshoot**
- **eco-entrepreneurs**
- **consumer awareness**

REFERENCES

Barr, S.; Gilg, A. W.; Ford, N. J. Differences between Household Waste Reduction, Reuse and Recycling Behaviour: A Study of Reported Behaviours, Intentions, and Explanatory Variables. *Environ. Waste Manag.* **2001,** *4*(2),69–82.

Chawla, K. S.; Rajaram, V. R. Reduce, Reuse, Recycle. In *Integration of Nature and Technology for Smart Cities;* Springer International Publishing: Cham, 2016, pp 269–282.

Dietz, T.; Rosa, E. A.; York, R. Driving the Human Ecological Footprint. *Front. Ecol. Environ.* **2007,** *5*(1), 13–18.

Dincer, I. Renewable Energy and Sustainable Development: A Crucial Review. *Renewable Sustainable Energy Rev.* **2000,** *4*(2), 157–175.

Eriksson, C. Can Green Consumerism Replace EnvironmentalRegulation?—A Differentiated-Products Example. *Resour.Energy Econ.* **2004,** *26*(3), 281–293.

Herring, H. Energy Efficiency—A Critical View. *Energy* **2006,** *31*(1), 10–20.

Hopewell, J.; Dvorak, R.; Kosior, E. Plastics Recycling: Challenges and Opportunities. *Philos. Trans. R. Soc. B. Biol. Sci.* **2009,** *364*(1526), 2115–2126.

Ishak, S.; Zabil, N. F. M. Impact of Consumer Awareness and Knowledge to Consumer Effective Behavior. *Asian Soc. Sci.* **2012,** *8*(13), 108.

King, A. M.; Burgess, S. C.; Ijomah, W.; McMahon, C. A. Reducing Waste: Repair, Recondition, Remanufacture or Recycle? *Sustainable Dev.* **2006,** *14*(4), 257–267.

Lundqvist, J.; de Fraiture, C.; Molden, D. *Saving Water: From Field to Fork: Curbing Losses and Wastage in the Food Chain.* Stockholm International Water Institute: Stockholm, 2008.

Menon, A.; Menon, A. Enviropreneurial Marketing Strategy: The Emergence of Corporate Environmentalism as Market Strategy. *J. Mark.* **1997,** *61*(1), 51–67.

Min, H.; Galle, W. P. Green Purchasing Strategies: Trends and Implications. *J. Supply Chain Manage.* **1997,** *33*(3), 10.

Morgan, L. R.; Birtwistle, G. An Investigation of Young Fashion Consumers' Disposal Habits. *Int. J. Consum. Stud.* **2009,** *33*(2), 190–198.

Ninlawan, C.; Seksan, P.; Tossapol, K.; Pilada, W. The Implementation of Green Supply Chain Management Practices in Electronics Industry. *Proc. Int. Multiconf. Eng.Comput. Sci.* **2010,** *3,* 17–19.

Peattie, K.; Ratnayaka, M. Responding to the Green Movement. *Ind. Mark. Manag.* **1992,** *21*(2), 103–110.

Ross, S.; Evans, D. The Environmental Effect of Reusing and Recycling a Plastic-Based Packaging System. *J. Cleaner Prod.* **2003,** *11*(5), 561–571.

Slater, D. *Handbook of Cultural Geography.* Sage Publications: London, 2003; pp 147–164.

Wackernagel, M.; Rees, W. E. *Our Ecological Footprint: Reducing Human Impact on the Earth(No. 9);* New Society Publishers: Canada, 1998.

Wackernagel, M.;Onisto, L.; Bello, P.; Linares, A. C.;Falfán, I. S. L.; Garcıa, J. M.; Guerrero M. G. S. National Natural Capital Accounting with the Ecological Footprint Concept. *Ecol. Econ.* **1999,** *29*(3), 375–390.

Wackernagel, M.; Schulz, N. B.; Deumling, D.; Linares, A. C.; Jenkins, M.; Kapos, V.; Monfreda, C.; Loh, J.; Myers, N.; Norgaard, R.; Randers, J. Tracking the Ecological Overshoot of the Human Economy. *Proc. Natl. Acad. Sci.* **2002,** *99*(14), 9266–9271.

White, K. K. *America Goes Green: An Encyclopedia of Eco-friendly Culture in the United States.* ABC–CLIO: Santa Barbara, CA, 2013; Vol. 1; www.footprintnetwork.org

Zsidisin, G. A.; Siferd, S. P. Environmental Purchasing: A Framework for Theory Development. *Eur. J. Purch. Supply Manag.* **2001,** *7*(1), 61–73.

CHAPTER 6

DETERMINANTS OF CONSUMER PURCHASE INTENTION FOR SOLAR PRODUCTS IN VARANASI CITY

OM JEE GUPTA[1,*] and ANURAG SINGH[2]

[1]*Institute of Management Studies, Banaras Hindu University, Varanasi, UP 221005, India, Mob.: +91 9716892156*

[2]*Institute of Management Studies, Banaras Hindu University, Varanasi, UP 221005, India, Mob.: +91 8004926090, E-mail: anuragbhadauria@gmail.com*

*Corresponding author. E-mail: omjee.soe@hotmail.com

6.1 INTRODUCTION

Eco-friendly products are "products that do not harm the environment whether in their production, use or disposal" (OECD, 2008). In other words, these products help to preserve the environment by significantly reducing the pollution, they could produce. Eco-products are also known as environment-friendly products or green products as they cause minimal harm to people and the environment (OECD, 2008). Green products are energy efficient, nontoxic, produces no air or water pollution, reusable, biodegradable, do not reduce ozone layer, made of recyclable, renewable resource, and unharmful compounds (Dincer, 2000). The green products include solar-powered, wind-powered, and hydro-powered products. Solar energy can be used to generate electricity which is further utilized to operate electrical appliances, vehicles. Solar energy product includes solar fountains, solar water heater, solar light/bulb, solar inverter, solar battery charger, and solar oven (Grupp, 2012; Lorenzo, 1994; Baumann et al., 2002). In other words, the home appliances, which can run with the help of solar power, are kept in this category.

This study is aimed at solar products, which is truly a green product, and is an inspiration for resource conservation. The consumers know that very use of solar products provides an enormous automatic solution to the problems which is dangerous for all species living on planet earth. Solar products usage benefits are not only limited to the person who is consuming but its benefitted effects are for the generation to come (Ottman et al., 2006). The consumer is aware that the use of solar products is beneficial to all living beings across the globe in many ways. It is recyclable, biodegradable, keeps environment protected, and runs through renewable energy sources. Even after there are enormous benefits of solar products, consumers in India are reluctant to use the same.

Many organizations including government and manufacturer of solar products, society, associations, charity, NGOs are continuously taking the initiatives to promote the solar products usage in the country (Velayudhan, 2003). Advertisement in print and electronic media promoting solar product usage can be seen on regular basis. The solar product usage pattern is being promoted with a message of better living through the clean and sustainable environment. Even after the gigantic promotion of solar products usage, the growth of the solar product market is very slow. The research says that the solar product purchasing and usage can only be promoted after the brain mapping of the consumers (Medeiros and Ribeiro, 2013). This study is also a step forward in this direction and tries to investigate the key factors, which influences the purchase intention of consumers for solar products.

6.1.1   GROWING SOLAR PRODUCTS USAGE PATTERN AT A GLANCE

The consequences of climate change have attracted concern from all across the world because they have resulted in melting icecaps in the polar areas, high level of temperature variability, unseasoned rainfall, and rainstorms all across the globe. Because of high variation in unexpected natural disasters the environmental scientists have now agreed to call it a most important addressable issue of the 21st century. Henceforth, it has also got the attention of "political diaspora" around the world (UNFCCC, 2016). The IPCC 4th assessment report (AR4) eloquently states that the topic today is not to discuss the presence of climate change but the discussion of the matter should be "How much change we are committed to, and how fast we can make it happen"? The report of IPCC also states that although this is a long-term problem but it can be handled properly with the help of short-term action plan. The issue of climate change cannot be handled by one country. The collective efforts (making strong policies, development of technology, adoption of technology, and transfer of technology across the countries without any hurdles) of every section of society can help in reducing the adverse effects of climate change (UNEP, 2009). Every section, includes business houses, agencies for environment protections government, policy makers, individuals, and NGOs.

In view of the concern for the protection and sustainability of the environment, the business houses are associating themselves directly or indirectly with green initiatives. Business houses show themselves as they are associated actively with the green social movement. For the purpose, they are taking the help of advertisements as well. To show their concern and consciousness for the protection of the environment, conservation of natural resources, recycling, and reduced wastage of products, the business houses spread the information and create the awareness for the environment. The continuous persuasion of business houses to use green products for the sustainability of the environment has increased the awareness level of customers and has motivated them to purchase and use green products.

On the other side, several international agencies like United Nations Environment Program (UNEP) have taken initiatives to develop the base through policies and cooperation so as to deal with climate change. UNEP is taking all the initiatives in providing information and generating the capacity to deal with climate change. It is also helping through financing for renewable energy investments in the form of solar panels, wind, and thermal systems. They are also creating the platform where they can engage investors and project developers. Most of the efforts are for small and micro level enterprise development especially in Asia, Africa, and Latin American countries. This is considered as a leading project of UNEP. It is pertinent to mention that the initiatives of international agencies on climate change have enormously contributed in generating awareness of green products across the globe.

The efforts of the government for the promotion of the solar products cannot be neglected as the initiatives of many governments across the globe are successful in generating the awareness, and the usage of solar products among the society members. The government of member countries is successful by waking the society to save the environment and making it sustainable for the future generation. To generate the awareness among the society and to make the society sensitive for the environment, the govern-ment has initiated many awareness campaigns (Solar Energy Corporation of India Limited, 2016). Government initiatives have led the society to realize that the use of conventional energy will be a great hazard to the environment hence it will be better to shift to renewable energy source products. Now, the individuals have realized that the alternative to conventional energy, renewable energy products are available such as solar energy, which can satisfy their needs by reducing the carbon emission, strengthening the economy, and making individuals energy independent.

On the other side, the government throughout the world has also realized the usage of solar energy in countries and industrial energy generation. Therefore, the government of many countries has decided to use renewable energy in industries rather than relying on any other forms of energy which are running out. Germany, Italy, Spain, and the United States are among the top 10 users of the solar energy panels in industries (Renewable Energy Corporation, 2016). These countries are meeting their maximum energy consumption targets through solar energy products. Germany has not only installed thousands of solar panels but is also planning to use renewable energy in a massive way by 2050. These countries are the top buyer of solar energy panels for several years and are expected to continue going forward in the same direction during the upcoming years.

As far as the scope of renewable products market across the globe is concerned, though India does not occupy the place in top 10 markets, but is ready for the extensive use of the solar product in Indian industries and consumer usage. The National (2014) says that the Indian market has the required constituents to attract renewable energy companies and consumers at one platform. It is also being said that because of the availability of high solar radiance and the power deficit, this sector can attract more foreign investors to Indian market (Medeiros and Ribeiro, 2013).

India has taken an initiative to electrify 60 million homes by solar power by the year 2022. Electrifying 60 million homes by solar power is the government goal to cover 40% of its uses from the nonfossil fuels by 2030 (Aljazeera, 2016). Though it is the unprecedented target, India has taken in its hand but is achievable. Thinking about electrifying a huge number of villages through solar energy shows the willingness of the country in curtailing global warming or climate change which ultimately translates into more viable society blended with green environment altogether. In view of the goal of the government, there is a huge scope for the companies working or willing to work in the area of renewable energy sources.

Along with the Indian government, the consumers are also thinking and working to propagate and accept the solar energy run products, which is beneficial for an individual and environment (The Economic Times, 2017). In other words, it can be said that the Indian consumers are getting ready to accept and use the solar products in coming time. Therefore, this study endeavors to investigate the objective considered in this research for the promotion of renewable energy run products in India.

6.2   LITERATURE REVIEW

The environment-friendly products are termed as the green products (Chan, 2001). The growth of green products are a result of undergoing and developing concern for the global and local pollution, climate change, depleting natural resources, overflowing of wastes, and so forth (Srivastava, 2007). The environment-friendly products are very helpful in avoiding the pollution and harm to the environment from the product use (Ramayah et al., 2010). It also helps in reducing the harm to self and to others, when using these products.

There are several factors which help in building the mood of a consumer and also influence the intention to buy green products (Ramayah et al., 2010). The green product purchase is said to be an ethical and accountable behavior of the consumer which intends to give sustainability to the society and does no harm to nature (Gadenne et al., 2011). Thinking about environment protection is one of the important reasons for consumers to purchase environment-friendly products (Gadenne et al., 2011). This thinking emerges due to the role of green products in improving the quality of the environment and sustainable life (Escalas and Bettman, 2005). Keeping the issue of environment protection in mind the companies designs the products which are less harmful to the environment (Papadopoulos et al., 2010). Most of the consumer behavior models address the environmental concern of consumption pattern and motivates to alter the conventional behavior of consumption in order to sustain along with clean environment (Hansen, 2009). The consumers boycott products of those companies who do not follow environmental regulations and do not produce environment-friendly products (Laroche et al., 2001). Ajzen (1991) said that the consumers' attitude toward the purchase of green products is due to favorable or unfavorable intention.

The purchase intention of green products can be described to support green companies (Albayrak et al., 2013; Schlegelmilch et al., 1996), adopting sustainable consumption practices (Gadenne et al., 2011), and likely to spend more for green products (Essoussi and Linton, 2010). A consumer feels responsible toward the environment and society who purchases green products; they also feel that nonuse of green products may cause the hazard to the environment (Moisander, 2007). Consumer feels that the green products are not harmful to nature and also give the meaningful quality of life, that is why the search and purchase of green products are increasing day by day (Escalas and Bettman, 2005).

Majority of the studies on green products are being undertaken either in European or the American context. Many researchers are now focusing to understand consumer behavior in Asian context (Lee, 2008, 2009; Gurau et al., 2005; Yam-Tang and Chan, 1998). Researchers, for example, Singh, 2013, have argued that India is a big potential market for green products. Most of the studies so far concentrated on modified food choices, organic food, and other green products (Anand, 2011; Chakrabarti, 2010; Knight and Paradkar, 2008). The available literature is silent on consumer intention for solar products.

The literature review reveals that many researchers around the world tried to study the purchase intention of green products, but have not deliberated to find the determinants, which influences the purchase intention of solar products. Finding the gap, this research tries to identify the determinants of consumer purchase intention for solar products in Varanasi city.

6.3   RESEARCH METHODOLOGY

6.3.1   STATEMENT OF PROBLEM

The availability of solar products is widespread in the market. This includes the products such as solar panel, street lights, solar lamps, pump sets, and many more. Government is taking enormous efforts to generate the demand for solar products. To generate the demand for solar energy run products, the government is giving the subsidies. Even then substantial individual demand is missing. Although the awareness about solar products is increasing, still it is felt that consumers are reluctant in buying these products. The consumer knows the benefits of solar products but the operational knowledge is missing. That is why the actual demand for these products is not at par with the desired level.

The information reveals that the awareness, benefits, and knowledge of solar product among the consumer is not at par, hence, the big percentage of consumers are not interested in purchasing the solar products (Sushma, 2016). No/Less interest of the customers to purchase the solar products is an issue which is creating the challenge for demand. In other words, it can be said that the available initiatives of the government and marketers have failed in generating the demand for the solar products. The demand can be generated by unlocking the consumers' mind and studying the factors considered by consumers to purchase the solar products. Understanding the challenge of solar product demand, this research proposes to figure out the determinants of consumer purchase intention for solar products in Varanasi city.

6.3.2   OBJECTIVES

The main aim of the study is to find out the determinants of consumer purchase intention for solar products in Varanasi city. To study the main aim of the research, this research endeavors to:

- Investigate the factors influencing the intention of the consumer to purchase solar products.
- Draw the managerial implication of those factors in the context of Varanasi city.

6.3.3   RESEARCH DESIGN

This study has followed the exploratory cum descriptive research design. The study starts by using the exploratory method. All the respective available literature was reviewed and then researcher framed the objectives accordingly. The objectives were selected keeping the pilot study in mind, which helped the researcher in its formulation. Then, the researcher employed the descriptive method to undertake further study.

The researcher has distributed 150 questionnaires to the respondents. But some of them due to some reasons did not fill the complete questionnaire. And, because of which the collected data after sorting it comes to 104 responses. Then, this data (of 104) respondents was tabulated and analyzed in SPSS 20.0 software. The demographic data shows that out of the total 104 respondents 85 were male and 19 were female. The male respondents were 81.73% and female respondents were 18.26%. The respondents belong to age group of 25–50 years. The internal consistency measure of the research instrument was tested by reliability analysis.

To facilitate the study in a structured way, the data was collected from both primary and secondary sources. The respective data which is already available in the body of literature was being investigated. All Secondary sources such as research papers published in journals, conference proceed-ings, Indian Government websites, international agencies websites, news articles, and books were used to conduct the literature review and to develop the questionnaire. The items were taken from previous studies (Kumar et al., 2015; Norazah, 2016; Lee, 2008). After the modification of the items as per the requirement of the study, the items were finalized and were devel-oped in the form of a questionnaire. Hence, a well-structured questionnaire containing 15 items was developed to collect the consumer's responses. The researcher has used purposive sampling technique to collect the data for the study. The primary source was used to collect the responses from the solar product users of Varanasi. The final items of the questionnaire are as follows:

1. I feel proper information will bring a positive change in the mind-set of people.
2. I like to have more education about solar products.
3. I feel people are less aware of its advantages.
4. I feel more information about solar products will lead to its increased uses.
5. I feel meaningful in supporting environmental protection.
6. I feel that the ingredients of solar products are not harmful to nature and animals.
7. I feel good in using solar products as it avoids the burning of fossil fuel.
8. I feel good to contribute to building clean environment for future generation.
9. I feel solar products lead to no harm.
10. I like to share my experience with solar products with my friends.
11. I like to learn more about environmental issues.
12. I feel supporting solar products makes me more socially attractive.
13. I feel better that I am a user of solar products.
14. I feel in leaving behind good precedence for future.
15. I feel owning a social cause builds a reputation.

The primary data was collected in five-point Likert scale from (1 = strongly disagree, 2 = disagree, 3 = neutral, 4 = agree, 5 = strongly agree).

The data was coded and was crammed in SPSS 20.0 software. Further, the exploratory factor analysis technique was followed to figure out the determinants influencing the purchase intention of Varanasi consumers for solar products.

6.4 ANALYSES AND FINDINGS

The analysis was started after the tabulation of the data in SPSS 20.0 software. The reliability of the questionnaire was calculated, which was followed by exploratory factor analysis. The findings are then elaborated subsequently.

6.4.1 RELIABILITY ANALYSIS

The descriptive statistics of the variables and reliability estimates are shown in Table 6.1 given below. Reliability analysis revealed Cronbach's α value for the questionnaire is 0.768. This shows that the data collected for the purpose is reliable for the study. Below table also shows 15 items taken for the study.

TABLE 6.1   Reliability Statistics.

Cronbach's α	Cronbach's α based on standardized items	No of items
0.768	0.764	15

6.4.2 FINDINGS AND DISCUSSION

The scale constituted on 15 items was analyzed using principal component analysis and varimax rotation technique. Table 6.2 illustrates the value of Kaiser-Meyer-Olkin (KMO) test of sampling adequacy and Bartlett's test of sphericity. The result of Bartlett's test of sphericity is 0.000 and the KMO value is 0.710, meeting the assumption for factorability (Field, 2013). The KMO test result with 0.5 or above is considered good for the study. And if the KMO test is below 0.5 then it is unacceptable. The literature says that if the KMO test result goes upwards it is always good for the study and fulfills the conditions of conducting exploratory factor analysis. Hence, exploratory factor analysis is considered well fit for the analysis (Hair et al., 2010).

TABLE 6.2   KMO and Bartlett's Test.

Kaiser-Meyer-Olkin measure of sampling adequacy	0.71
Approx. chi-square	696.943
Bartlett's test of sphericity df	105
Sig	0.000

Bartlett's measure test reveals the significance of the study. We need some relationship between the items while conducting factor analysis. So that the items converge into factors. Bartlett test is significant ($p<0.001$) of the data for the study, therefore, it can be concluded that the factor analysis would be appropriate for the study.

6.4.3   FACTOR EXTRACTION

Principal component analysis as extraction method and varimax rotation method with Kaiser Normalization technique was used for factor analysis. The number of factors was determined using eigenvalue greater than 1 and a screen plot for the emerged model was also plotted (see Fig. 6.1). Below given Table 6.3 illustrates the eigenvalue for each item. This process distinguishes certain 4 factors with the eigenvalue more than 1. SPSS has identified 15 items initially. This table also shows that after extraction, 4 factors are being identified, where factor explains the percentage of variance (1st factor=25.29%, 2nd factor=18.73%, 3rd factor=13.92%, and 4th factor=11.05%). Cumulatively all the 4 factors combined together explains 68.99% of the data.

Figure 6.1 portrays the screen plot by showing the dimension of all the items and pointing out 4 factors with eigenvalue more than 1. Rest of the factors, that is, 11 items are below the eigenvalue 1. As per the factor analysis, these 4 factors are adequate enough to explain all the 15 items considered for the study. Each extracted factor has subsumed all the correlated items into it.

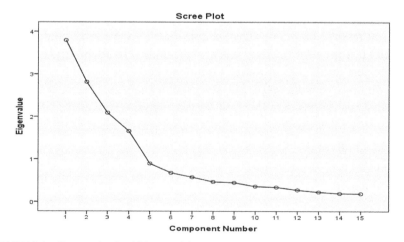

FIGURE 6.1   Screen plot for 15 items of the survey.

TABLE 6.3 Total Variance Explained.

Component	Initial eigenvalues			Extraction sums of squared loadings			Rotation sums of squared loadings		
	Total	% of variance	Cumulative %	Total	% of variance	Cumulative %	Total	% of variance	Cumulative %
1	3.794	25.293	25.293	3.794	25.293	25.293	3.340	22.264	22.264
2	2.810	18.734	44.027	2.810	18.734	44.027	2.644	17.624	39.888
3	2.089	13.924	57.951	2.089	13.924	57.951	2.253	15.020	54.908
4	1.657	11.048	68.998	1.657	11.048	68.998	2.114	14.090	68.998
5	0.896	5.971	74.969						
6	0.676	4.505	79.474						
7	0.575	3.834	83.307						
8	0.467	3.111	86.418						
9	0.448	2.987	89.405						
10	0.359	2.396	91.801						
11	0.341	2.275	94.076						
12	0.275	1.835	95.911						
13	0.227	1.516	97.427						
14	0.196	1.304	98.731						
15	0.190	1.269	100.000						

Extraction method: Principal component analysis.

6.4.4 FACTOR ROTATION ANALYSIS

Table 6.4 illustrates that all the 4 factors loaded by 15 items are considered for the study. In this table, there is no overlapping of factor loading that means the loading of particular items is only considered for one extracted factor. All the item loadings to the extracted factors are in the range of 0.729–0.869. The constructs of all the 15 items to their subsequent loadings demonstrate a significant contribution to the respective factors, that is, extracted factors. Factor 1 gets loadings from 5 items, Factor 2 gets loadings from 4 items, Factor 3 gets loadings from 3 items, and finally, Factor 4 gets loadings from 3 items. This only can be demonstrated using varimax technique under rotation method. This matrix helps in eliminating the dual loadings from a single item. Rotation method improves the loadings of each item on the extracted factors, and at the same time minimizes the loading on rest the other extracted factors.

We see the extracted factors produced by principal component analysis using varimax rotation method with Kaiser Normalization technique. Looking at their properties, we can make some genuine common proposition which carries weight for all the respective items.

TABLE 6.4 Rotated Component Matrix.

	Component			
	1	2	3	4
I feel that the Ingredients of solar products are not harmful for nature and animals	0.847			
I feel solar products lead to no harm	0.836			
I feel good in using solar products as it avoids the burning of fossil fuel	0.824			
I feel good to contribute in building clean environment for future generation	0.799			
I feel meaningful in supporting environmental protection	0.748			
I feel more information about solar products will lead to its increased uses		0.822		
I like to have more education about solar products		0.801		
I feel proper information will bring a positive change in mind-set of people		0.767		
I feel people are less aware about its advantages		0.748		
I feel owning a social cause builds reputation			0.869	
I feel better that I am a user of solar products			0.831	
I feel in leaving behind good precedence for future			0.729	

TABLE 6.4   *(Continued)*

	Component			
	1	**2**	**3**	**4**
I like to learn more about environmental issues				0.835
I like to share my experience of solar products with my friends				0.831
I feel supporting solar products makes me more socially attractive				0.764
Extraction method: Principal component analysis.				
Rotation method: Varimax with Kaiser normalization				
a. Rotation converged in 5 iterations				

6.5   MARKETING IMPLICATIONS

This research illustrates the significant implications for the marketers and policymakers as it carries the information about the consumer's mind-set. Understanding the customers' mind-set is always a top priority for the marketers and policymakers. All the marketing efforts are centered on customers, and that forms the core of marketers and policymakers concern (HBR, 2014). In other words, this research talks about the factors influencing the consumer purchase intention for solar products in Varanasi. By understanding and bringing the factors into practice, the marketers and policymakers will be able to serve consumers in a better way and can generate better demand for solar products. This would be helpful for existing companies serving in this field and will also help new entrants (companies) to formulate their strategy accordingly. Each extracted factor has been given an adequate name followed by the subsequent constituents of respective factor (with factor loadings) below.

6.5.1   FACTOR 1: CONCERN FOR PROTECTION

Among the four distinguished factors, Factor 1 in Table 6.5 which emerges as the most important factors in this study. Factor 1 gets loadings from five statements. The name of this factor cannot be other than "concern for protection." All the five statements talk about the protection of nature, animals, avoiding burning fossil fuels, clean environment, and shows concern for future generation. This shows the pragmatic views of the consumer from which we can observe that the consumer is not only interested in buying green products for his benefits but also interested in contributing something to the society by not using the devices run by conventional energy sources. They also feel that using renewable energy sources will contribute to developing a clean and sustainable environment, and not cause harm to nature and will give better future to next generations. Therefore, to have a requisite demand for solar products, marketers need to showcase the products with the vicarious benefits, so that the customer can take the benefits into consideration before purchasing.

TABLE 6.5   Constituents of Factor 1.

I feel that the Ingredients of solar products are not harmful for nature and animals (0.847)
I feel solar products lead to no harm (0.836)
I feel good in using solar products as it avoids the burning of fossil fuel (0.824)
I feel good to contribute in building clean environment for future generation (0.799)
I feel meaningful in supporting environmental protection (0.748)

Marketers must keep this factor in mind that today's consumers are more concerned about these facts. This will help them in understanding consumers' mind in regard to solar products.

6.5.2   FACTOR 2: AVAILABILITY OF INFORMATION AND KNOWLEDGE

Another important factor, that is, Factor 2 in Table 6.6, talks about the "availability of information and knowledge." This extracted factor talks about the consumers who are interested to have more information about these products, and that potential customers also not properly informed about the solar products. This shows that the majority of potential consumers are less aware of the benefits of these products. If the marketer provides information and the knowledge about the solar products and its benefits to the environment, it will be of great use to the customers. As the existing customer considers the information of solar product use and its benefits as an important factor, hence, providing the same may generate demand for the solar products. To provide the information and knowledge to the customers, marketers can conduct a workshop and proceed with informative advertisements. Proper availability of information and knowledge will produce a substantial result in the subsequent future as it has been observed that people, due to lack of knowledge and no past experience of its benefits, do not purchases solar products. Marketers can start a campaign to counter this underpinning by providing some incentive from their side. This could be through advertisements, direct communication to the existing and potential customers. This will motivate people to come out from their existing zone to green zone.

TABLE 6.6 Constituents of Factor 2.

I feel more information about solar products will lead to its increased uses (0.822)
I like to have more education about solar products (0.801)
I feel proper information will bring a positive change in mind-set of people (0.767)
I feel people are less aware of its advantages (0.748)

6.5.3 FACTOR 3: SOCIAL REPUTATION AND APPEAL TO OTHERS

The next important factor, Factor 3 in Table 6.7, deals with the "social reputation and appeal to others." This factor demonstrates the feel of social reputation by the use of solar product usage in the society. This section of consumer views demonstrates that the one who uses these products holds a reputational position in the society. Consumers do think that they are perceived as an environmentally conscious member of the society. This also leaves a good precedence for the future generation to follow.

In other words, this factor portrays, that the usage of solar products builds the reputation of people in the society and the same appeals to others to do the same. People are interested in attaching themselves with the social causes. Marketers are advised to collaborate themselves with those programs which build and promote the image of the solar product users. Further, the marketer should use the social image of the solar product users to appeal the potential solar product users.

TABLE 6.7 Constituents of Factor 3.

I feel owning a social cause builds reputation (0.869)
I feel better that I am a user of solar products (0.831)
I feel in leaving behind good precedence for future (0.729)

6.5.4   FACTOR 4: EXPERIENCE AND RESPONSIBILITY

The analysis has extracted the 4th factor in Table 6.8, that is, "experience and responsibility." The result reveals that the consumers are always interested to protect the environment through the experience of renewable energy run solar products. Solar product users are always the attraction of society and also feel self-responsible by sharing their experience of solar products use with potential users. The research result also reveals that the customers are self-motivated to share the solar product experience with others, as they feel the use of the solar product will not only benefit them but also will benefit the next generation. Hence, the marketers should understand and take the proper initiative to utilize experienced consumers (willing to share their experience with others) in motivating the potential customers through marketing programs. The existing consumers are displaying their interest to share the experience with others. They also, as a user of solar products, demonstrate that they are the responsible citizen of the country. All the existing and potential solar products producing companies keep all these facts in mind when they go to market their products.

TABLE 6.8   Constituents of Factor 4.

I like to learn more about environmental issues (0.835)
I like to share my experience of solar products with my friends (0.831)
I feel supporting solar products makes me more socially attractive (0.764)

6.6   CONCLUSION

Many international treaties are conforming to the fact that the global warming or climate change is no longer tolerable in modern society. The central government is also trying to battle emission of fossil fuel from every end. The government through Prime Minister Ujjawala Yojna is providing the LPG (nonrenewable energy) connection for the rural poor and is giving the subsidy. That is, providing substantial health benefits to the housewives. It is pertinent to mention that the government is equally serious to promote the solar energy products by offering the subsidy to rural and urban people through different government and nongovernment agencies (Ministry of New and Renewable Energy, 2016). The government of India is taking the help of many organizations in designing and developing products which are environment-friendly and are less harmful to people at large. People are showing interest to purchase and use these solar energy run products.

This study has explored 15 independent items and sought the responses from the existing users of solar products in Varanasi. Taking the aim to explore the items influencing the purchase intention of solar products into consideration, the research revealed that the four separate factors emerge and get the loadings from all the 15 items. The factor carries weight in the form of application which needs to be included in the strategies of the marketers. Once the marketer tries and includes all these factors into their marketing program, then, these things will show in their results. Hence, the marketers are advised that they should include these responses of existing consumer into their strategy in order to improve the market for solar products and to satisfy the consumers' demand. It is very obvious that if the marketer considers these 4 factors into their strategy and takes the proper initiatives to practice, the market and demand for the solar products can be improved in future. This research paper is unique of its kind which has brought the consumer's side in the forefront. Every marketing strategy must be formulated keeping the consumer at the center.

As no study is an exception to the limitation. This study also has some limitations. The study was conducted in the Varanasi city only. The result has the capacity to be emulated for the similar cities only. Therefore, it may not be prudent to use its results in cities which have a distinct type of consumer base. The very important limitation of the study is that any statement of price has not been taken into consideration. The same may be one of the important factors to influence the intentions of the customer to purchase the solar product. In a country like India, where we have large growing middle class, the price may have been an important determinant.

Limitations also suggest the scope for future researches. Therefore, similar kind of studies can be conducted in metro cities and in rural areas. It would be interesting to understand how consumer thinks about these products in the rural area. This will be an exciting opportunity for the researcher to conduct similar kind of study in rural area of Varanasi. This will be a step ahead in reducing the limitations stated above. The research conducted in metro cities and rural areas might produce some glaring revelation about determinants of consumer intentions to purchase and uses of solar products.

KEYWORDS

- **green products**
- **solar products**
- **consumer psyche**
- **factors**
- **purchase intention**

REFERENCES

Ajzen, I. The Theory of Planned Behavior. *Organ. Behav. Hum. Decis. Process* **1991**, *50*(2), 179–211.

Albayrak, T.; Aksoy, S.; Caber, M. The Effect of Environmental Concern and Scepticism on Green Purchase Behaviour. *Mark. Intell. Plann.* **2013**, *31*(1), 27–39.

Aljazeera. India Unveils The World's Largest Solar Power Plant. 2016, http://www.aljazeera.com/news/2016/11/india-unveils-world-largest-solar-power-plant-161129101022044.html (accessed Nov 26, 2016).

Anand, R. A Study of Determinants Impacting Consumers Food Choice with Reference to the Fast Food Consumption in India. *Soc. Bus. Rev.* **2011**, *6*(2), 176–187.

Baumann, H.; Boons, F.; Bragd, A. Mapping the Green Product Development Field: Engineering, Policy and Business Perspectives. *J. Cleaner Prod.* **2002**, *10*(5), 409–425.

Chakrabarti, S. Factors Influencing Organic Food Purchase in India—Expert Survey Insights. *Br. Food J.* **2010**, *112*(8), 902–915.

Chan, R. Y. Determinants of Chinese Consumers' Green Purchase Behavior. *Psychol. Mark.* **2001**, *18*(4), 389–413.

Dincer, I. Renewable Energy and Sustainable Development: A Crucial Review. *Renewable Sustainable Energy Rev.* **2000**, *4*(2), 157–175.

Escalas, J. E.; Bettman, J. R. Self-Construal, Reference Groups and Brand Meaning. *J. Cons. Res.* **2005**, *32*(3), 378–389.

Essoussi, L. H.; Linton, J. D. New or Recycled Products: How Much are Consumers Willing to Pay? *J. Consum. Mark.* **2010**, *27*(5), 458–468.

Field, A. *Discovering Statistics Using IBM SPSS,* 4th ed.; SAGE Publications India Pvt. Ltd: New Delhi, 2013.

Gadenne, D.; Sharma, B.; Kerr, D.; Smith, T. The Influence of Consumers' Environmental Beliefs and Attitudes on Energy Saving Behaviour. *Energy Policy* **2011**, *39*(12), 7684–7694.

Grupp, M. *Time to Shine: Applications of Solar Energy Technology;* John Wiley and Sons: Hoboken, New Jersey, 2012.

Gurau, C.; Ranchhod, A. International Green Marketing: A Comparative Study of British and Romanian Firms. *Int. Mark. Rev.* **2005**, *22*(5), 547–561.

Hair, J. F.; Black, W. C.; Babin, B. J.; Anderson, R. E; Tatham, R. L. *Multivariate Data Analysis: A Global Perspective.* Pearson Education Inc.: New Jersey, NJ, 2010.

Hansen, M. G. *Environmental Engagement and Product Knowledge Among Consumers of Electric Light Bulbs in Albany, California*; published in ES196 May 2009, Senior Research Seminar, Environmental Sciences Group Major, University of California at Berkeley: Berkeley, CA, 2009; pp 1–12.

HBR. Decision-Driven Marketing. 2014, https://hbr.org/2014/07/decision-driven-marketing (accessed Jan 03, 2017).

IPCC. 2016, https://www.ipcc.ch/activities/activities.shtml (accessed Dec 28, 2016).

Knight, J.; Paradkar, A. Acceptance of Genetically Modified Food in India: Perspectives of Gatekeepers. *Br. Food J.* **2008,** *110*(10), 1019–1033.

Kumar, P.; Ghodeswar, M. Factors Affecting Consumers 'Green Product Purchase Decisions. *Mark. Intell. Plann.* **2015,** *33*(3) 330–347.

Laroche, M.; Tomiuk, M.; Bergeron, J.; Barbaro-Forleo, G. Targeting Consumers Who Are Willing to Pay More for Environmentally Friendly Products. *J. Consum. Mark.* **2001,** *18*(6), 503–520.

Lee, K. Opportunities for Green Marketing: Young Consumers. *Mark. Intell. Plann.* **2008,** *26*(6), 573–586.

Lee, K. Gender Differences in Hong Kong Adolescent Consumers' Green Purchasing Behaviour, *J. Consum. Mark.* **2009,** *26*(2), 87–96.

Lorenzo, E. Ed. *Solar Electricity: Engineering of Photovoltaic Systems;* Earthscan/James and James: Spain, 1994.

Medeiros, J. F.; Ribeiro, J. L. D. Market Success Factors of Sustainable Products. *Independent J. Manage. Prod.* **2013,** *4*(1), 188–207.

Ministry of New and Renewable Energy. Solar Scheme. 2016, http://mnre.gov.in/related-links/jnnsm/introduction-2/ (accessed Dec 29, 2016).

Moisander, J. Motivational Complexity of Green Consumerism. *Int. J. Consum. Stud.* **2007,** *31*(4), 404–409.

Norazah, M. S. Green Product Purchase Intention: Impact of Green Brands, Attitude, and Knowledge. *Br. Food J.* **2016,** *118*(12) 2893–2910.

OECD. Promoting Sustainable Consumption, GOOD PRACTICES IN OECD COUNTRIES, 2008, https://www.oecd.org/greengrowth/40317373.pdf (accessed Oct. 22, 2016).

Ottman, J. A.; Stafford, E. R.; Hartman, C. L. Avoiding Green Marketing Myopia: Ways to Improve Consumer Appeal for Environmentally Preferable Products. *Env. Sci. Policy Sustainable Dev.* **2006,** *48*(5), 22–36.

Papadopoulos, I.; Karagouni, G.; Trigkas, M.; Platogianni, E. Green Marketing: The Case of Greece in Certified and Sustainably Managed Timber Products. *Euro. Med. J. Bus.* **2010,** *5*(2), 166–190.

Ramayah, T.; Lee, J. W. C.; Mohamad, O. Green Product Purchase Intention: Some Insights from a Developing Country. *Res. Conserv. Recycl.* **2010,** *54*(12), 1419–1427.

Renewable Energy Corporation. Benefits of Solar, 2016, http://renewableenergysolar.net/benefits-of-solar/ (accessed Nov 25, 2016).

Schlegelmilch, B. B.; Bohlen, G. M.; Diamantopoulos, A. The Link between Green Purchasing Decisions and Measures of Environmental Consciousness. *Eur. J. Mark.* **1996,** *30*(5), 35–55.

Singh, G. Green: The New Colour of Marketing in India. *ASCI J. Manage.* **2013,** *42*(2), 52–72.

Solar Energy Corporation of India Limited. Major Government Initiatives, 2016, http://www.seci.gov.in/content/govt_initiatives.php (accessed Nov. 20, 2016).

Srivastava, S. K. Green Supply Chain Management: A State of the Art Literature Review. *Int. J. Manage. Rev.* **2007,** *9*(1), 53–80.

Sushma. Why More People Now Own Their Home's Solar Panels Instead of Lease Them, 2016, www.marketwatch.com (accessed Dec 29, 2016).

The Economic Times. Why India Might Not Achieve its 2020 Renewable Energy Targets, 2017, http://economictimes.indiatimes.com/industry/energy/power/why-india-might-not-achieve-its-2020-renewable-energy-targets/articleshow/56389839.cms (accessed Jan 07, 2017).

The National. India Bids to Harness More Green Power, 2014, http://www.thenational.ae/business/energy/india-bids-to-harness-more-green-power (accessed Nov 20, 2016).

UNEP. Climate Change Strategy, 2009, http://www.unisdr.org/files/8356_UNEPCCS TRATEGYweb1.pdf (accessed Dec 22, 2016).

UNEP. India Opens the World's Biggest Solar Farm And A Chemicals Plant Using CO_2 to Make Baking Powder, 2016, http://www.bbc.com/news/business-38391034 (accessed Dec 26, 2016).

UNFCCC, Nationally Determined Contributions (NDCs), 2016, http://unfccc.int/paris_agreement/items/9485.php (accessed Nov. 26, 2016).

Velayudhan, S. K. Dissemination of Solar Photovoltaics: a Study on the Government Programme to Promote Solar Lantern in India. *Energy policy,* **2003,** *31*(14), 1509–1518.

Yam-Tang, E. P. Y.; Chan, R. Y. K. Purchasing Behaviours and Perceptions of Environmentally Harmful Products. *Mark. Intell. Plann.* **1998,** *16*(6), 356–362.

ENVIRONMENTALLY CONSCIOUS CONSUMER BEHAVIOR AND GREEN MARKETING: AN ANALYTICAL STUDY OF THE INDIAN MARKET

PRADEEP KAUTISH*

College of Business Management, Economics and Commerce, Mody University of Science and Technology, NH-11, Lakshmangarh-332311, Sikar, Rajasthan, India, Tel.: +91 9001097506

Corresponding author. E-mail: pradeep.kautish@gmail.com

7.1 INTRODUCTION

The emergence of global environmental problems as the critical business issue is well evident in rising awareness of the relationship between industrialized societies and environment deterioration (Stern et al., 1992). Everyday life activities and business aspects such as technology, product purchases, consumption, and consumerism are getting affected by environmental movement (Zimmer et al., 1994). The human civilization has abused the natural resources beyond its recovering capacity (Ottman, 1993), adversely altered the ecosystem, and extinguished the flora and fauna on the planet to its level (Kleiner, 1991; Yam-Tang and Chan, 1999). With advent to proliferation of the environmental theme, during the last few years, it has also been observed that Indian consumers have also started paying attention to their existing patterns of consumption and subtly they have become more concerned about the impact of consumption on environment (Jain and Kaur, 2006; Nair and Menon, 2008). It has been widely investigated across markets that human activities have altered the natural ecosystem so in order to make natural resources available for future generations. There is a greater need to achieve more sustainable forms of economic development (Antil, 1984; Bloom, 1995; Burgess, 2003; Griskevicius et al., 2010; Kautish, 2016). According to consumer studies across the globe, they clearly state that 30–40% of current environmental degradation is due to the consumption activities of private households which is a substantial market for green practices (Gronhoj, 2006; Kautish and Dash, 2016; Straughan and Roberts, 1999). The concern for environment protection dominant the marketing theme in many forms since its inception (Drumwright, 1994; Ottman et al., 2006) and then after public concern picked the steam up as a result of environment protection promotional campaigns of the corporations in big way, which geared up the market in this direction (Charter and Polonsky, 1999; Fineman, 1996; Stern et al., 1995). This is also an established fact that the modern fiercely urbanized and very fast life style of last few decades, which got fortified with postindustrial revolutionary phase has done irreparable damage to the environment on earth in a drastic way (Diamantopoulos et al., 2003). There is a need to investigate this topic in less affluent societies like India. India stands next only to China in terms of population, accounting for about 16% of the world population (Chan, 1996; Jain and Kaur, 2006). Even in the emerging economies like India, people have come to realize the importance of environment protection for next generation and which is well evidenced by emergence of environmental issues in last decade (Nair and Menon, 2008). Even if a fraction of the Indian consumers goes green, they can contribute significantly to the cause of environmental protection. Indian marketers and policy makers too can benefit by employing segmentation-based green marketing strategies (Jain and Kaur, 2003). But no significant study exists in the country to guide the decision-makers in this respect. This is a good evidence of a paradigmatic business opportunity shift in marketing orientation toward environment (Lee, 2008; Moisander and Pesonen, 2002; Spangenberg, 2004) and the driving force responsible for such a shift is owing to growing fundamental awareness in the society about green behavior (Rugman and Verbeke, 2000; Schultz, 2000; Segun et al., 1998). The last couple of decades have contributed unprecedented growth in the proliferation of consumer awareness about environmental deterioration (Anderson and Cunningham, 1972; Antil, 1984; Smart, 2010). The companies offering such products are interested in finding the consumer behavioral determinants in order to formulate marketing strategies for their successful businesses (Crane, 2000; Kinnear et al., 1974; Wheatly, 1993).Beginning in the 1990s, a very significant amount of research has been conducted on environmentally concerned consumer behavior for green products across developed countries (Ling-Yee, 1997; Schlegelmilch et al., 1996) but very few are from developing countries (Furman, 1998). In India too, some researches were conducted to establish the fact that Indian consumer's awareness is reasonably good and companies can tap the markets; this segment has great potential (Bhate, 2002; Jain and Kaur, 2003; Jain and Kaur, 2004; Jain and Kaur, 2006; Kautish and Soni, 2012; Punyatoya, 2014). Many consumer behavioral dimensions and factors emerged in the context to drive choice set in purchasing environment-friendly products (Nair and Menon, 2008). Growing concerns about the environment has prompted marketers to consider the influencing factors and consumer perceptions of a firm's offering purchase decision; the issue has been explored in new lights with early theories of behavioral understanding as well to give fresh perspective to the existing body of knowledge about consumer behavior (Balderjahn, 1988; Kalafatis et al., 1999). These factors can be grouped into values, beliefs/knowledge, needs and motivations, attitudes, and demographics (Wagner, 2003). Moreover, a number of intervening variables affect consumers' intention to pay more for an environment-friendly product, grouped into eco-labels and consumer backlash. The major issue, that is, who is concerned and who is not helps companies target their consumer offerings with adequate information about tastes and preferences (Henriques and Sadorsky, 1996; Straughan and Roberts, 1999). Tanner and Kast (2003) have conducted a study on Swiss consumers on determinants of green purchases which promotes sustainable consumption; Burgess (2003) tried to figure out feasibility continuum of sustainable consumption. Similar study has been conducted by Vermeire and Verbeke (2006) which explored the sustainable food consumption pattern with consumer attitude and intention gap model which is one of the most appropriate citations on environmental concern (EC) from green consumer behavior.

7.2 THEORY AND FRAMEWORK

Over the years, marketer's EC and consumer interests in environment-friendly products were instrumental in bringing varied terms in marketing discipline, that is, ecological marketing (Henion, 1976), environmental marketing (Coddington, 1993; Nair and Menon, 2008), green marketing (Charter and Polonsky, 1999; Ottman et al., 2006), socially responsible marketing (McDonald and Oates, 2006), and sustainable marketing (Fuller, 1999). Webster (1975) demarcated the socially conscious consumer in altogether different way "a consumer who takes into account the public consequences of his private consumption or who attempts to use his or her purchasing power to bring about social change." The environmentally conscious consumer is convinced that he/she can actually change the status quo of continued environmental deterioration (Follows and Jobber, 2000; Henion, 1976). Van Liere and other researchers made a distinction between the "traditional" socially concerned consumer and the environmentally conscious consumer. Mohr et al. (2001) defined the socially responsible consumer behavior based on the well-researched concept of corporate social responsibility which is one approach to emphasize ECs (Sen and Bhattacharya, 2001). Many of the consumers want companies to become more socially responsible through greater emphasis on environmentally responsible approach, and a number of corporations are responding to these desires of the consumers which are healthy sign for environmentalists (Bhate, 2002; Bodur and Sarigollu, 2005; Chan, 1999). Van Liere and Dunlap (1981) found attitudinal, demographic, socioeconomic, and leisure activity profiles were delineated. Attitudes toward specific behaviors were found to be the best predictors of behavior, followed by general attitudes, education, and locus of control. They hypothesized that the environmentally concerned consumer would possess certain demographic characteristics, and more so, psychological characteristics to a significantly greater degree than the (traditional) socially concerned consumer (Krause, 1993; Roberts and Bacon, 1997). Though number of corporations has become socially responsible because they firmly believe, others may be environmental friendly soon only when they will find it a financially viable option (Bansal and Roth, 2000; Berger and Corbin, 1992). The economic fall outs of ECs have contributed further to it in terms of business growth (Henion, 1976; Orsato, 2006; Princen, 2002; Sanne, 2002). However, this is a challenging dimension as research studies have revealed this fact that consumers do not always purchase environment-friendly or green products (Schultz, 2000). That is why the other modalities such as innovativeness have been empirically tested for environmentally conscious consumer behavior (ECCB) (Englis and Phillips, 2013; Jansson et al., 2010). To better understand consumers, marketers' innovative approach can provide new dimensions to the green market opportunities. The previous research studies have clearly evidenced that results often contradicted with another study on the matter of predicting the environmental concerned consumer behavior. As a phenomenon, it is very difficult to understand because so many factors play a vital role in it such as New Environmental Paradigm (Albrecht et al., 1982; Bamberg and Moser, 2007; Dunlap et al., 2000). Brown and Wahlers (1998) conducted an exploratory study on possible segments of environmentally conscious consumers and their purchasing motivations. Sen and Bhattacharya (2001) further extended the work on the same theme with exhaustive consumer reactions to it. In addition to the consumer standpoint, the environmental issue has been reviewed from the point of view of corporate strategy too which shows its importance for survival trajectory as well for the corporations in years to come (Crane, 2000; Gabriel and Tim, 1995; Ginsberg and Bloom, 2004; Hart, 1997; McDonald and Oates, 2006; Orsato, 2006), and in marketing arena advertising creative strategy has also been reviewed for green consumers (buyer characteristics) in order to get edge on promotional campaigns for environment-friendly products (D'Souza et al. 2006; Roberts, 1996; Shrum et al., 1995). Significant number of firms has initiated critically well insightful procedural steps to become environment-friendly through R&D on products which profitably meet demand of environmentally conscious consumers in coming years (Cornwell and Schwepkar, 1995; Prakash, 2000; Smart, 2010). ECCB includes many facets like reading labels on product packets and using biodegradable products with recyclable and/or refillable packaging and so forth. In addition to this it includes post-purchase behavior as well as returning bottles/cans, recycling papers/diapers/tissue papers, and sorting the trash according to the product categories (Laroche et al., 2001). There are many environmental facets normal public also know very well and in some cases people have advance knowledge as well that is shrinking natural resources for mankind, overwhelming landfill sites, pollution, the depletion of the ozone layer, and greenhouse effect and so forth. According to this theoretical perspective which we derive out of literature, human functioning (an individual's behavior) is uniquely determined by each of these three factors; a dynamic interplay of personal (knowledge, willingness), behavioral [concern, perceived consumer effectiveness (PCE)], and environmental influences (past behavior). Consumer knowledge about environment is considered a relevant and significant construct that affects how consumers gather and organize information (Alba and Hutchinson, 1987), how much information related to product or brand is used in decision making (Punyatoya, 2014), and how consumers evaluate products and services (Schwepker and Cornwell, 1991). The role of product knowledge/education affecting purchase decision is of primary importance (Hines et al., 1987; Kautish, 2015). Therefore, it seemed reasonable to assume that highly educated people more readily see the relation between environmental issues and themselves. Laroche et al. (2001) have pointed out that the education of the consumer is seen as an appropriate method for increasing perceived convenience and establishing credibility in terms of being environment-friendly.

H_{1a}: Environmental knowledge (EK) has a significant and positive effect on EC.

H_{1b}: EK has a significant and positive effect on willingness to be environmental friendly (WEF).

H_{1c}: EK has a significant and positive effect on PCE.

The underlying question marketers' face is whether or to what extent, EC is predictive of ECCB (Ellen et al., 1991). Other ECCB dimensions are providing services to environment protection initiatives, becoming an active member of environmental groups, and using public transportation system to curb pollution hazards. There is a plethora of research studies conducted on varied demographical trends of ECCB (Chan, 1999; Chan, 1996; Follows and Jobber, 2000). Some more surveys proposed that people's attitudes reveal quite a bit of environment concern, suggesting general environmentalist attitude is becoming more and more prevalent (Balderjahn, 1988; Berger and Corbin, 1992; Bohlen et al., 1993; Chan, 1999; Gill et al., 1986). Behavioral change is contingent upon the value of the reinforcement (Henion and William, 1976). Bodur and Sarigollu (2005) conducted a consumer cluster analysis based on behaviors toward the environment, and three distinct segments were identified: actively concerned; passive concerned; and unconcerned. Van Liere and Dunlap (1981) provided the first of its kind measures for EC.

H_{2a}: EC has a significant and positive effect on PCE.

H_{2b}: EC has a significant and positive effect on WEF.

Many paths breaking researches have been conducted on different forms of environment concerns like climate change, recycling behavior (McCarthy and Shrum, 1994), global warming, pollution, and ozone damage (Charter and Polonsky, 1999; Coddington, 1993). In addition to this, the implications of ecological awareness have been studied with bottom line effect of greening from the point of consumer and marketers as well even advertising contribution in it (Forte and Lamont, 1998; Zinkhan and Carlson, 1995). The research conducted by Balderjahn (1988) and Picket et al. (1993) found that people who performed a specific ECCB were likely to be different from those that performed another.

H_{3a}: Past environment-friendly behavior (PEB) has a significant and positive effect on EC.

H_{3b}: PEB has a significant and positive effect on PCE.

H_{3c}: PEB has a significant and positive effect on WEF.

Broadly the researches fall into three streams such as descriptive studies of consumers environmental information, knowledge, concerns, attitudes, and behavior in particular countries (Bodur and Sarigollu, 2005; Joonas et al., 2005; Murphy et al., 1978), across national contexts (Bhate, 2002; Lee and Holden, 1999), or across different demographic segments of given population (Chan, 1999; Ellen et al., 1991; Hines et al., 1987). It is important to understand the demographic characteristics of different consumer groups, the marketing researches attempt to establish, identify, and predict environmental behavior and/or intensions through demographic variables (Kautish and Soni, 2012). Gill et al. (1986) defined EC as a general or global attitude with indirect effects on behavior through behavioral intentions. Zimmer et al. (1994) described EC as "a general concept that can refer to feelings about many different green issues." The notion of a general attitude which precedes more specific attitudes, intentions, and behaviors seem to have their own predictors (Balderjahn, 1988; Lee, 2008; Picket et al., 1993). Though the level of consciousness and concern about the environment is proven to be high in many countries at the same time, it doesn't translate automatically into pro-environmental behavior (Kollmuss and Agyeman, 2002). Cleveland et al. (2005) studied pro-environmental behavior with the locus of control dimension to understand psychological side of the consumer behavior, pro-environmental was studied with value orientation by Karp (1996) for marketing implications and reviewed the research with the theory of planned behavior. The term "PCE" was first described and defined by Kinnear et al. (1974), as the conviction of the environmentally concerned consumer that he can actually change the status quo of continued environmental deterioration. PCE is demarcated as a measure of the subject's judgment in the ability of individual consumers to affect environmental resource problems (Antil, 1984), and is similar to the concept of internal locus of control (Kinnear et al., 1974). PCE further refined and explored by Ellen et al. (1991) in motivating environmentally conscious consumers. PCE was initially considered a measure or element of the attitude itself and consequently was modeled as a direct predictor of environmentally conscious behavior (Antil, 1984; Berger and Corbin, 1992; Kinnear et al., 1974; Webster, 1975). PCE represents an evaluation of the self in the context of the issue and it is a domain specific belief that the efforts of an individual can make a difference in the solution to a problem (Ellen et al., 1991). To the extent that PCE is conceptualized as unique and separate from the attitude itself, it can likewise be modeled separately and it may function as more than just a direct predictor of behavior (Berger and Corbin, 1992).

H_4: PCE has a significant and positive effect on ECCB.

Jansson et al. (2010) reflected upon green consumer behavior with eco-innovation adoption framework. Kuhn and Jackson (1989) evaluated the stability of factor structures in the measurement of public environmental attitudes which is the most important facets to understand environmental consciousness. Some empirical studies provide evidence that consumers' difficulty in locating environmentally directed products is rooted partly in lack of information (Brown and Wahlers, 1998). These researchers further state that environmentally concerned consumers, on the contrary, feel serious about the environment, are aware and informed, and take responsibility for preserving the environment. The central proposition to the ideation of environment-friendly behavior connotation is a willingness to perform it in practice because until and unless one is willing to execute what he/she is thinking the whole idea comes to an end. As a rational human being we all understand that our different choices contributes directly or indirectly to the environmental deterioration but how many of us willingly take steps to curb, it is a matter of debate. There is a strong belief among consumer behavior researchers and environmentalists that by paying attention to issues like purchasing environment-friendly products and other environmental measures, consumers can contribute significantly to mitigate the situation around us.

H_5: WEF has a significant and positive effect on ECCB.

A thorough review of literature highlighted little evidence of consistency in conceptualization in terms of theory building, measurement of the construct with adequate reliability, validity issues and so forth. In recent time, much measurement scale has been developed to get insights on numerous issues related to environment-friendly behavior in varied contexts. Later, the developed scales have been reexamined by different authors in different consumer class, markets, and countries as well in order to understand the dynamic underpinnings. In the early 1990s, Noe and Snow (1990) analyzed the new environmental paradigm and tried to develop measurement scale for the same. In the mid-1980s, Geller and Lasley (1985) reexamined the existing new environmental paradigm scale. In the early 2000s, the new environmental paradigm scale had been challenged for its relevance (Laronde and Jackson, 2002). Balderjahn (1988) used causal modeling as a methodological construct to investigate the efficacy of a number of predictors such as personality, demographic, attitude, etc. in explaining the variations in behavioral dimensions. The behavioral dimensions were home insulation, energy curtailment, ecologically responsible buying, and using the products

EC. The results of the study suggested that each behavior pattern had its own "cluster of predictors" and no generalizable set of predictors could serve to integrate the dimensions investigated. Picket et al. (1993) employed nine-items, a self-report measure to assess the respondent's level of ECCB. The construct labeled CONSERV measured the extent to which consumers engage in traditional recycling behavior, pro-environmental dispositional behavior, and public resource conservation behavior through regressing the CONSERV scale on demographic and psychological variables, and they concluded that demographics did not significantly explain variation in the dependent variable but psychological variables were found signifi-cant. Mohr et al. (1998) developed and tested a comprehensive measure of skepticism toward environmental claims in marketers' communication which was unique because of its application, to some extent similar scale has been developed by Gatersleben (2002) as well for environmental behavior. Various studies concluded that the combination of items that focus on a wide range of substantive issues into a single measure may mask the true relation-ship between potential dimensions of the measure and ECCB (Brown and Wahlers, 1998; Cornwell and Schwepker, 1995; Henion and William, 1976). Carrigan and Ahmed (2005) were more concerned towards ethical dimen-sion of consumer purchase behavior; similar approach has been reiterated by Shaw and Shiu (2003) through multivariate modeling for decision making process. In a very recent attempt, Webb et al. (2008) revisited the measure-ment of socially responsible consumption paradigm.

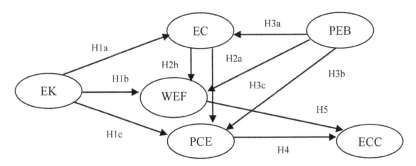

FIGURE 7.1   Conceptual model.

7.3   METHODOLOGY

The potential consequences of random or systematic measurement error can be quite serious in application-oriented research (Fornell and Larcker,

1981). Random error can be statistically controlled by developing multiple measures of constructs to reduce the potential influence of random error in measurement (Terblanche and Boshoff, 2008). In view of this, the authors have utilized multiple measures to develop ECCB scale and the scale items have been modified as per the requirements, that is, language adaptation, reworded in such a way to render the questionnaire comprehensive, altering the sequence of the items, and so forth. To develop the scale to measure ECCB, Churchill's (1979) paradigm for developing better multi-item measures was used. In order to rationalize the measure, the integrated framework was used to identify the items in measurement scale (Malhotra et al., 2012). A list of 28 questions was developed and pretested. The scale was pretested on a sample of 50. Reliability analysis produced a coefficient alpha of >0.80, which confirms the internal consistency of the instrument (Churchill, 1979; Robinson et al., 1991). Where necessary, items were reverse coded and then recoded to maintain consistency. Some of the questions in the initial questionnaire were edited because few respondents raised doubts while the survey was administered. The frequency distributions from the pretest statistics showed that answers to some questions were skewed to one side of the scale or the other, while others were more normally distributed. This reveals that the questions ask about a variety of environmentally concerned behaviors ranging from those that are commonly performed to those that are commonly performed by a few. Also, included among the environmental behavior questions were social desirability questions. A common problem often mentioned in the environmental literature is respondents over-reporting their environmental behavior because acting environmentally is the socially desirable thing to do (Crowne and Marlowe, 1964). If survey respondents in this study, demonstrate a social-desirability response set it may not be possible to conclude anything from their survey responses because they are not true responses. The present study was conducted with the help of online service provider specialized in the survey instrument. A cover letter described the objectives of the study and assured confidentiality. The survey was voluntary, anonymous, and structured. A total of 325 questionnaires were received but some of them were incomplete so it was decided to utilize 305 fully completed questionnaires for final analysis, which were representative samples of consumers between 21 and 50 years of age. Data collection was completed by the summer of 2016. The scale adapted ECCB was a 7-point Likert-type scale with 1= "almost never" and 7= "very frequently" (Roberts and Bacon, 1997). For the present research, author developed a self-administered questionnaire which was specifically designed on the basis of some of the most prominent existing reliable and validated scales (Antil, 1984; Bohlen et al., 1993; Bloom, 1995; Lee and Holden, 1999; Abdul-Muhmin, 2007; Zabkar and Hosta, 2013) including environment knowledge (six items) adapted from Abdul-Muhmin (2007), environment concern (12 items) adapted from Antil (1984), Bohlen et al. (1993) and Zabkar and Hosta (2013), PEB (four items) adapted from Bloom (1995) and Abdul-Muhmin (2007) and perceived psychological consequences of environment-friendly behavior (three items) adapted from Minton and Rose (1997) and Abdul-Muhmin (2007), and willingness to perform environment-friendly behavior (three items, adapted from Abdul-Muhmin 2007). Data was collated with statistical tests and results of the same that are available in tables. To assess the temporal stability of the instrument, the questionnaire was presented to 25 participants selected at random from the sampling frame and after 2 weeks, the same questionnaire was readministered to the same participants. The responses on both administrations were used to assess the test-retest reliability of the instrument. In fact, all constructs produced a significant correlation coefficient ($p < 0.01$); thus, confirming the temporal stability of the instrument.

7.4 DATA FINDINGS AND ANALYSIS

Once the initial measurement construct was developed, pretesting was carried to ensure the instrument's content validity. Missing values were substituted by "trend at a point" (Tabachik and Fidell, 1996). Diagnostic tests were undertaken; extreme values were ascertained to be $<0.05\%$ of observations; normality was confirmed, and no pattern was observed in the distribution of error terms. In addition, psychometric assessment of measurement scales was also undertaken: scale dimensionality through factor analysis and to find out the sampling adequacy for the sample Kaiser-Meyer-Olkin (KMO) measure of sampling adequacy was performed. High KMO value (0.871) as shown in Table 7.1 which is much above the threshold limit shows the acceptance of factor analysis for present research.

TABLE 7.1   KMO and Bartlett's Test of Sphericity.

KMO and Bartlett's Test		
Kaiser-Meyer-Olkin measure of sampling adequacy		**0.871**
Bartlett's test of sphericity	Approx. chi-square	7019.961
	df	328
	Sig.	0.000

The results of Bartlett's test of sphericity was also significant (sig=0.000), which indicate that there was a significant relationship among the construct variables, and analysis could be conducted between these factors as data was sufficiently appropriate (Hair et al., 2006). To confirm the internal consistency of the scale, reliability statistics were also run (Table 7.2).

TABLE 7.2 Reliability Estimates and Factor Loadings.

Factors		Scale items	Factor loading	No. of items	Cronbach's α
Environmental knowledge	1	KNOWELDGE1	0.881	6	0.881
	2	KNOWELDGE2	0.776		
	3	KNOWELDGE3	0.631		
	4	KNOWELDGE4	0.852		
	5	KNOWELDGE5	0.821		
	6	KNOWELDGE6	0.754		
Environmental concern	1	CONCERN1	0.815	12	0.923
	2	CONCERN2	0.832		
	3	CONCERN3	0.845		
	4	CONCERN4	0.706		
	5	CONCERN5	0.696		
	6	CONCERN6	0.812		
	7	CONCERN7	0.807		
	8	CONCERN8	0.902		
	9	CONCERN9	0.723		
	10	CONCERN10	0.842		
	11	CONCERN11	0.834		
	12	CONCERN12	0.768		
Past-environmental behavior	1	PAST1	0.880	3	0.803
	2	PAST2	0.850		
	3	PAST3	0.806		
Perceived-environmental consequences	1	PERCEIVED1	0.861	3	0.861
	2	PERCEIVED2	0.870		
	3	PERCEIVED3	0.913		
Willingness to perform environmentally	1	WILL1	0.842		
	2	WILL2	0.931	4	0.892
	3	WILL3	0.862		
	4	WILL4	0.862		

Extraction method: Principal component analysis. Rotation method: Varimax with Kaiser normalization.

[a]Rotation converged in three iterations.

Principal component analysis and scale reliability were assessed using item-total correlations and Cronbach's α. Results for the same are available in tables. Construct validity of the measurement was demonstrated by the significant correlations between each of the 28 items of ECCB adapted from different scales. Cronbach's α for the construct was above 80% indicating a good reliability. The result of the factor analysis discovered that two factors were extracted through varimax rotation (Table 7.3 and 7.4). These factors were found to explain 69.31% of the variability. To ensure error-free analysis, items with low reliability (<0.50) were removed from the data analysis (Nunnally and Bernstein, 1994). Factor Analysis can identify the structure of a set of variables and provides a process for data reduction (Hair et al., 2006). To assess the construct validity of the instrument, exploratory factor analysis was used and all the hypotheses found to be significant from the study.

7.5   DISCUSSION

This chapter reviews a wide range of literature on EC by synthesizing consumer marketing domain and perspectives of corporations on the issue. It broadly explores what has already been written so far on this important theme that makes people buy green products and the factors which can help business start looking at environment-friendly products as growing opportunity (Ellen, 1994). This exploratory research not only provides descriptive information but also suggests areas for further study in the near future. The current research used few consumption issues for understanding the EC. There is a huge scope for conducting research on green marketing using a comprehensive scale which examines the multi-dimensional attributes of it. Notwithstanding these important findings, we must also be cautious with regard to generalizations because the study has been conducted in only one city of the country. In conclusion, it may be mentioned that the direct linkage between EK, concern, perceived environmental consequences, and willingness has wide-ranging implications for the various environmental stakeholders, namely consumers, business and marketing, educational institutions, government, and environmental organizations, in their quest to preserve and protect the earth's physical and natural environment.

TABLE 7.3 Total Variance Explained.

Component	Initial eigenvalues			Extraction sums of squared loadings			Rotation sums of squared loadings		
	Total	% of Variance	Cumulative %	Total	% of Variance	Cumulative %	Total	% of Variance	Cumulative %
1	7.911	28.252	28.252	7.912	28.252	28.252	7.821	27.932	27.931
2	3.953	14.113	42.365	3.952	14.113	42.365	3.881	13.856	41.788
3	2.987	10.669	53.034	2.986	10.668	53.033	3.076	10.986	52.773
4	2.445	8.733	61.768	2.445	8.733	61.767	2.403	8.582	61.357
5	2.112	7.544	69.311	2.112	7.544	69.310	2.227	7.954	69.311
6	0.988	3.534	72.844						
7	0.693	2.479	75.323						
8	0.611	2.184	77.508						
9	0.554	1.980	79.488						
10	0.519	1.853	81.342						
11	0.501	1.789	83.130						
12	0.468	1.670	84.801						
13	0.432	1.544	86.344						
14	0.424	1.522	87.866						
15	0.388	1.385	89.252						
16	0.346	1.232	90.482						
17	0.337	1.202	91.683						
18	0.299	1.067	92.750						
19	0.286	1.021	93.773						

TABLE 7.3 *(Continued)*

Component	Initial eigenvalues			Extraction sums of squared loadings			Rotation sums of squared loadings		
	Total	% of Variance	Cumulative %	Total	% of Variance	Cumulative %	Total	% of Variance	Cumulative %
20	0.274	0.978	94.747						
21	0.250	0.893	95.640						
22	0.236	0.843	96.483						
23	0.206	0.735	97.218						
24	0.199	0.710	97.928						
25	0.187	0.668	98.596						
26	0.173	0.622	99.217						
27	0.110	0.394	99.612						
28	0.108	0.389	100.000						

Extraction method: Principal component analysis.

TABLE 7.4   Rotated Component Matrix.

	Rotated component matrix[a]				
	Component				
	1	2	3	4	5
KNOW1		0.882			
KNOW2		0.778			
KNOW3		0.632			
KNOW4		0.872			
KNOW5		0.831			
KNOW6		0.764			
CONC1	0.820				
CONC2	0.804				
CONC3	0.844				
CONC4	0.709				
CONC5	0.693				
CONC6	0.808				
CONC7	0.800				
CONC8	0.930				
CONC9	0.752				
CONC10	0.849				
CONC11	0.835				
CONC12	0.786				
PAST1					0.883
PAST2					0.856
PAST3					0.809
PERC1				0.864	
PERC2				0.872	
PERC3				0.918	
WILL1			0.847		
WILL2			0.908		
WILL3			0.865		
WILL4			0.860		

Extraction method: Principal component analysis.

Rotation method: Varimax with Kaiser normalization.

[a]Rotation converged in three iterations.

7.6 CONCLUSION

Though the assumption of multivariate normality for behavioral variables was violated by the data; however, no improper random effect was found, and the sample size is sufficient so one can assume that the parameter estimates and their associated errors will be approximately multivariate normally distributed. The present study examined the environmentally conscious consumption behaviors of residents of a small town in Rajasthan only; thus, research can be undertaken on the much larger scale and data may be collected from across different cities of the country in order to understand the vital market intricacies. The population sampled was primarily people, who were relatively known to the green issues. It is possible that there could be differences in the value orientation between the sample and the general population. In addition, the study is constrained by virtue of being cross-sectional in nature; a longitudinal study would shed light on the potential mutuality of cause and effect among the variables of interest. Furthermore, the study relies on scales that were developed during different stages of the environmental movement and may need closer examination. Future research should focus on development and validation of improved measures with domain specific scales for other constructs. Many of the scales cover a plethora of issues which are not at all necessarily applicable in a different context (Alwitt and Pitts, 1996). The study does not intend to provide any final verdict but to lay the foundation on consumer perception on green consumption in the Indian market. In terms of academic research, this research forms the basis for further in-depth studies on this important area. Still, the result provided valuable insights and well researched empirical evidence on the state of consumption behavior on environmental issues. The state of Rajasthan in India has a growing economy and predominantly has a promising young population as a potential market. This coupled with increased education and changing demographics like income, age, and educational level, etc. give immense opportunities to the marketers (Stern et al., 1993). EC is fast transforming the way the products are served to the target consumers. These conclusions could assist policy makers at both the national level in formulating public environmental policies. The green consumption is no longer restricted to major cities of the country but has reached across the country (Ramirez et al., 2015). The research study indicates that environment-friendly products may be used as a consumption strategy for improving the environment (Mendleson and Polonsky, 1995). The implications derived from the present study have an impact in a number of marketing areas. The results indicate that consumers in India have a positive attitude toward the EC, they demand green products, and they attempt to figure out the impact of their consumption on environmental well-being. The present work is an initial attempt to understand the results of environmental marketing practices of companies for Indian consumers. Given the narrow demographic characteristics of the sample, one should be cautious when interpreting the results for the general population even in another part of the state of Rajasthan. Generalizability could be increased by applying the model to products used by a broader range of consumers. The present study primarily focused on fewer environmentally concerned variables mainly because it was conducted in a developing country. The environmentally concerned variables such as recycling as a construct have not been included as it was not prominent due to contextual limitations. Future research should continue to delve into this issue as to what other factors drive environmentally concerned behavior among consumers so that we can have a comprehensive model to understand the phenomenon for wider applications.

KEYWORDS

- environment
- environment friendly
- green marketing
- consumer behavior
- Indian market

REFERENCES

Abdul-Muhmin, A. G. Explaining Consumers' Willingness to be Environmentally Friendly. *Int. J. Consum. Studies* **2007,** *31*(3), 237–247.

Alba, J. W.; Hutchinson, J. W. Dimensions of Consumer Expertise. *J. Cons. Res.* **1987,** *13*(4), 411–454.

Albrecht, D.; et al. The New Environmental Paradigm Scale. *J. Environ. Educ.* **1982,** *13*(3), 39–43.

Alwitt, L. F.; Pitts, R. E. Predicting Purchase Intentions for an Environmentally Sensitive Product. *J. Consum. Psychol.* **1996,** *5*(1), 49–64.

Anderson, W. T. Jr.; Cunningham, W. H. C. The Socially Conscious Consumer. *J. Mark.* **1972,** *36*(3), 23–31.

Antil, J. H. Socially Responsible Consumers: Profile and Implications for Public Policy. *J. Macro. Mark.* **1984,** *4*(2), 18–39.

Balderjahn, I.; Personality Variables and Environmental Attitudes as Predictors of Ecologically Responsible Consumption Patterns. *J. Bus. Res.* **1988,** *17*(1), 51–56.

Bamberg, S.; Moser, G. Twenty Years After Hines, Hungerford, and Tomera: A New Meta-Analysis of Psycho-Social Determinants of Pro-Environmental Behavior. *J. Environ. Psychol.* **2007,** *27*(1), 14–25.

Bansal, P.; Roth, K. Why Companies Go Green: A Model of Ecological Responsiveness. *Acad Manage. J.* **2000,** *43*(4), 717–736.

Berger, I. E.; Corbin, R. M. Perceived Consumer Effectiveness and Faith in Others as Moderators of Environmentally Responsible Behavior. *J. Public Policy Mark.* **1992,** *11*(2), 79–100.

Bhate, S. One World, One Environment, One Vision: Are We Close to Achieving This? An Exploratory Study of Consumer Environmental Behavior Across Three Countries. *J. Consum. Behav.* **2002,** *2*(2), 169–184.

Bloom, D. E. International Public Opinion on the Environment. *Science* **1995,** *269*(5222), 354–358.

Bodur, M.; Sarigollu, E. Environmental Sensitivity in a Developing Country: Consumer Classification and Implications. *Environ. Behav.* **2005,** *37*(1), 487–510.

Bohlen, G. M.; et al. Measuring Ecological Concern: A Multi-Construct Perspective. *J. Mark. Manage.* **1993,** *19*(4), 415–430.

Brown, J. D.; Wahlers, R.G. The Environmentally Concerned Consumer: An Exploratory Study. *J. Mark. Theory Practice* **1998,** *6*(2), 39–47.

Burgess, J. Sustainable Consumption: Is It Really Achievable? *Consum. Policy Rev.* **2003,** *13*(3), 78–84.

Carrigan, M.; Ahmad, A. The Myth of the Ethical Consumer—Do Ethics Matter in Purchase Behavior? *J. Consum. Mark.* **2005,** *18*(7), 560–577.

Chan, T. S. Concerns for Environmental Issues and Consumer Purchase Preferences: A Two-Country Study. *J. Int. Consum. Mark.* **1996,** *9*(1), 43–55.

Chan, R. Y. K. Environmental Attitudes and Behavior of Consumers in China: Survey Findings and Implications. *J. Int. Consum. Mark.* **1999,** *11*(4), 25–52.

Charter, M.; Polonsky, M. J. Greener Marketing: A Global Perspective on Greening Marketing Practice. Greenleaf: Sheffield, **1999.**

Churchill, G. A. A Paradigm for Developing Better Measures of Marketing Constructs. *J. Mark. Res.* **1979,** *16*(1), 54–73.

Cleveland, M.; et al. Shades of Green: Linking Environmental Locus of Control and Pro-Environmental Behaviors. *J. Consum. Mark.* **2005,** *22*(4/5), 198–213.

Coddington, W. Environmental Marketing: Positive Strategies for Reaching the Green Consumer. McGraw- Hill: New York, **1993.**

Cornwell, B. T.; Schwepker, C. H. Ecologically Concerned Consumers and Their Purchase Behavior, Environmental Marketing. Haworth Press Inc.: New York, **1995.**

Crane, A. Facing the Backlash: Green Marketing and Strategic Reorientation in the 1990S. *J. Strategic Mark.* **2000,** *8*(3), 277–296.

Crowne, D. P.; Marlowe, D. The Approval Motive. John Wiley and Sons: New York, **1964.**

Diamantopoulos, A.; et al. Can Socio-Demographics Still Play a Role in Profiling Green Consumers? A Review of the Evidence and an Empirical Investigation. *J. Bus. Res.* **2003,** *56*(6), 465–480.

Drumwright, M. E. Socially Responsible Organizational Buying: Environmental Concerns as a Noneconomic Buying Criterion. *J. Mark.* **1994,** *58*(3), 1–19.

D'Souza, C.; et al. An Empirical Study on the Influence of Environmental Labels on Consumers. *Corporate Commun. Int. J.* **2006,** *11*(2), 162–173.

Dunlap, R. E. V. L. et al Measuring Endorsement of the New Ecological Paradigm: A Revised NEP Scale. *J. Soc. Issues* **2000,** *56,* 425–442.

Ellen, P. S. Do We Know What We Need to Know? Objective and Subjective Knowledge Effects on Pro-Ecological Behaviors. *J. Bus. Res.* **1994,** *30*(1), 43–52.

Ellen, P. S.; et al. The Role of Perceived Consumer Effectiveness in Motivating Environmentally Conscious Behaviors. *J. Public Policy Mark.* **1991,** *10*(2), 102–117.

Englis, B. G.; Phillips, D. M. Does Innovativeness Drive Environmentally Conscious Consumer Behavior? *Psychol. Mark.* **2013,** *30*(2), 160–172.

Fineman, S. Constructing the Green Manager. The Dryden Press: London, **1996**.

Follows, S. B.; Jobber, D. Environmentally Responsible Purchase Behavior: a Test of a Consumer. *Eur. J. Mark.* **2000,** *34*(5/6), 723–746.

Fornell, C.; Larcker, D. F. Structural Equation Models with Unobserved Variables and Measurement Error: Algebra and Statistics. *J. Mark. Res.* **1981,** *18*(3), 382–388.

Forte, M.; Lamont, B. The Bottom Line Effect of Greening: Implications of Ecological Awareness. *J. Academy Manage. Exec.* **1998,** *12*(1), 89–91.

Fuller, D. *Sustainable Marketing: Managerial Ecological Issues.* Sage. Thousand Oaks: CA, **1999**.

Furman, A. A Note on Environmental Concern in a Developing Country: Results from an Istanbul Survey. *Environ. Behav.* **1998,** *30*(4), 520–534.

Gabriel, Y.; Tim, L. The Unmanageable Consumer: Contemporary Consumption and its Fragmentation. Sage Publications: London, **1995**.

Gatersleben, B.; et al. Measurement and Determinants of Environmentally Significant Behavior. *Environ. Behav.* **2002,** *34*(3), 335–362.

Geller, J. M.; Lasley, P. The New Environmental Paradigm Scale: A Reexamination. *J. Environ. Educ.* **1985,** *17*(1), 9–12.

Gill, J. D.; Crosby, L. A.; Taylor, J. R. Ecological Concern, Attitudes and Social Norms in Voting Behavior. *Publ. Opin Q.* **1986,** *50*(4), 537–554.

Ginsberg, J. M.; Bloom, P. N. Choosing the Right Green Marketing Strategy. *MIT Sloan Manage. Rev.* **2004,** *46*(1), 79–84.

Griskevicius, V.; et al. Going Green to Be Seen: Status, Reputation, and Conspicuous Conservation. *J. Pers. Soc. Psychol.* **2010,** *98*(3), 392–404.

Gronhoj, A. Communication about Consumption: A Family Process Perspective on "Green" Consumer Practices. *J. Consum. Behav.* **2006,** *5*(6), 491–504.

Hair, J. F. Jr.; et al. Multivariate Data Analysis. Pearson Education: New Delhi, **2006**.

Hart, S. L. Beyond Greening: Strategies for a Sustainable World. *Harv. Bus. Rev.* **1997,** *75*(1), 66–76.

Henion, K. E. Ecological Marketing. Columbus. Grid, Inc.: Ohio, 1976.

Henion, K. E. II; William, H. W. The Ecologically Concerned Consumer and Locus of Control. In *Ecological Marketing*; Henion, K., Thomas, K. Eds.; American Marketing Association: Chicago, IL, **1976,** 131–144.

Henriques, I.; Sadorsky, P. The Determinants of an Environmentally Responsive Firm: An Empirical Approach. *J. Environ. Econ. Manage.* **1996,** *30*(3), 381–395.

Hines, J. M.; et al. Analysis and Synthesis of Research on Responsible Environmental Behavior. *J. Environ. Educ.* **1987,** *18*(2), 1–8.

Li-Tze, Hu.; Bentler, P. M. Cut off Criteria for Fit Indices in Covariance Structure Analysis: Conventional Criteria Versus New Alternatives. *Struct. Equation Model. Multidiscip. J.* **1999,** *6*(1), 1–55.

Jain, S. K.; Kaur, G. Strategic Green Marketing: How Should Business Firms Go About Adopting It? *Indian J. Commerce.* **2003,** *55*(4), 1–16.

Jain, S. K.; Kaur, G. Green Marketing: An Indian Perspective. *Decision* **2004,** *31*(2), 168–209.

Jain, S. K.; Kaur, G. Role of Socio-Demographics in Segmenting and Profiling Green Consumers: An Exploratory Study of Consumers in India. *J. Int Consum. Mark.* **2006,** *18*(3), 107–145.

Jansson, J.; et al. Green consumer Behavior: Determinants of Curtailment and Eco Innovation Adoption. *J. Consum. Mark.* **2010,** *27*(4), 358–370.

Joonas, K.; et al. An Investigation of the Environmental Beliefs and Attitudes of Business Students in Mexico and the U.S. *Southwest Bus. Econ. J.* **2005,** *14*(1), 321–345.

Kalafatis, S. P.; et al. Green Marketing and Ajzen's Theory of Planned Behavior: A Cross-Market Examination. *J. Int. Consum. Mark.* **1999,** *16*(5), 441–460.

Karp, D. G. Values and Their Effect on Pro-Environmental Behavior. *Environ. Behav.* **1996,** *28*(1), 111–133.

Kautish, P. Empirical Study on Understanding of Consumer Behavioral Factors for Marketing of Environmental Friendly Products. *IMR Manage. Speak.* **2015,** *8*(2), 1–12.

Kautish, P. Volkswagen AG: Defeat Device or Device Defeat? *IMT Case J.* **2016,** *7*(1), 19–30.

Kautish, P.; Dash, G. Environmentally Concerned Consumer Behavior: Evidence from Consumers in Rajasthan. *J. Modell. Manage.* **2017,** *12*(4), 712–738.

Kautish, P.; Soni, S. The Determinants of Consumer Willingness to Search for Environmental-Friendly Products: A Survey. *Int. J. Manage.* **2012,** *29*(2), Part. 2, 696–711.

Kinnear, T. C.; et al. Ecologically Concerned Consumers: Who are they? *J. Mark.* **1974,** *38*(2), 20–24.

Kleiner, A. What Does It Mean To Be Green? *Harv. Bus. Rev.* **1991,** *69*(4), 38–47.

Kollmuss, A.; Agyeman, J. Mind the Gap: Why do People Act Environmentally and What are the Barriers to Pro-Environmental Behavior? *Environ. Educ. Res.* **2002,** *8*(3), 239–260.

Krause, D. Environmental Consciousness: An Empirical Study. *Environ. Behav.* **1993,** *25*(1), 126–142.

Kuhn, R. G.; Jackson, E. L. Stability of Factor Structures in the Measurement of Public Environmental Attitudes. *J. Environ. Educ.* **1989,** *20*(3), 27–32.

Lalonde, R. J.; Jackson, E. L. The New Environmental Paradigm Scale: Has it Outlived its Usefulness? *J. Environ. Educ.* **2002,** *33*(4), 28–36.

Laroche, M.; et al. Targeting Consumers Who are Willing to Pay More for Environmentally Friendly Products. *J. Consum. Mark.* **2001,** *18*(6), 503–520.

Lee, K. Opportunities for Green Marketing: Young Consumers. *Mark. Intell. Plann.* **2008,** *26*(6), 573–586.

Lee, J. A.; Holden, S. J. S. Understanding the Determinants of Environmentally Conscious Behavior. *Psychol. Mark.* **1999,** *16*(5), 373–392.

Ling-Yee, L. Effect of Collectivist Orientation and Ecological Attitude on Actual Environmental Commitment: the Moderating Role of Consumer Demographics and Product Involvement. *J. Int. Consum. Mark.* **1997,** *9*(4), 31–53.

Malhotra, N. K.; et al. One, Few or Many? An Integrated Framework for Identifying the Items in Measurement Scales. *Int. J. Mark. Res.* **2012,** *54*(6), 835–862.

McCarthy, J. A.; Shrum, L. J. The Recycling of Solid Wastes: Personal Values, Value Orientations, and Attitudes About Recycling as Antecedents of Recycling Behavior. *J. Bus. Res.* **1994,** *30*(1), 53–62.

McDonald, S.; Oates, C. J. Sustainability: Consumer Perceptions and Marketing Strategies. *Bus. Strategy Environ.* **2006,** *15*(3), 157–170.

Mendleson, N.; Polonsky, M. J. Using Strategic Alliances to Develop Credible Green Marketing. *J. Consum. Mark.* **1995,** *12*(2), 4–18.

Minton, A. P.; Rose, R. L. The Effects of Environmental Concern on Environmentally Friendly Consumer Behavior: an Exploratory Study. *J. Bus. Res.* **1997,** *40*(1), 37–48.

Mohr, L.; et al. The Development and Testing of a Measure of Skepticism Toward Environmental Claims in Marketers Communication. *J. Consum. Aff.* **1998,** *31*(1), 30–55.

Mohr, L. A.; et al. Do Consumers Expect Companies to Be Socially Responsible? The Impact of Corporate Social Responsibility on Buying Behavior. *J. Consum. Aff.* **2001,** *35*(1), 45–72.

Moisander, J.; Pesonen, S. Narratives of Sustainable Ways of Living: Constructing the Self and the Other as a Green Consumer. *Manage. Decis.* **2002,** *40*(4), 329–342.

Murphy, P. E.; et al. Environmentally Concerned Consumers—Racial Variations. *J. Mark.* **1978,** *42*(4), 61–66.

Nair, S. R.; Menon, C. G. An Environmental Marketing System- a Proposed Model Based on Indian Experience. *Bus. Strategy Environ.* **2008,** *17*(8), 467–479.

Noe, F. P.; Snow, R. The New Environmental Paradigm and Further Scale Analysis. *J. Environ. Educ.* **1990,** *21*(4), 20–26.

Nunnally, J. C.; Bernstein, I. H. Psychometric Theory. McGraw Hill: New York, **1994.**

Orsato, R. J. Competitive Environmental Strategies: When Does it Pay to Be Green? *Calif. Manage. Rev.* **2006,** *48*(2), 127–144.

Ottman, J. A. Green Marketing, NTC Business Books: Lincolnwood, IL, 1993.

Ottman, J. A.; et al. Avoiding Green Marketing Myopia: Ways to Improve Consumer Appeal for Environmentally Preferable Products. *Environ. Sci. Policy Sustainable Dev.* **2006,** *48*(5), 21–36.

Picket, G. M.; et al. Is There a General Conserving Consumer? A Public Concern. *J. Public Policy Mark.* **1993,** *12*(2), 234–243.

Prakash, A. Greening the Firm: The Politics of Corporate Environmentalism. Cambridge University Press: Cambridge, 2000.

Princen, T. Consumption and Its Externalities: Where Economy Meets Ecology. MIT Press: London, 2002.

Punyatoya, P. Linking Environmental Awareness and Perceived Brand Eco-friendliness to Brand Trust and Purchase Intention. *Global Bus. Rev.* **2014,** *15*(2), 279–289.

Ramirez, E.; et al. Concrete and Abstract Goals Associated with the Consumption of Environmentally Sustainable Products. *Eur. J. Mark.* **2015,** *49*(9/10), 1645–1665.

Roberts, J. A. Green Consumers in the 1990s: Profile and Implications for Advertising. *J. Bus. Res.* **1996,** *36*(2), 217–231.

Roberts, J. A.; Bacon, D. R. Exploring the Subtle Relationships between Environmental Concern and Ecologically Conscious Consumer Behavior. *J. Bus. Res.* **1997,** *40*(1), 79–89.

Robinson, N. M.; et al. Criteria for Scale Selection and Evaluation. In *Measures of Personality and Social Psychological Attitudes;* Robinson, J. P., Shaver, P. R., Wrightsman, L. S., Eds.; Academic Press: San Diego, CA, 1991; pp 1–16.

Rugman, A.; Verbeke, A. Environmental Regulations and the Global Strategies of Multinational Enterprises. Routledge: London, 2000.

Sanne, C.; Willing Consumers or Locked In? Policies for Sustainable Consumption. *Ecol. Econ.* **2002,** *43*(2/3), 127–140.

Schultz, W. P Emphasizing with Nature: the Effects of Perspective Taking on Concern for Environmental Issues. *J. Soc. Issues.* **2000,** *56*(3), 391–406.

Schlegelmilch, B. B.; et al. The Link between Green Purchasing Decisions and Measures of Environmental Consciousness. *Eur. J. Mark.* **1996,** *30*(5), 35–55.

Segun, C.; et al. Towards a Model of Environmental Activism. *Environ. Behav.* **1998,** *30*(5), 628–652.

Sen, S.; Bhattacharya, C. B. Does Doing Good Always Lead to Doing Better, Consumer Reactions to Corporate Social Responsibility. *J. Mark. Res.* **2001,** *38*(1), 225–243.

Shaw, D.; Shiu, E. Ethics in Consumer Choice: a Multivariate Modeling Approach. *Eur. J. Mark.* **2003,** *37*(10), 1485–1498.

Schwepkar, C. H. Jr; Cornwell, T. B. An Examination of Ecologically Concerned Consumers and Their Intentions to Purchase Ecologically Packaged Products. *J. Public Policy Mark.* **1991,** *10*(2), 77–101.

Shrum, L. J.; et al. Buyer Characteristics of the Green Consumer and Their Implications for Advertising Strategy. *J. Advertising.* **1995,** *24*(2), 71–82.

Smart, B. *Consumer Society: Critical Issues and Environmental Consequences.* Sage Publications: London, UK, 2010.

Spangenberg, J. H. The Society, its Products and the Environmental Role of Consumption. Edward Elgar Publishing Limited: Cheltenham, 2004.

Stern, P. C.; et al. Value Orientations, Gender, and Environmental Concern. *Environ. Behav.* **1993,** *25*(5), 322–348.

Stern, P. C.; et al. The New Ecological Paradigm in Social-Psychological Context. *Environ. Behav.* **1995,** *27*(5), 723–743.

Stern, P. C.; et al. Global Environment Change: Understanding the Human Dimensions. National Academy Press: Washington, D. C., 1992.

Straughan, R. D.; Roberts, J. A. Environmental Segmentation Alternatives: A Look At Green Consumer Behavior in the New Millennium. *J. Consum. Mark.* **1999,** *16*(6), 558–575.

Tabachik, B. G.; Fidell L. S. *Using Multivariate Statistics.* Harper Collins: New York, 1996.

Tanner, C.; Kast, S. Promoting Sustainable Consumption: Determinants of Green Purchases by Swiss Consumers. *Psychol. Mark.* **2003,** *20*(10), 883–902.

Terblanche, N. S.; Boshoff, C. Improved Scale Development in Marketing. *Int. J. Mark. Res.* **2008,** *50*(1), 105–119.

Van Liere, K. D.; Dunlap, R. E. Environmental Concern: Does it Make a Difference How It's Measured? *Environ. Behav.* **1981,** *13*(6), 651–676.

Vermeire, I.; Verbeke, W. Sustainable Food Consumption: Exploring the Consumer Attitude Behavioral Intention Gap. *J. Agric. Environ Ethics.* **2006,** *19*(1), 169–194.

Wagner, S. *Understanding Green Consumer Behavior: A Qualitative Cognitive Approach.* Routledge: London, 2003.

Webb, D. J.; et al. A Re-Examination of Socially Responsible Consumption and Its Measurement. *J. Bus. Res.* **2008,** *61*(2), 91–98.

Webster, F. E. Jr. Determining the Characteristics of the Socially Conscious Consumer. *J. Cons. Res.* **1975,** *2*(3), 188–196.

Wheatly, M. Green Business, Making it Work for Your Company. Pitman Publishing: London, 1993.

Yam-Tang, E. P. Y.; Chan, R. Y. K. Purchasing Behaviors and Perceptions of Environmentally Harmful Products. *Mark. Intell. Plann.* **1999,** *16*(6), 365–382.

Zabkar, V.; Hosta, M. Willingness to Act and Environmentally Conscious Consumer Behavior: Can Prosocial Status Perceptions Help Overcome the Gap? *Int. J. Consum. Studies.* **2013,** *37*(3), 257–264.

Zimmer, M. R.; et al. Green Issues Dimensions of Environmental Concern. *J. Bus. Res.* **1994,** *30*(1), 63–74.

Zinkhan, G. M.; Carlson, L. Green Advertising and the Reluctant Consumer. *J. Adv.* **1995,** *24*(2), 1–6.

CHAPTER 8

EMPIRICALLY EXAMINING GREEN BRAND ASSOCIATIONS TO GAIN GREEN COMPETITIVE ADVANTAGE THROUGH GREEN PURCHASING INTENTIONS

NITIKA SHARMA[1,*] and MADAN LAL[2]

[1]*Department of Commerce, University of Delhi, New Delhi, India, Mob.: +91 9873118030*

[2]*Department of Commerce, University of Delhi, New Delhi, India, Mob.: +91 9415204753, E-mail: madanfms@gmail.com*

Corresponding author. E-mail: nitikasharma28@gmail.com

8.1 INTRODUCTION

Conceptually, there is a common agreement that consumers at present are conscious regarding green. Consumers with green consciousness exhibit higher green purchasing decisions (Sharma and Kesharwani, 2015; Sharma and Sharma, 2013; Schlegelmilch et al., 1996), leading to the creation of what has come to be known as green competitive advantage. Companies are trying to underline the concept of green marketing to attract customers by undertaking sustainable initiatives in their corporate goals, operations, and management. Hence, green marketing by companies is used by featuring "eco," "environment-friendly," "green," "earth-friendly," and "sustainable" As a result both practitioners and academicians are paying much attention to examine the concept of green purchasing intentions (GPI) in green marketing through different lens, as such, to gain prominence in the competitive market. The study has comprehensive literature which states that green consumerism leads to green buying decisions, behavior, and/or intentions via sociodemographic variables, psychographic variables (Sharma and Sharma, 2013; Akehurst et al, 2012; Diamantopoulo et al, 2003; Straughan and Roberts, 1999), environmental knowledge, environmental motivation, and environmental value and attitude (Sharma and Kesharwani, 2015). Further, Chen (2010) proposed novel constructs of the green brand image, green brand equity, and green trust (GT) and found that there is a positive and significant relationship between these constructs and buying intentions. In addition, the study affirmed that brand associates/attributes positively affect consumers' buying intentions. Furthermore, many researchers have also reported that brand associates have a positive influence on consumer's choice, preferences, and intention of buying (Chaudhuri and Holbrook, 2001; Yoo et al., 2000; Agarwal and Rao, 1996). Although previous studies have paid significant attention to examine the pertinent issues in brand image, equity and/or association, it may be asserted that broader dimension of green brand association (GBA) and GPI remain under-researched. The present study argues that the exploration of GBA has been limited explored with GPI. Hence, by empirically investigating the extent to which or how, GBA influences consumers' intentions of green purchasing, the present chapter addresses the important research gap to gain a green competitive advantage.

Further, several studies have purported that brands association can be classified into three major categories: attributes, benefits, and attitudes (Keller, 1993; Keller, 1998); and these classifications help in determining the amount of leverage a brand has (Tauber, 1988) such as uniqueness, trustworthiness, or dullness (Biel, 1991). Such associates may generate values such as brand knowledge, awareness, and/or associations which differentially affect consumer's response. Moreover, the perception of knowledge and awareness influence trust among customers (Peters et al., 1997). Indeed, strong and positive brand associations strengthen the brand (James, 2005) and beget the trust (Delgado-Ballester and Munuera-Alemán, 2005). Moreover, it is argued that trust reduces uncertainty and expects a brand to display opportunistic behaviors and these influences purchase intentions (Chong et al., 2003). This chapter may offer an important insight into a subjective way in which a consumer derive trust from the brand associations. Moreover, the relationship between trust and brand association, particularly in environment context, has not straightforwardly focused. Hence, another contribution of this study is to extend the brand research by examining two novel constructs—GBA and GT. Therefore, an argument is premised within this chapter to explore the relationship between the GBA and GT to achieve a green competitive advantage.

In addition, Chen and Chang (2012) have examined that GPI are affected by GT, as trust is an antecedent of customer purchase intentions. However, the relationships of these two constructs such as GT and GPI have not been investigated via GBA. Hence, the main research question of the present study is "how to enhance purchase intentions in environmental context via establishing greener brand associates and generating environmental trust among customers?" Therefore, this chapter intended to help marketers by empirically exploring the premised relationship between GBA and green purchasing behavior through GT. Hence, this study summarizes the literature on three novel constructs, that is, GBA, GT, and GPIs and extending the research of GPIs into a new framework. Followed by an overview of research methodology, the study would undertake an empirical test to verify the relationship among GBA, GT, and GPIs. The present study posits a new framework of GPIs in compliance with brand research to explore new environmental trends to help the companies in increasing their GPI by proposing four hypotheses. Subsequently, the descriptive statistics, the reliability of the measurement, factor analysis, correlation coefficients between the constructs, discriminant validity, convergent validity, and the results of structural equation modeling (SEM) are done. The model was analyzed using Hayes' (2008) mediation model. In the end, the present study draws the conclusions and mentions the discussions about the findings, implications, research limitations, and possible directions for future research.

8.2   LITERATURE REVIEW AND FRAMEWORK

8.2.1   THE POSITIVE EFFECT OF GREEN BRAND ASSOCIATION (GBA) ON GREEN TRUST (GT) AND GREEN PURCHASE INTENTIONS (GPAS)

Aaker (1996) stated that brand associations are anything about the likableness of a brand, and this likableness provides cues for information retrieval (Osselaer and Janiszewski, 2001). Thus, these informational nodes linked to the brand node in memory contain the meaning of the brand for consumers. Brand associations are called as the "heart and soul of the brand," for example, set of brand associations with McDonald are convenient, consistent in quality, hassle free, and provide fast service (Aaker, 1996). Moreover, certain associations are owned by brands, for example, "safety" is associated with Volvo brand; "photography" attribute is associated with Kodak (Aaker, 2012; Aaker and Biel, 1993). Indeed, Keller (1998, 1993) summarized brand associations in three classifications, that is, attributes, benefits, and attitudes; where attributes are the descriptive features that characterize a brand; benefits are the personal values a consumer can get out of the brand; and attitude involves the overall evaluation of the brand. Hence, it is argued that brand association drives the choice of the consumers for the brand (Aaker, 1996). Based on the above definition, the present study proposes a construct, "GBA," and offers one of the possible definitions of the term as "a set of environmental benefits/value attached to the brand attributes." Green brand attributes include environmentally sound production processes, responsible product uses, or product elimination, which consumers compare with other conventional products (Peattie, 1995). Hence, the study can safely assert that GBA may create a greener picture in consumer's mind and differentiate the products from the competitors. Certainly, Hartmann and Apaolaza Ibáñez (2006) emphasized that marketers should accentuate the environmental credentials of the green brand to cultivate environmental consciousness. It is worth noting that consumer's environmental consciousness motivates them to adopt environmentally sustainable behavior and consumers are placing increasingly greater emphasis on environment-friendly products (Sharma and Kesharwani, 2015; Sharma and Sharma, 2013). More recently, researchers have been relating the concepts like the green brand image, green brand equity, and green brand satisfaction with GPIs/decisions (Chen and Chang, 2012; Chen, 2009). As such an environmental holistic view, marketers should communicate the target audience regarding the significant impact of buying green products on the welfare of the environment (Laroche et al., 2001) to generate self-involvement toward protection of the environment. Consequently, the GBA can be significant to academicians as well as practitioners for several reasons. One, it can help the consumer in differentiating the green brand. Second, it can help in consumer process and retrieving environmental information regarding the brand. Third, if green associations are positive, they will create beneficial attitudes and may create an intention to buy the green product. Therefore, GBA is likely to lead to GPI. Thus:

H1: GBA positively affects GPIs.

A novel study of Chen (2010) found that green brand image has a positive and significant relationship with the GBA and GT, and brand image can be defined as a set of perceptions of consumers about a brand resonated by brand associations (Keller, 1993). In addition, the study of Hartmann and Apaolaza Ibáñez (2006) discussed the positive effect of green branding on evoking the significant emotional response among the consumers by illustrating the case of Spanish green energy brand *Iberdrola Energía Verde* (green energy), and the British conventional energy supplier, BP. It is worth noting that maintaining GBA through divergent ways as a brand strategy may create affirmative consumers' response, as green information offers environmentally sound attributes which makes consumers feel better using those green products. It may be asserted that brand comprising the attributions of environmental intentions in relation to consumer's environmental interests and needs may beget GT. It may, therefore, be conjectured that GBA has a positive relationship with GT which helps in GPIs.

H2: GBA positively affects GT.

Trust is the fundamental determinant to stimulate green purchase as it has been shown that both the purchase decision (Gefen and Straub, 2000) as well as purchasing intention (Schlosser et al., 2006) are influenced by consumer's trust. The definition of trust across disciplines varies, as psychologists define the term as a tendency to trust others (Rotter, 1971). Social psychologists define trust as cognition about the trustee (Rempel et al., 1985). Sociologists define the term as a characteristic of the institutional environment (McKnight et al., 2002). However, within the divergent description of trust, there are similarities in the terms of how each discipline conceptualizes the construct of trust. McKnight and Chervany (2001a) broadly conceptualized the term "trusting intentions" as the "willingness to depend, or intends to depend, on the other party with a feeling of relative security, in spite of the lack of control over that party, and even though negative consequences are possible." It is well accepted that trust-related behavior helps consumers to overcome the perception of uncertainty and risk involved in purchase intentions (McKnight et al., 2002). Conceptually, trust includes many attributes like honesty, competence, commitment, credibility, expertise, reliability, integrity, ability, predictability, and benevolence (Bhattacherjee, 2002; McKnight and Chervany, 2001b). Hence, trust owes its evolution in purchasing intentions' context as attributes of trust demonstrates greater purchase intention under the conditions of higher trust (Chong et al., 2003). In environmental context, Chen (2010) argued that GT would lead to GPIs and defined the novel construct "GT" as "willingness to depend on a product, service, or brand based on the belief or expectation resulting from its credibility, benevolence, and ability about its environmental performance." Netemeyer et al. (2005) defined GPIs as the possibility of buying green products due to consumer's environmental needs. As such, in line with willingness to depend on brand for better environmental performance, consumers may exhibit positive GPI. It may be, therefore, postulated that there exists an association between GT and GPIs in GBAs' context.

H3: GT positively affects GPIs.

As the study conceptualizes GBA effecting GT and GT effecting GPIs, it throws up another question; does GT play a mediating or moderating role, if at all, between the GBA and GPIs. In the case of mediating variables, there is a casual relationship between antecedent and consequent variables and in case of moderating variables; a variable has a contingent effect on antecedent and consequent variables. Lin et al. 2003 defined trust as the intention to accept vulnerability based on positive expectations of the integrity and capability of another one. Therefore, rather than being present or absent (i.e., moderating variable) trust is tacitly to be present in overt form is forming GPIs. If GBA presumably affect GT toward intentions to purchase green, then GT is more likely to assume the role of a mediator between the GBA and GPIs. Thus,

H4: GT has a mediating effect on GBA and GPIs.

8.3 METHODOLOGY

8.3.1 DATA COLLECTION AND SURVEY INSTRUMENT

The present chapter has developed the quantitative study to test the relationship among GBA, GPI, and GT. The questionnaire survey was applied to verify the hypotheses and research framework. The questionnaires were mailed, floated on the internet through social media and also data were collected personally. The sample size of 227 respondents was considered from Delhi and NCR. A reasonable degree of randomness could be assumed because of the absence of systematic effort in the selection of the respondents causing convenience biasness which cannot be ruled out. However, the limitation is acceptable given the exploratory nature of the study. 31% of the respondents were in the age group of 18–27 years and 29% in 28–35 years. 22% in 36–40 years and rest belong to age group between 41–46 years. Gender-wise distribution of the sample showed that 44% were females and 56% were males. 30% respondents were graduates and rest were post graduate or above.

8.3.2 MEASURES

To study the impact of GBA on GPIs, data were collected for three variables: GPI, GBA and GT. The measurement of the instrument items in this study was done by means of five-point Likert scale from "1–5" rating from strongly disagree to strongly agree.

1. Green brand associations (GBA): To study the impact of GPIs through GBA, two advertisements where shown to the respondents carrying green information. Then GBA was measured by four attributes of the advertisement shown, that is, dependable, genuine, environment-friendly and committed. The Cronbach's alpha coefficient of measurement was analyzed to measure the internal reliability and the Cronbach's alpha of GBA was 0.880 which is considered as good internal consistency (Malhotra, 2004).

2. Green trust (GT): This study refers to Chen's (2010) research to measure the GT construct. GT was measured by four items. Each item was measured on a five point Likert scale ranging from "strongly agree" to "strongly disagree." The Cronbach's alpha of the instrument was 0.862 which is quite good. The following questions were asked from the respondents after showing them the advertisements:

   i. I feel that green product's environmental reputation is generally reliable.
   ii. I feel that green product's environmental claims are generally trustworthy.
   iii. Green product's environmental concern meets my expectations.
   iv. Green product keeps promises and commitments for environmental protection.

3. Green purchasing intentions (GPI): This chapter refers the research of Chen and Chang (2012) to administer the items for GPIs. The study included three items and each item was measured on a five point Likert scale ranging from "strongly agree" to "strongly disagree." The Cronbach's alpha of the instrument was 0.843 which is quite good. The following questions were asked from the respondents after showing them the advertisements:

   i. I intend to purchase this product because of its environmental concern.
   ii. I expect to purchase this product in the future because of its environmental performance.
   iii. Overall, I am glad to purchase this product because it is environment-friendly.

In order to test the construct validity of GBA, GPI, and GT the study utilizes the factor analysis and the grouping presented in Table 8.1. Further to verify the hypothesis, the present study employed SEM using AMOs 21.0

and used mediation model of Hayes (2008) to obtain empirical results. Two levels were investigated in the current study that is the measurement model through SEM and the structure model through mediation model of Hayes (2008). The results are presented in the following section.

8.4 RESULTS

8.4.1 THE RESULTS OF MEASUREMENT MODEL

To determine the validity of the instruments, discriminant and convergent validity was conducted with the help of principal component analysis (PCA) and confirmatory factor analysis (CFA). The factor analysis was conducted on 227 cases and there was no cross loading above 0.4 and total extracted variance was 0.75. The Varimax factor rotation converged in five iterations and relatively simple factor structure emerged. Three factors are shown in principal component analysis (PCA)-employing varimax rotation as per Table 8.1 and each loading is more than 0.7 which is considered to be excellent except GT4 variable. Appropriateness of sampling adequacy for factor analysis is indicated by significant ($p<0.01$) Kaiser-Mayer-Olkin (KMO) value 0.794.

TABLE 8.1 Exploratory Factor Analysis (Principal Components Method with Varimax Rotation).

	Component		
	1	2	3
GT1		0.831	
GT2		0.876	
GT3		0.696	
GT4		0.788	
GPI1			0.851
GPI2			0.835
GPI3			0.766
GA1	0.871		
GA2	0.818		
GA3	0.837		
GA4	0.854		

Also, factor structure within GBA, GPI, and GT was examined. Conver-gent validity; CFA was conducted using SEM. As a first step model, fit of measurement model was examined using standard indices. It was found that goodness of fit statistic (GFI) was 0.92, Normed fit index (NFI) was 0.921, comparative fit index (CFI) 0.944, and root mean square error of approximation (RMESA) was 0.079. All the parameters indicate the pres-ence of goodness of fit. Convergent validity was tested using two criteria. First, lambda coefficients of each item should be greater than 0.70 and significant at 5% level of significance (Gefen et al., 2000). Second, each path loading should be twice the standard error (Anderson and Gerbing, 1998). It may be noted that if intercorrelation among the latent variables is less than 0.6 that would further confirm the presence of discriminant validity among latent variables (Carlson et al., 2000). It can be observed that all the lambda coefficients are above 0.69 (0.69–0.92). The path coeffi-cients are significant and path loading are greater than twice the respective standard error. The correlation between three constructs is less than 0.60, positive, and significant. Thus, discriminant and convergent validity are confirmed.

8.4.2   THE RESULTS OF THE MODEL

Data are analyzed using mediation model which focuses on the estima-tion of the indirect effect of X on Y through an intermediary mediator variable M causally located between X and Y (i.e., a model of the form $X \rightarrow M \rightarrow Y$) (Hayes, 2015), where X is the input variable, Y is output and M is the mediating variable. The method is useful in studies whose goal is to investigate causal relationship between two variables X and Y while it is postulated that X exerts its influence on Y through another intervening variable(s) M. Hayes (2008) suggests that the causal model may be presented as shown in Figure 8.1. It can be seen from Figure 8.1 that it is a causal system in which at least one causal antecedent X variable is projected as influencing an outcome Y through a single inter-vening variable M. Such a model traces two pathways which influence Y by direct and indirect effect. In direct effect, pathway leads from X to Y without passing through M. In indirect effects, the pathway of X to Y leads through M. Hayes (2008) has shown that direct effect as well as indirect effect of X and Y can be estimated, with the help of two Ordinary Least Squares (OLS) equations using Y and M as consequent variables. There are two consequent variables forming two equations, and these equations can be estimated by conducting OLS regression analyses using SPSS or by using PROCESS.sps in SPSS developed by Andrew F. Hayes. Using Hayes (2008) notations the equations are:

$$M = i_1 + aX + e_M \qquad (8.1)$$

$$Y = i_2 + \grave{c}X + bM + e_Y \qquad (8.2)$$

where a and $\grave{c}$ respectively estimate direct effect of X on M and Y. Hayes (2008) has shown that can quantify indirect effect of X on Y through M by multiplying a and b obtained in Equations 8.1 and 8.2 above. Thus, one unit change in X will lead to ab units change in Y because of effect of X on M which, in turn, influences Y. It is interesting to note that total effect of X on Y can be estimated by running OLS regression:

$$Y = i_3 + cX + eY \qquad (8.3)$$

where c is measure of total effect. It may be noted that sum of direct and indirect effects constitutes total effect. Thus,

$$c = \grave{c} + ab \qquad (8.4)$$

This implies

$$ab = c - \grave{c} \qquad (8.5)$$

Equation 8.5 provides an interesting insight into the nature of indirect effect (Hayes, 2008) that can obtain indirect, mediation effect by subtracting or eliminating effect of X on Y while holding M constant from the total effect of X on Y. In respect of statistical inference regarding the significance of coefficients a, b, $\grave{c}$, c, and ab. Hayes (2008) has shown through literature review as well as empirically the advantage of employing bootstrapping approach for generating the sampling distribution of these coefficients. The study reports the confidence interval of these coefficients as well as p-values based on such sampling distribution employing PROCESS. SPS software developed by Hayes. (It is available on his website www. afhayes.com)

Table 8.2 presents, the results of three OLS regressions estimating the impact of GBA on GPI (total effect), GT and influence of GT on GBA (indirect effect), while Figure 8.1 shows diagrammatic presentation where all the three paths estimated are significant using Hayes' (2008) mediation model.

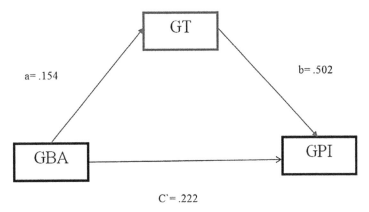

FIGURE 8.1   Statistical diagram of mediation model of GPI.

It can be observed from the table that coefficient of GBA GT (model 1) is significant ($p<0.05$), thereby, supporting hypothesis 3. R² at 0.029 may not be called substantial. In order to verify hypothesis 1, a regression was run to examine the total effect of GBA. The regression results of GPI as a function of GBA showed slope coefficient. 299 is positive and significant with standard error 0.056, p value of.0000, R²=0.111 and R=0.334. Note that the OLS slope coefficient represents total effect of GBA on GPI (it is c of Eq. 8.3 specified in results section). It may also be mentioned here that the direct effect of GBA on GPI is also positive and significant (Eq. 8.2) with slope coefficient. 222, standard error 0.049 and p value 0.0000. It implies that GBA affects GPI significantly even if GT is held constant. Thus, there is sufficient evidence to support hypothesis 1. Eventually, brand attributes have considerable effects on purchasing intentions of the consumers.

TABLE 8.2   Model Coefficients for the Green Purchasing Intentions (GPIs) Study.

Antecedent	Consequent					
	GT			GPI		
	Coefficient	SE	p	Coefficient	SE	p
GBA	0.154	0.059	0.009	0.222	0.049	0.000
GT	–	–	–	0.502	0.055	0.000
Constant	3.1906	0.2176	0.000	1.415	0.251	0.000
	R2=	0.029			R2=	0.3526
	F (1226)=	6.895	p=0.009	F (2225)=	81.476	p=0.000

In model 2, it is worth noting that there is a substantial increase in R^2 of model 2 (0.353) as compared to R^2 of model 1 (0.29). It implies that if consumers acknowledge GT to the total effect equation, the R^2 of the model goes up significantly $(p < 0.01)$[1]. This substantial increase in explained variation lends credence to the hypothesis 4 that GT has a partial mediating role in the proposed model. Model 2 shows that there is a positive and significant relationship between GT and GPI at 5% level of significance, supporting hypothesis 2. It shows, prima facie, better relevance of model 2 relatively than model 1. The findings affirm that one, model 2 has a better model specification compared to model 1 due the modest increase of R^2. Two, GBA explains greater variation in GPI when GT also is taken into account. It implies that it not only the environmental attributes which intend to lead GPI among consumers, but addition of willingness to depend on the brand to fulfill their environmental needs increases the intentions to purchase green products. Hence, this study demonstrates that presence of GT is a vital determinant in GBA for affecting purchase intentions of consumers.

The indirect effect between the two coefficients, for example, GPIs and GBA through GT was examined and indicated by ab which was found to be 0.0771 $(p < 0.05)$.

It demonstrates the partial mediation effect in the study that GPI is more powerful when green attributes of brand operates through of GT. It may, therefore, be asserted that consumers will have higher GPI, environmental consciousness and rejection of products/practices damaging the society owing to the impact of GBA through GT rather than the direct relationship between GBA and GPI. Thus hypothesis 4 is supported. Table 8.3 summarizes the support to various hypothesis proposed in this study.

TABLE 8.3   Support to Different Hypotheses.

Hypothesis	Proposed effect	Coefficient	Results
H1	+	0.299*	H1 is supported
H2	+	0.154*	H2 is supported
H3	+	0.502*	H3 is supported
H4	+	0.077*	H4 is supported

*$p < 0.01$

[1]The significance of difference of two R^2 was calculated using the F test proposed by Gujarati, Porter and Gunasekar (2009, p.260)

8.5 LIMITATIONS AND FUTURE SCOPE

One of the limitations of this study is randomness in the sampling. It might adversely affect the generalizability of the results. Moreover, sample size is confined to Delhi and NCR. There is a need to examine the subject over a sample extending to other states of the country. Second, in this study GBA evaluated only on the two advertisements shown to the respondents and not specifying any particular brand or segment due to the exploratory nature of the study. Future research can focus on the different and particular brand/ segment to compare the results of this study or for conclusive results. Third, only a few variables of GBA have been investigated. For further research, researchers can add more attributes related to trustworthiness of the brand. Fourth, factors and dynamics of perception toward brand associations, trust, and intentions change over period of time. Hence, future researchers should consider longitudinal study.

8.6 CONCLUSION

Potentially, green marketing is still in the genesis phase of research and practice besides reaching to its zenith, as researchers and practitioners may explore the concepts in green marketing using different lens. By converging the multidisciplinary view of brand management and green marketing, the present chapter proffers an approach to understand the purchasing intensions of environmentally friendly products through two constructs namely GBA and GT. Eventually, when discussing the purchasing intentions dimensions of environment-friendly products, scholars have drawn the multidisciplinary literature to reveal the understanding of how consumers subjectively seek intentions to purchase green through GT, green satisfaction, green brand equity, green values, green knowledge, green attitude, and/or green consciousness. Indeed, this study develops a new research framework of GBA-influencing GPI through GT to acknowledge other perspective of GPIs. The empirical study embodies that green brand association is positively related to GT. In addition, it has been observed that GPI is partially mediated by GT. Thus, this chapter presents a rich conclusion among multiplicity of green marketing literature. All results in the present study are supporting posited hypothesis.

Four wider conclusions can be drawn in the study. First, this study combines the concepts of green marketing and brand associations to develop a research framework of GPIs. Second, when respondents were asked regarding brand attributes after showing them the advertisements to understand the subjective view of consumers, they associated the brands as green and trustworthy. Third, there is positive relationship between GBA and GPIs and this relationship is partially mediated by GT. Fourth, this chapter extends the research of consumer buying behavior and green branding by analyzing the impact of GBA in the field of green marketing.

The practical contributions in this study verify that influence of environmental attributes have an impact on consumer's GPI. It may be purported that if companies would like to enhance their purchase intentions for their environmental products, they should underline the concept of GBA and try to utilize the environmental attributes of brand to build long term environmental strategies. Second, in green marketing, companies should focus on conviction aspect of consumers to influence GPIs through GT. Third, companies need to make strategies to enhance GT among consumers because present study reported the significant mediation effect of GT. Many respondents believe that green products in showed advertisements give the impression of being trustworthy and meeting their environmental expectations. It may be affirmed that trustworthiness attribute in advertisement may enhances GT of the brand and leads to commitment and reliability. Hence, to stimulate consumer's GT organizations should improve the green products' dependability, consistency, and assurance in branding strategies.

It is hoped that this chapter has put forward a case of examining GPIs of consumers derive through GBA and GT. In a managerial context, manager may affix more trustworthiness features in the advertisements of green products to enhance GPI. However, to gain green competitive advantage companies should avoid greenwash as this leads to green distrust among consumers (Chen and Chang, 2013).

KEYWORDS

- **green brand associations**
- **green trust**
- **green purchase intentions**
- **green competitive advantage**
- **mediation model**

REFERENCES

Aaker, D. A. *Building Strong Brands;* The Free Press: New York, 1996.

Aaker, D. A. Brand Extensions: The Good, the Bad and the Ugly. *Sloan Manage. Rev.* **2012,** *31*(4).

Aaker, D. A.; Biel, A. L. The Dual Structure of Brand Associations. In *Brand Equity and Advertising: Advertising's Role in Building Strong Brands;* Farquhar, P. H., Herr, P. M., Eds.; Psychology Press: New York, **1993**; Vol. 1, p 263.

Agarwal, M. K.; Rao, V. R. An Empirical Comparison of Consumer-Based Measures of Brand Equity. *Mark. lett.* **1996,** *7*(3), 237–247.

Akehurst, G.; Afonso. C.; Gonçalves H. M. Re-Examining Green Purchase Behaviour and the Green Consumer Profile: New Evidences. *Manage. Decis.* **2012,** *50*(5), 972–988.

Anderson, J. C.; Gerbing, D. W. Structural Equation Modeling in Practice: A Review and Recommended Two-Step Approach. *Psychol. Bull.* **1988,** *103*(3), 411.

Bhattacherjee. A. Individual Trust in Online Firms: Scale Development and Initial Test. *J. Manage. Inf. Syst.* **2002,** *19*(1), 211–241.

Biel, A. L. The Brandscape: Converting Brand Image into Equity. *Admap* **1991,** *26*(10), 41–46.

Carlson, D. S.; Kacmar, K. M.; Williams, L. J. Construction and Initial Validation of a Multidimensional Measure of Work–Family Conflict. *J. Vocational Behav.* **2000,** *56*(2), 249–276.

Chaudhuri, A.; Holbrook, M. B. The Chain of Effects from Brand Trust and Brand Affect to Brand Performance: The Role of Brand Loyalty. *J. Mark.* **2001,** *65*(2), 81–93.

Chen, Y. S. The Drivers of Green Brand Equity: Green Brand Image, Green Satisfaction, and Green Trust. *J. Bus. Ethics* **2010,** *93*(2), 307–319.

Chen, Y. S.; Chang. C. H. Enhance Green Purchase Intentions: The Roles of Green Perceived Value, Green Perceived Risk, and Green Trust. *Manage. Decis.* **2012,** *50*(3), 502–520.

Chen, Y. S.; Chang, C. H. Greenwash and Green Trust: The Mediation Effects of Green Consumer Confusion and Green Perceived Risk. *J. Bus. Ethics* **2013,** *114*(3), 489–500.

Chong, B.; Yang, Z.; Wong, M. Asymmetrical Impact of Trustworthiness Attributes on Trust, Perceived Value and Purchase Intention: A Conceptual Framework for Cross-Cultural Study on Consumer Perception of Online Auction. *In Proceedings of the 5th International Conference on Electronic Commerce.* ACM. Sept, 2003; pp 213–219

Delgado-Ballester, E.; Luis Munuera-Alemán, J. Does Brand Trust Matter to Brand Equity? *J. Prod. Brand Manage.* **2005,** *14*(3), 187–196.

Diamantopoulos, A.; Schlegelmilch, B. B.; Sinkovics, R. R.; Bohlen, G. M. Can Socio-Demographics Still Play a Role in Profiling Green Consumers? A Review of the Evidence and an Empirical Investigation. *J. Bus. Res.* **2003,** *56*(6), 465–480.

Gefen, D.; Straub, D.; Boudreau, M. C. Structural Equation Modeling and Regression: Guidelines for Research Practice. *Commun. Assoc. Inf. Syst.* **2000,** *4*(1), 7.

Hartmann, P.; Apaolaza Ibáñez, V. Green Value Added. *Mark. Intell Plann.* **2006,** *24*(7), 673–680.

Hayes, A. F. *Introduction to Mediation, Moderation, and Conditional Process Analysis: A Regression-Based Approach.* Guilford Press: New York City, NY, 2008.

Hayes, A. F. An Index and Test of Linear Moderated Mediation. *Multivar. Behav. Res.* **2015,** *50*(1), 1–22.

James. D. Guilty Through Association: Brand Association Transfer to Brand Alliances. *J. Consum. Mark.* **2005,** *22*(1), 14–24.

Keller, K. L. Conceptualizing, Measuring, and Managing Customer-Based Brand Equity. *J. Mark.* **1993,** *57,* 1–22.

Keller, K. L *Strategic Brand Management. Building, Measuring and Managing Brand Equity;* Prentice Hall: Englewood Cliffs, NJ, 1998.

Laroche, M.; Bergeron, J.; Barbaro-Forleo, G. Targeting Consumers who are Willing to Pay More for Environmentally Friendly Products. *J. Consum. Mark.* **2001,** *18*(6), 503–520.

Lin, N. P., Weng, J. C.; Hsieh, Y. C. Relational Bonds and Customer's Trust and Commitment—A Study on the Moderating Effects of Web Site Usage. *Serv. Ind. J.* **2003,** *23*(3), 103–124.

Malhotra, N. *Marketing Research.* Pearson Education: Englewood Cliffs, NJ, 2004.

McKnight, D. H.; Chervany, N. L. Trust and Distrust Definitions: One Bite at a Time. In *Trust in Cyber-Societies;* Springer Berlin Heidelberg; 2001a; pp 27–54.

McKnight, D. H.; Chervany, N. L. What Trust Means In E-Commerce Customer Relationships: An Interdisciplinary Conceptual Typology. *Int. J. Electron. Commer.* **2001b,** *6*(2), 35–59.

McKnight, D. H.; Choudhury, V.; Kacmar, C. Developing and Validating Trust Measures for E-Commerce: An Integrative Typology. *Inf. Syst. Res.* **2002,** *13*(3), 334–359.

Netemeyer, R. G.; Maxham, J. G., III; Pullig, C. Conflicts in the Work–Family Interface: Links to Job Stress, Customer Service Employee Performance, and Customer Purchase Intent. *J. Mark.* **2005,** *69*(2), 130–143.

Peattie, K. *Environmental Marketing Management: Meeting the Green Challenge;* Pitman: London, 1995.

Peters, R. G.; Covello, V. T.; McCallum, D. B. The Determinants of Trust and Credibility in Environmental Risk Communication: An Empirical Study. *Risk Anal.* **1997,** *17*(1), 43–54.

Rempel, J. K., Holmes, J. G.; Zanna, M. P. Trust in Close Relationships. *J. Personal. Soc. Psychol.* **1985,** *49*(1), 95.

Rotter, J. B. Generalized Expectancies for Interpersonal Trust. *Am. Psychol.* **1971,** *26*(5), 443.

Schlegelmilch, B. B.; Bohlen, G. M.; Diamantopoulos. A. The Link between Green Purchasing Decisions and Measures of Environmental Consciousness. *Eur. J. Mark.* **1996,** *30*(5), 35–55.

Schlosser, A. E., White, T. B.; Lloyd, S. M. Converting Web Site Visitors into Buyers: How Web Site Investment Increases Consumer Trusting Beliefs and Online Purchase Intentions. *J. Mark.* **2006,** *70*(2), 133–148.

Sharma, N.; Kesherwani, S. Encouraging Green Purchasing Behavior by Increasing Environmental Consciousness. In *Reinventing Marketing for Emerging Markets;* Das, J. K., Zameer, A., Narula, A., Tripati, R., Eds.; Bloombury Publishing: India, 2015; pp 288–301.

Sharma, N.; Sharma, C. S. Encouraging Green Purchasing Behavior Through Green Branding. *Bus. Anal.* **2013,** *34*(2), 65–76.

Straughan, R. D.; Roberts, J. A. Environmental Segmentation Alternatives: A Look at Green Consumer Behavior in the New Millennium. *J. Consum. Mark.* **1999,** *16*(6), 558–575.

Tauber, E. M. Brand leverage-Strategy for Growth in a Cost-Control World. *J. Advertising Res.* **1988,** *28*(4), 26–30.

Van Osselaer, S. M.; Janiszewski, C. Two Ways of Learning Brand Associations. *J. Consum. Res.* **2001,** *28*(2), 202–223.

Yoo, B.; Donthu, N.; Lee, S. An Examination of Selected Marketing Mix Elements and Brand Equity. *J. Acad. Mark. Sci.* **2000,** *28*(2), 195–211.

PART II
Sustainability Aspects of Green Marketing

CHAPTER 9

SUSTAINABLE GREEN MARKETING: A TREND OF CONSUMERISM

HARSH TULLANI[1,*] and RICHA DAHIYA[2]

[1]Department of Management Studies, SRM University, Delhi NCR, Sonepat, Haryana 131029, India, Mob.: +91-9034666131

[2]Department of Management Studies, SRM University, Delhi NCR, Sonepat, Haryana 131029, India, Mob.: +91-8901291933, E-mail: richa_dahiya18@yahoo.co.in

*Corresponding author. E-mail: harshtullani21@gmail.com

9.1 INTRODUCTION

The beginning of 21st century faced environmental challenges such as resource conservation, climate changes, and global warming, which are directly related to offshoots of served business practices that have a detrimental impact on the economy, environment, and society. All the resources are depleting day by day and their sustainability needs to be replenished urgently. This chapter is intended to investigate the impacts of green marketing strategy on the firms' performance through an analysis of the green innovation and green promotion. Marketing is central to global society, and when harnessed responsibly can encourage us to recycle, reuse, buy Fairtrade, eat healthily, drink sensibly, save energy, and support good causes. Marketing serves an important function in promoting economic development around the world, raising living standards in many countries (Fisk, 2001). The commercial potential following the recent economic crisis to further the goals of sustainability and provide ethical products and services remains considerable (Carrigan and de Pelsmacker, 2009). In the race for economic development, governments, society, and business have created environments in which marketing has flourished to serve consumer and societies' needs. The term "sustainable" becomes a popular notion in recent literature. It discusses the basic premise, that businesses have a responsibility to satisfy the human needs and wants while preserving nature. The current ecological challenges require the managers to formulate strategies that control pollutions and preserve the natural resources. Millar et al. (2012) stated that integrating sustainability into the organization is still a challenge and difficult for many leaders. Hence, many industries are adopting a green business strategy to ensure sustainable growth that encompasses the green characteristic of their business operation. Although on a voluntary basis, more and more companies are taking this initiative and it has become one of the main agendas and is important strategic concerns for companies. The remarkable growth of new green industries has attracted positive attention from the marketers.

This chapter provides an understanding of the term sustainability and inclusivity in context of green marketing. It also discusses how green innovation and green promotions are major tools of green marketing that have led to firm's performance resulting in the sustainable green market. The outcome of this chapter is a model that shows a mutual interrelation between green innovation, green promotion, and firm's performance in a sustainable and inclusive green market.

9.2   PURPOSE OF THE STUDY

Green marketing is still in nascent stage in India. "Preserving environment" is a major concern of not only India but of the entire world. Moreover, sensitivity and consciousness of current government led by Prime Minister Shri Narender Modi toward *Swachh Bharat Abhiyan* "clean India campaign" has given a momentum to green India and green marketing. Whereas companies in India, now, show more concern for the environment and consumers also have an inclination toward the purchase of environment-friendly products. So, green marketing is emerging as the latest buzzword of marketing.

Green marketing is done according to the new rules also affects how a corporation manages its business and brands and interacts with all of its stakeholders who may be affected by its environmental and social practices (Ottman, 2011). This chapter is intended to investigate the impact of green marketing strategy on the firms' performance through an analysis on the green innovation and green promotion practices as it investigates the impact of green innovation and green promotion on firm performance with the emergence of sustainable and inclusive market in the society. This qualitative chapter is syntheses of all the available literature along with an analysis of some selected cases on sustainable green market propose a conceptual model which can further be empirically tested to make projections for future. This is a pioneer research in this area and will formulate a building block for future researches.

9.3   TURN THROUGH GREEN MARKETING

Green marketing is also known in different terminologies such as environmental marketing, ecological marketing, or sustainable marketing. Polonsky (1994) stated that green marketing comprises product modification, changes to the production method and process, packaging, and modifying advertising. Juwaheer et al. (2012) recommended that effective green marketing strategy should be further developed on green branding, packaging, labeling, and advertising to create demand for the green products. Chamorro and Bañegil (2006) stated that the objective of the green marketing is to lessen the impact on the natural environment during the process of planning and implementations of products or services, price, place, and promotion, and Mourad and Ahmed (2012) pointed out that the goal of green marketing is to create profit and maintain the social responsibility. Interestingly, Gordon et al. (2011) supported that green marketing that includes ranging from production to post-purchasing service with the goal to balance the company's profit and protect the environment as well. Sarkar (2012) also agreed that green marketing encompasses a broad range of activities including product modification, changes to the production process, packaging changes, remodeling, and stylizing as well as modifying advertising. In general, green marketing can be applied to the much broader concept; consumer goods, industrial goods, and services.

9.4   GREEN MARKETING AND SUSTAINABILITY

Sustainability has been defined as "the consumption of goods and services that meet basic needs and quality of life without jeopardizing the needs of future generations" (OECD, 2002). As Cooper (2005) indicates, this may be interpreted in a number of ways, but principally sustainability is about limiting the throughput of resources while making the best use of those resources available. Sustainability is now a mainstream issue, evidenced by the growing interest shown in sustainable issues (Bandura, 2007; Fitzsimmons, 2008). Sustainability is a relatively recent notion, it is worth mentioning that the term of sustainability has so far been encompassed by countless definitions and has been applied along with a number of various notions such as sustainable marketing. According to Peattie (2001, pp 131–132), the concept of sustainability is significant because: it pulled together issues relating to the physical environment, society, and economy and recognized their interdependence. Previously, these were presented as separate agendas between which interests were traded off (in particular, environmental protection and economic growth were presented as a choice). The concept was widely discussed and, at least in principle, adopted as a strategic aim by the majority of the world's governments and major corporations. For the term of sustainability, there is no universal definition of sustainable development, through its foundations include the fact that it is oriented to economic and social progress without affecting the natural equilibrium.

As economies implode and nefarious business decisions are exposed, there has been considerable criticism that marketers have been complicit through their encouragement of unsustainable patterns of consumption (De Graaf et al., 2005). It is evident that "extensive environmental damage has been caused by continuous consumption, marketing, manufacturing, processing, discarding, and polluting" (Saha and Darnton, 2005, p 117), leading to suggestions that "a sustainable future is not achievable while disregarding the key contributors to ecological degradation population growth and high consumptive lifestyles" (Bandura, 2007, p 32). Marketing "negatives" are effect of wider societal and structural conditions and not necessarily a direct cause of unsustainable practices, just as consumption is a collection of social practices that influence, and are influenced by, lifestyle choices, social norms, societal structures and institutions (Connolly and Prothero, 2003; Jackson, 2005). While historically governments have been reluctant to curb the material consumption of their citizens (Cohen, 2001), there is a growing impatience with the less sustainable outcomes from marketing activity. While arguments about wealth disparities are not new, engendering good relationships with developing markets are critical to the future of the consumer industry, while the credibility of products becomes increasingly linked to their ability to advance well-being rather than harm (Nair, 2008; Charles, 2010). It appears that the hegemony of a dominant social paradigm and the role of marketing within it, it has played a part in environmental decline (Kilbourne and Carlson, 2008). Herein, lies the crux of the marketing problem. Currently, marketing does exactly what it is supposed to do, that is, selling more goods, encouraging consumption, and making profits. It is not inherently managed to deliver sustainability, thus, its potential to do so is often overlooked.

Marketing's functionality emanates not only from structural conditions but also from features within the discipline's fundamental principles and practices. The dominance of a managerialism, functionalist theoretical approach to marketing often means less focus on the effect of green marketing as a social institution of significant importance and impact with undoubted implications on sustainability. Yet, recent concerns and pressures have begun to shift the landscape. According to Porter's (1985) value chain has been used to conceptualize the positive image created by a firm for its customers, and, through complex exchanges for a range of organizational stakeholders. Polonsky et al. (2003) "harm chain" consider how those exchanges may also result in the generation of harm. These often unintentional harms or "moral externalities" (Gowri, 2004), as they are defined, have received only limited attention within the marketing literature (cf. Desmond and Crane, 2004; Fry and Polonsky, 2004a). Fry and Polonsky (2004b, p 1209) identify that while typically firms engage in marketing activities with outcomes beneficial to both the firm and its stakeholders, an increasing number of situations occurs where successful marketing activities impact on society in an "unanticipated negative manner." This has generated reviews of green marketing in order to allow it to play a viable role in sustainable development (van Dam and Apeldoorn, 1996; Desmond and Crane, 2004; Peattie and Peattie, 2009). To achieve this it has become clear that green marketing itself needs to become sustainable. Several marketing academics have begun exploring a sustainable green marketing paradigm (Cooper, 2005; Fuller, 1999; Peattie and Peattie, 2009). Sustainability would, therefore, become a central component of green marketing thought and practice, one that carries a convincing business case from a consumer perspective. Consumers report an increased disposition toward ethical behaviors; a recent Boston Consulting Group study of 9000 consumers in nine countries (Manget et al., 2009) concluded that green and ethical issues are a significant factor, influencing where consumers shop and what they buy.

9.5 GREEN MARKETING AND INCLUSIVITY

Inclusion is a state of being valued, respected, and supported. It is about focusing on the needs of every individual and ensuring the right conditions are in place for each person to achieve his or her full potential. Inclusion should be reflected in an organization's culture, practices, and relationships that are in place to support a diverse workforce.

In simple terms, sustainability is the mix; inclusion is getting the mix to work well together. The notion of inclusivity is for additional focus and innovation in the way companies do business. It involves creating new forms of employment, new markets, and affordable products and services. This spurs economic growth and encourages entrepreneurship (Bonnell and Veglio, 2011). Inclusive businesses are entrepreneurial initiatives that are economically profitable and environmentally and socially responsible. Underpinned by a philosophy of creating mutual value, inclusive businesses contribute to improving the performance of employees and increase the productivity of the firm. Inclusivity is fundamental to our culture and core values at Accenture. We believe that no one should be discriminated because of their differences such as age, disability, ethnicity, gender, religion, or sexual orientation. Accenture also believes that government laws, regulations, and business practices should uphold the principles of diversity, inclusion, and equality. While laws may vary in the countries where Accenture operates, we remain committed to an inclusive and diverse workplace—where people can feel comfortable, be themselves and, as a result, be productive and leads to the positive performance of the firm. To optimize talent performance, diversity and inclusion practices should be included within all of an organization's policies, practices, and training. We also know that there are times when specialized diversity and inclusion strategy is exactly what is needed in the organization to meet the challenges.

It is strongly believed that a diverse workforce and sustainability in an inclusive environment will improve individual and organizational performance and result in better value to customers, clients, taxpayers, and other stakeholders.

9.6   SUSTAINABLE GREEN MARKET—CASE ANALYSIS

Green marketing has been described as a race without a finish line. Every organization wants to be in a race to achieve idealistic goals. Corporate conservationism acts as a series of small steps on a learning curve of green innovations and promotion in their products to create green consumerism.

9.6.1   PHILIPS

At Philips, innovation efforts are closely aligned with business strategy. It does so through cooperation among research, design, marketing, strategy, and business in interdisciplinary term along with the innovation chain, from front-end-to first-of-a-kind product development. Philips life light, the new zero-energy, solar-powered LED lighting range designed for homes in off-grid rural and semi-urban communities, as part of Philips drive to deliver innovations that are locally relevant.

9.6.2 LG

South Korean consumer durables major LG Electronics recently introduced LG LED E60 and E90 series monitors in the Indian market. The LED E60 and E90 series monitors are packed with the eco-friendly features such as 40% less energy consumption than the traditional LCD monitors. Also, the use of hazardous materials such as halogen or mercury is kept to minimal in this range. Globally, LG launched a range of eco-friendly products "Eco-Chic" such as the platinum coated two-door refrigerator and washing machine with steam technology.

9.6.3 HCL

Last year, HCL launched its range of eco-friendly notebooks, HCL ME 40. HCL claims that this was India's first polyvinyl chloride (PVC) free and eco-friendly notebook. This notebook is completely free from PVC and other harmful chemicals. Further, Bureau of Energy Efficiency has given HCL's eco-friendly products a five-star rating, and they also meet REACH (REACH is the European Community Regulation on chemicals and their safe use) standards and are 100% recyclable and toxin free.

9.6.4 SAMSUNG ELECTRONICS

Samsung Electronics, a leading brand in display products, is continuing to evolve LED monitor technology by offering environment-friendly monitors with its range of 50 and 30 series LED monitors. Samsung products have an eco-friendly LED backlight, which contains few or no environmentally hazardous substances such as mercury or lead. Additionally, it uses about 40% less energy. Also, Samsung's touch of color (ToC) technology doesn't use paints, sprays or glues, ensuring they contain no volatile organic compounds (VOCs), making recycling simpler and safer.

Its other eco-feature allows users to adjust a monitor's brightness based on different energy consumption levels with four preset energy-saving options to choose from. Samsung has also launched its advanced range of split ACs, which claim to save up to 60% energy.

9.6.5   VOLTAS

Air-conditioners, refrigerators, and plasma or LCD TVs are going green with a vengeance. Next in the line is, Voltas from the Tata Group. In 2007, Voltas initiated the "Green" range of air-conditioners, following which the government made it mandatory for home appliances to have energy star ratings. Energy Star is an international standard for energy efficient consumer products that originated in the United States. Thus, devices carrying the star logo such as computer products and peripherals, kitchen appliances and other products, use about 20–30% less energy than the set standards.

9.6.6   PANASONIC

Panasonic has an ECONAVI range of air conditioners and LCD screens which is once again based on energy conservation. ECONAVI home appliances use sensor and control technologies to minimize energy consumption based on a family's lifestyle. For instance, a door-opening sensor and lighting sensor allows the refrigerator to learn the time periods when the family typically doesn't use—when they are sleeping or away from home. The refrigerator goes into sleep mode accordingly. Globally, Panasonic is aiming to become top green innovation company in the electronics industry by 2018 and is laying a lot of emphasis on eco-friendly products.

9.6.7   VIVANTA BY TAJ HOTEL, RESORTS, AND PALACES

Earth-friendliness is the hallmark of the Vivanta line, just as it is at all Taj hotels. The parent company follows guidelines established at a United Nations Earth Summit and endorsed by nearly 200 countries. These green benchmarks are monitored by a leading worldwide certifier, Green Globe. Taj aims to bring the total of its Vivanta hotels to 30+ in the next 2 years, totaling 5000+ guest rooms. So, from LCDs to clothing, every brand is giving you an opportunity to ride the green wave. Go ahead, make a choice and join the green revolution.

9.6.8   STARBUCKS UTILIZES FACEBOOK FOR CAMPAIGN

Facebook is a social media that is capable enough to provide the necessary facilities to its users, it is not surprising that Starbucks chose to do green marketing campaign in Facebook. Starbucks actually did not only choose Facebook account but they also use a Twitter, Google+, and Instagram.

9.6.9   AMAZON

Its marketing strategy is designed to increase customer traffic to their websites, drive awareness of products and services they offer, promote repeat purchases, develop incremental product and service revenue opportunities, and strengthen and broaden the Amazon.com brand name. They save the environment by adopting green means of advertising which consists primarily of online advertising through associates program, sponsored search, portal advertising, e-mail campaigns, and other initiatives.

9.6.10   KFC

KFC's new green restaurant the latest store was built using elements that follow the leadership in energy and environmental design certification process created by the U.S. Green Building Council. "This new KFC Green restaurant is a part of our E3 initiative, which looks at economically responsible ways of saving energy and being environmentally aware."

9.6.11   STATE BANK OF INDIA—GREEN IT @ SBI

State Bank of India (SBI) entered into green service known as "Green Channel Counter." SBI is providing many services like paperless banking, no deposit slip, no withdrawal form, no checks, no cash transactions. All these transactions are done through SBI shopping and ATM cards.

9.7   OUTCOME DISCUSSION

Green marketing and what is increasingly being called "sustainable branding" is the topic of discussion today. According to the new rules of green marketing, effectively addressing the needs of consumers with a heightened environmental and social consciousness cannot be achieved with the same assumptions and formulae that guided consumer marketing in the post-war era. Times have changed. A new paradigm has emerged, requiring new strategies with a holistic point of view and green innovative product, and service offering as green promotions.

"The literature enables to provide results on the role of green innovation and green promotion as a marketing strategy. Polonsky (1994) stated that green marketing is a tool to promote green products to satisfy customers' needs and wants. Based on the literature and case analysis, it can be concluded that green innovation and green promotion will lead to enhance firm's performance and which will, in turn, lead to emergence of sustainable green market."

9.7.1 PROPOSED MODEL

The proposed model shows that green innovation is induced by competitors, government, customer, supplier, employees, and technology in the organization and green promotion has five tools, that is, green advertising, direct marketing, sales promotion, public relations, personal selling for generating awareness about the product, providing knowledge related to goods and services which create sustainability and inclusivity in the market and enhance product and customer value in the market which further leads to firm performance (socially, economically, and environmentally) in a positive direction by increasing their profitability and productivity in the sustainable green market (Fig. 9.1).

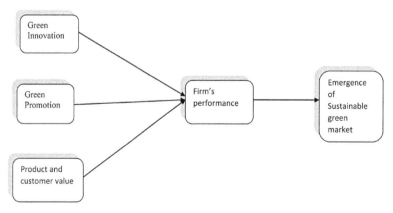

FIGURE 9.1   Proposed model.

Source: Author's compilation

9.8   GREEN INNOVATION

Green innovation is a synonym for the successful production, assimilation, and exploitation of novelty in the economic and social spheres. It offers new solutions to problems and, thus, makes it possible to meet the needs of both the individual and society. As a driving force, it points firms toward ambitious long-term objectives.

According to the definition proposed by the Organisation for Economic Co-operation and Development (OECD) in its "Frascati Manual," it involves the transformation of an idea into a marketable product or service, a new or improved manufacturing or distribution process, or a new method of social service. The term, thus, refers to the process. On the other hand, when the word "innovation" is used to refer to the new or improved product, equipment or service which is successful on the market, the emphasis is on the result of the process.

Green innovation is at the heart of the spirit of enterprise: practically all new firms are born from a development which is innovative, at least in comparison to its existing competitors on the market. Innovation also means anticipating the needs of the market, offering additional quality or services, organizing efficiently, mastering details, and keeping costs under control.

Green innovation is defined as the modification in the product or process with the concern of environment. Chen et al. (2006) suggest that green innovation consists of technical improvements or new administrative practices that improve the environment, firm performance, and the competitive advantage of an organization.

We intended to identify the green innovation practices and examine the effects of stakeholders on the adoption of green innovation practices and the consequences on the environment and the performance of companies. We viewed each stakeholder as a factor exerting pressure on the companies and driving the companies toward better environmental practices.

9.8.1   PRESSURE FROM COMPETITORS

Companies usually react and respond to the actions of their competitors. When competitors adopt new environmental practices, companies in the same industry will feel pressured to reevaluate their current status regarding environmental responsibility and to decide whether to increase and/or improve the implementation of environmental practices (Christmann, 2004).

In general, companies need to be aware of their competitors' offerings and industry norms to ensure that their innovation capabilities are similar to those of the rest of the industry.

9.8.2   GOVERNMENTAL PRESSURE

This item is related to the stringency of government regulations and the degree to which future regulation and its effects on business could be predicted. Regulatory changes and enforcement of these changes by the government affect companies' actions regarding environmental management and sustaining their business (Huang et al., 2009). Additionally, to compete globally, companies need to follow both global and local regulations to protect the environment. The policies of the government affect the working of business as they provide their own rules and regulations to regulate with different bodies and with change in a political party or government; it will lead to change in the working of government.

9.8.3   CUSTOMER PRESSURE

Companies have great impact regarding environmental practices due to customers' pressure for going ecological. Customer expectations have become one of the most important factors influencing companies' environmental practices. More and more customers now have strong concerns about the environment and prefer to purchase environment-friendly products. Customers may refuse to buy products that damage the environment, which encourages companies to create green products.

9.8.4   PRESSURE FROM SUPPLIERS

Suppliers affect the cost, lead time, development risks, and market availability of manufacturers. Pujari (2006) pointed out that a firm's green innovation is largely determined by "upstream" environmental impacts, meaning that supplier materials and components could influence the quality, design, and competitiveness of a company's products. Sometimes, suppliers may refuse to supply products to firms that they believe damage the environment. Geffen and Rothenberg (2000) noted that through unique partnerships with suppliers, companies can improve their environmental performance, indicating that supplier involvement plays an important role in the firm's innovation.

9.8.5   EMPLOYEE CONDUCT

Top managers recognize the importance of environmental protection and their company's responsibility to influence strategic planning with regard to environmental management. Strong recognition of and attention to environmental factors by management should yield better innovation and performance (Huang, 2009). Additionally, a company's future direction with regard to environmental practices depends heavily on whether the management team encourages employees to actively participate in environmental management initiatives and on management's own commitment to green practices (Fergusson, 2006). Companies will have difficulty in accomplishing environmental goals if employees do not support their policies (Zhu et al., 2008). Thus, companies need to provide employees with training on environmental issues to involve appropriate employees and to enhance their commitment to environment-friendly practices (Reinhardt, 1999).

9.8.6   TECHNOLOGY

Reducing global greenhouse gas emissions and protecting environmental assets will require innovation and the large-scale adoption of green technologies. Without innovation, it will be very difficult and very costly to sustain current growth while addressing major environmental issues such as climate change. Consequently, OECD governments and emerging economies are giving priority to R&D activities and incentives for the diffusion and adoption of green technologies in the society.

9.9   GREEN PROMOTION

Green promotion is the communication tools and tactics that a company uses to promote and market their product. There are many ways to communicate a company's products and benefits. Branding is the cornerstone of the communications platform.

Green marketing likewise involves promoting the material of a business. Sales promotions, direct marketing, public relations, and green advertising are some of the means of conveying to the customers the core message of greenness; the latter two approaches are the most extensively used platforms for projecting the green outlook of a firm. Going green, which aims to bridge the business and the community, can sometimes require a major public relations initiative. Publicizing products and rationalizing their features and prices are two goals that green advertising could achieve (Arseculeratne and Yazdanifard, 2014). For companies, green promotion programs are an effective means of informing stakeholders about their environmental preservation efforts, commitment, and achievements. From the tactical perspective, firms can undertake programs that are designed to reduce the detrimental environmental effects of their marketing communication efforts (Kotler, 2011). For instance, Dell uses roughly 50% recycled paper in its direct mail catalogs, and it has directly linked all of its printed promotional materials to carbon-offsetting programs. From the strategic perspective, firms can use green promotion tactics to communicate the environmental benefits of their goods and services, such as advertising environmental claims, publicizing environmental efforts, and integrating environmental claims into product packaging. For instance, to communicate the environmental impact of its products, Timberland introduced a green index rating system (Ottman, 2011). In the United Kingdom, Procter and Gamble successfully deployed a campaign to reduce the washer temperatures of consumers, which benefitted from the technological advancements of Ariel; in the process, annual savings of 60,000 t of carbon dioxide were realized (Belz and Peattie, 2009).

A green marketing plan is focused on the target market and made up of five key elements. They are green advertising, public relations, personal selling, direct marketing, and sales promotion. These are used as tools to communicate to the target market and produce organizational sales goals and profits. Besides that, the government also uses the green advertising to promote the green environment concept to increase the level of awareness to protect the environment. Abd Rahim et al. (2012) and Ann et al. (2006) recommended that the environmental labels should be aware and stay visible in order to influence purchase from the consumer. In general, it can be concluded that green promotional strategy must be able to communicate the relationship between products or services with the environment, serve as a campaign to promote a green lifestyle, and finally can enhance the corporate image of the companies. The tremendous growth of green brands in the market due to the consumer demand has encouraged the green marketing to appear in a more strategic way (Taghian, 2005). Thus, there will be a parallel rise in an environmental promotional scheme in order to increase the level of awareness among consumers and release the corporate image in the mind of the customer.

Same as green innovation we intended to examine green promotion along with its major components and their effect on firm's performance.

9.9.1   GREEN ADVERTISING

Green advertisement can be defined as the "paid form of nonpersonal presentation and promotion of idea, goods, or services by an identified sponsor in an eco-friendly way." It is an impersonal presentation where a standard or common message regarding the merits, price, and availability of product or service is given by the producer or marketer. One of the most tangible of these criteria can be found in a study conducted by Banerjee et al. (1995). Green advertising is defined as any advert that meets one or more of the following criteria:

1. Explicitly or implicitly addresses the relationship between a product/ service and the biophysical environment.
2. Promotes green lifestyle with or without highlighting a product/ service.
3. Presents corporate image of environmental responsibility.

The green advertisement builds pull effect as it tries to pull the product by directly appealing to the customer to buy it, which increases its profitability and level of performance in the market. It is always felt that advertising increases the cost of product or service but advertising is considered economical as compared to other promotional techniques because it reaches masses and if we calculate the cost per customer it is very low or nominal.

9.9.2   DIRECT MARKETING

Direct marketing refers to sales communications delivered directly to individual customers through e-mail, direct mail, and telemarketing. The goal is to use information about individuals in order to present them with messages relevant to their needs and interests. Direct marketing enables the organization to communicate with the customers in a more personalized way than advertising, such as greeting them with a letter or telephoning them directly. Telemarketing, direct mail, catalogs, and coupon mailers are all examples of direct-marketing techniques. Successful direct marketing depends on whether you can acquire and maintain a database of your target market.

9.9.3   SALES PROMOTION

Sales promotion refers to the short-term use of incentives or other promotional activities that stimulate the customer to buy the product. Sales promotion techniques are very useful because they bring short and immediate effect on sale, stock clearance is possible with sales promotion. Sales promotion techniques (rebates, discounts, refunds, contest, lucky draws) induce customers as well as distribution channels. And sales promotion techniques help to win over the competitor.

9.9.4   PUBLIC RELATIONS

Public relations evaluate public attitudes; identify the policies and procedures of an organization with the public interest to earn public understanding and acceptance. The public does not mean only customers, but it includes shareholders, suppliers, intermediaries, customers, and so forth. The firm's success and achievement depend upon the support of these parties, for example, the firm needs the active support of middlemen to survive in the market, and it must have good relations with existing shareholders who provide capital. The consumers' group is the most important part of the public as the success of business depends upon the support and demand of customers only.

9.9.5   PERSONAL SELLING

Personal selling means selling personally. This involves face to face interaction between seller and buyer for the purpose of sale; the personal selling does not mean getting the prospects to desire what seller wants but the concept of personal selling is also based on customer satisfaction. When the seller and buyer come together this may improve the relationship between the customer and seller. Salespersons normally make friendly relations with the customers it's a win-win type of situation for a firm which builds a familiar relation among customer and firm.

9.10   PRODUCT VALUE/CUSTOMER VALUE

Green innovation and promotion are simply about harnessing good ideas and seeing them through their development and implementation across a company. In particular, managing customer value is about understanding different value perspectives, components, and dynamics. Potential economic values play an important role in the various conceptualizations of customer values (Peppers and Rogers, 2005). Such ideas of green promotions and green innovations can increase revenue, reduce costs, improve competitive advantage, improve company sustainability and inclusivity—and in doing so lift staff productivity and morale through seeing their ideas come to firm by which they can create their customer value by meeting the need of market and providing quality and eco-friendly product which leads to product value in the market.

9.11   FIRM'S PERFORMANCE

Performance is a critical concern for companies. The major drivers of firm performance are resources that are unique, invaluable, and difficult to imitate and replace (Peppers and Rogers, 2005). Excellent firm performance is likewise at the core of competitive advantage. A number of scholars provide similar definitions of performance, but their criteria for measuring performance vary. Therefore, the research topic of a study should determine the performance measurement index to be used (Evans and Davis, 2005). The marketing literature indicates the advantages of pursuing green initiatives such as larger financial gains and market share, high levels of employee commitment, increased firm performance, and enhanced capabilities. Environmentally responsible actions likewise increase customer satisfaction and firm value and reduce threats to the company, thus, increasing firm valuations. However, green promotional initiatives also yield negative stock returns. At the same time, cost savings may be obtained from green practices. Firms that reduce pollution and inputs that may cause waste would obtain cost-saving advantages. According to the dictionary of sustainable management (2010), sustainability can be assessed in many ways including using social criteria (e.g., being socially desirable, acceptable, psychologically nurturing); using environmental criteria (e.g., being environmentally robust, generationally sensitive, and being capable of continuous learning), it can also be assessed by employing financial criteria. Therefore, this study used financial measures to compare the performance of small business organizations with a high level of green marketing versus organizations with a low level of green marketing. Moreover, since there is no best way to measure performance, the performance measurement in this study is in line with other studies that employed multi-item measures of performance (e.g., Venkatraman and Ramanujam, 1986; Tan and Litschert 1994; Lysonski and Pecotich, 1992). This is so because using multiple performance measures should yield a more accurate picture of overall performance (Walker and Ruekert, 1987). The performance should also include effectiveness which is commonly measured by items such as market share and relative sales growth (Walker et al., 1987). Relative market share in terms of the top three competitors rather than absolute market share can also be used to measure performance. Performance in this study was measured in terms of overall performance in relation to competitors, market share, and profitability.

As supported by Psomas et al. (2011), the firms can enhance their environmental performance and business efficiency by implementing the green innovation and promotions discussed in the literature which benefited firm by enhancing good corporate image, reducing waste, minimizing cost, increase customer satisfaction, increase productivity, better goodwill, increase market share, and of course profitability. Chen et al. (2006) believe that the investment of the green product innovation and green process innovation contribute a positive impact on the businesses. According to Johansson (2002), the product development project success is measured by the profit, sales, and market share. In the same way, Doran and Ryan (2012) concluded that an eco-innovation has a positive and significant impact on the firms' performance. Various study (Ann et al., 2006; Hwee Nga, 2009) discuss whether companies that implement the environmental strategy and certification of ISO 14001 environmental management system have a positive influence on the firms' performance, but despite the number of studies in this area, none are emphasized on the influence of green innovation and green promotion on the firms' performance in details. Punitha and Rasdi (2013) highlighted the benefits and advantage that the companies will achieve as they are involved in the green businesses, for instance, superior financial and market performance.

Therefore, it can be concluded that the firm's financial and nonfinancial state is the most common dimension used to measure the overall organizational performance which is highly influenced through green innovation and green promotional strategies of the firm.

9.12   CONCLUSION

Following an exhaustive review and selected case analysis, it can be concluded that the green marketing strategy contributes to the firms' profitability, competitive advantage, and encourage a greener pattern of consumption among consumers. The green innovation, particularly research constructs can be divided into green products and process including the innovation in technologies that are involved in waste recycling, green products design, and energy saving (Chen et al., 2006). Green innovation represents the concept of environmental protection into the design of the products. The second construct is a green promotion also referred as the communication that promotes the product and the services. Besides promoting the green advertising campaign, it should also have the characteristics to enhance the corporate image of social responsibility.

Thus, the success of the green innovation and green promotion is a success factor to influence the firms' performance and the results of this study show that green innovation practices have positive and significant effects on environmental performance, indicating that a firm that engages in green innovation will indeed observe better environmental performance. Through implementing green innovation practices, firms can fulfill governmental and industry requirements, decrease waste and pollution, protect the environment, and simultaneously increase their competitiveness. The model indicates that green innovation can have positive effects on firm's overall performance economically, socially, and environmentally.

Greenmarket has been an emerging issue worldwide driving companies to continuously enhance their green capabilities and implement innovative green practices to protect the environment and improve business performance. This study provides a basic framework of information and a conceptual model to researchers for future empirical studies. This chapter also provides a base to the managers and practitioners to revamp their strategies in the line with sustainable and inclusive green markets. Despite innovation and promotion from scholars and policymakers, several fundamental issues in green marketing such as the relationship between green marketing mix and firm performance, remain under-investigated.

Well, in this scenario, many corporate have taken green marketing further and as a part of their company strategy just to create a brand image, gain the attention of the consumers. More and more companies need to emerge and also facilitate to the environment. This chapter can also be viewed as a source of new opportunities to grow in today's highly competitive global environment. Enhancing a company's green innovation and green promotion capacity can provide a new strategic weapon for managers to maintain sustainability and inclusivity in emerging new level of the market.

KEYWORDS

- **sustainability**
- **inclusivity**
- **green innovation**
- **green promotion**
- **green marketing**

REFERENCES

Books

Belz, F. M.; Peattie, K. *Sustainability Marketing: A Global Perspective;* Wiley: Chichester, 2009.

Ottman, J. *The New Rules of Green Marketing: Strategies, Tools, and Inspiration for Sustainable Branding;* Berrett Koehler Publishers: San Francisco, CA, 2011.

Chapters in Edited Book

Peattie, K. Towards Sustainability: The Third Age of Green Marketing. *Mark. Rev.* **2001,** *2*(2), 129–146.

Online Journal Articles/Advance Online Publication Articles

Abd Rahim, M. H.; Ahmad Zukni, R. Z. J.; Ahmad, F.; Lyndon, N. Green Advertising and Environmentally Responsible Consumer Behavior: The Level of Awareness and Perception of Malaysian Youth. *Asian Soc. Sci.* **2012,** *8*(5). DOI: 10.5539/ass.v8n5p46.

Ann, G. E.; Zailani, S.; Abd Wahid, N. A study on the Impact of Environmental Management System (EMS) Certification Towards Firms Performance in Malaysia. *Manage. Environ. Qual. Int. J.* **2006,** *17*(1), 73–93. DOI: 10.1108/14777830610639459.

Arseculeratne, D.; Yazdanifard, R. How Green Marketing can Create a Sustainable Competitive Advantage for a Business. *Int. Bus. Res.* **2014,** *7*(1), 130.

Bandura, A. Impeding Ecological Sustainability through Selective Moral Disengagement. *Int. J. Innovation Sustainable Dev.* **2007,** *2*(1), 8–35.

Banerjee, S.; et al. Shades of Green: A Multidimensional Analysis of Environmental Advertising. *J. Advertising* **1995,** *24*(2), 21.

Bonnell, V.; Veglio, F. Inclusive Business for Sustainable Livelihoods. *Field Actions Sci. Rep.* **2011,** *5*, 1–5.

Carrigan, M.; De Pelsmacker, P. Will Ethical Consumers Sustain their Values in the Global Credit Crunch? *Int. Mark. Rev.* **2009,** *26*(6), 674–687.

Chamorro, A.; Bañegil, T. M. Green Marketing Philosophy: A Study of Spanish Firms with Ecolabels. *Corporate Soc. Responsib. Environ. Manage.* **2006,** *13,* 11–24. DOI: 10.1002/csr.

Charles, G. Ethics Come into Fashion. *Marketing,* 2010.

Chen, Y.-S.; Lai, S.-B.; Wen, C.-T. The Influence of Green Innovation Performance on Corporate Advantage in Taiwan. *J. Bus. Ethics* **2006,** *67,* 331–339.

Christmann, P. Multinational Companies and the Natural Environment: Determinants of Global Environmental Policy Standardization *Acad. Manage. J.* **2004,** *47,* 747–760.

Cohen, M. J. The Emergent Environmental Policy Discourse on Sustainable Consumption In *Exploring Sustainable Consumption: Environmental Policy and the Social Sciences;* Cohen, M. J., Murphy, J., Eds.; Amsterdam: Pergamon, 2001; pp 839–844.

Connolly, J.; Prother, A. Sustainable Consumption: Consumption, Consumers and the Commodity Discourse. *Consumption, Mark. Cult.* **2003,** *6*(4), 275–291.

Cooper, T. Slower Consumption: Reflections on Product Life Spans and the Throwaway Society. *J. Ind. Ecol.* **2005,** *9*(1/2), 51–67.

De Graaf, J.; Wann, D.; Naylor, T. H. *Affluenza: The All Consuming Epidemic,* 2nd ed.; Berrett-Koehler: San Francisco, CA, 2005.

Desmond, J.; Crane, A. Morality and the Consequences of Marketing Action. *J. Bus. Res.* **2004,** *57*(11), 1222–1230.

Doran, J.; Ryan, G. Regulation and Firm Perception, Eco-Innovation and Firm Performance. *Eur. J. Innovation Manage.* **2012,** *15*(4), 421–441. DOI: 10.1108/14601061211272367.

Evans, W. R.; Davis, W. D. High-Performance Work Systems and Organizational Performance: The Mediating Role of Internal Social Structure. *J. Manage.* **2005,** *31*(5), 758–775.

Fergusson, H.; Langford, D. A. Strategies for Managing Environmental Issues in Construction Organizations. *Eng. Constr. Archit. Manage.* **2006,** *13,* 171–185.

Fisk, G. Reflections of George Fisk: Honorary Chair of the 2001 Macro Marketing Conference. *J. Macromarketing* **2001,** *21*(2), 121–122.

Fitzsimmons, C. Make it Green and Keep them Keen, The Guardian, 21 January, 2008. http://www.guardian.co.uk/media/2008/jan/21/marketingandpr (accessed May 7, 2016).

Fry, M.; Polonsky, M. Introduction: Special Issue on Examining Marketing's Unintended Consequences. *J. Bus. Res.* **2004a,** *57*(11), 1209–1210. DOI: 10.1016/S0148-2963(03)00056-0.

Fry, M.; Polonsky, M. Examining the Unintended Consequences of Marketing. *J. Bus. Res.* **2004b,** *57,* 1303–1306. DOI: 10.1016/S0148–2963(03)00073–0.

Fuller, D. A. *Sustainable Marketing: Managerial–Ecological Issues*; Sage: Thousand Oaks, CA, 1999.

Geffen, C. A.; Rothenberg, S. Suppliers and environmental innovation: The Automotive Paint Process. *Int. J. Oper. Prod. Manage.* **2000,** *20,* 166–186.

Gordon, R.; Carrigan, M.; Hastings, G. A Framework for Sustainable Marketing. *Mark. Theory* **2011,** *11*(2), 143–163. DOI: 10.1177/1470593111403218.

Gowri, A. When Responsibility can't do it. *J. Bus. Ethics* **2004,** *54,* 33–50.

Huang, Y.-C.; Ding, H.-B.; Kao, M.-R. Salient Stakeholder Voices: Family Business and Green Innovation Adoption. *J. Manage. Organization* **2009,** *15,* 309–326.

Hwee Nga, J. K. The Influence of ISO 14000 on Firm Performance. *Soc. Responsib. J.* **2009,** *5*(3), 408–422. DOI: 10.1108/17471110910977311.

Jackson, T. *Motivating Sustainable Consumption: A Review of Evidence on Consumer Behaviour and Behaviour Change.* A Report to the Sustainable Development Research Network; Policy Studies Institute: London, 2005. http://admin.sd-research.org.uk/wp-content/uploads/2007/04/motivatingscfinal_000.pdf (accessed May 24, 2016).

Johansson, G. Success Factors for Integration of Ecodesign in Product Development: A Review of State of the Art. *Environ. Manage. Health* **2002,** *13*(1), 98–107. DOI: 10.1108/09566160210417868.

Juwaheer, T. D.; Pudaruth, S.; Noyaux, M. M. E. Analysing the Impact of Green Marketing Strategies on Consumer Purchasing Patterns in Mauritius. *World J. Entrepreneurship Manage. Sustainable Dev.* **2012,** *8*(1), 36–59. DOI: 10.1108/20425961211221615.

Kilbourne, W. E.; Carlson, L. The Dominant Social Paradigm, Consumption, and Environmental Attitudes: Can Macromarketing Education Help? *J. Macromarketing* **2008,** *28*(2), 106–121.

Kotler, P. Reinventing Marketing to Manage the Environmental Imperative. *J. Mark.* **2011,** *75*(4), 132–135.

Lysonski, S.; Pecotich, A. Strategic Marketing Planning, Environmental Uncertainty and Performance. *Int. J. Mark. Res.* **1992,** *9,* 247–488.

Manget, J.; Roche, C.; Munnich, F. *Capturing the Green Advantage for Consumer Companies*; Boston Consulting Group: Boston, MA, 2009.

Millar, C.; Hind, P.; Magala, S. Sustainability and the Need for Change: Organisational Change and Transformational Vision. *J. Organ. Change Manage.* **2012,** *25*(4), 489–500. DOI: 10.1108/09534811211239272.

Mourad, M.; Ahmed, Y. S. E. Perception of Green Brand in an Emerging Innovative Market. *Eur. J. Innovation Manage.* **2012,** *15*(4), 2012. DOI: 10.1108/14601061211272402.

Nair, C. The Last Word: Luxury Goods – Ethics Out of Fashion. Ethical Corporation, 2008. http://www.ethicalcorp.com/content.asp?ContentID¼6192 (accessed May 20, 2016).

OECD. *Towards Sustainable Household Consumption? Trends and Policies in OECD Countries;* OECD: Paris, 2002.

Peattie, K.; Peattie, S. Social Marketing: A Pathway to Consumption Reduction? *J. Bus. Res.* **2009,** *62*(2), 260–268.

Polonsky, M. J. An Introduction to Green Marketing. *Electron. Green J.* **1994,** *1*(2), 1091. DOI: 10.1016/j.neuron.2011.12.010 (Journal of field actions. 5, 1–5).

Polonsky, M. J.; Carlson, L.; Fry, M. L. The Harm Chain: A Public Policy Development and Stakeholder Perspective. *Mark. Theory* **2003,** *3*(3), 345–364.

Porter, M. E. *Competitive Advantage: Creating and Sustaining Superior Performance;* The Free Press: New York; 1985.

Psomas, E. L.; Fotopoulos, C. V.; Kafetzopoulos, D. P. Motives, Difficulties and Benefits in Implementing the ISO 14001 Environmental Management System. *Manage. Environ. Quality Int. J.* **2011,** *22*(4), 502–521. DOI: 10.1108/14777831111136090.

Pujari, D. Eco-innovation and New Product Development: Understanding the Influences on Market Performance. *Technovation* **2006,** *26,* 76–85.

Punitha, S.; MohdRasdi, R. Corporate Social Responsibility: Adoption of Green Marketing by Hotel Industry. *Asian Soc. Sci.* **2013,** *9*(17), 79–93. DOI: 10.5539/ass.v9n17p79.

Reinhardt, F.L. Bringing the Environment Down to Earth. *Harv. Bus. Rev.* **1999,** *77,* 149–158.

Saha, M.; Darnton, G. Green Companies or Green Con-Panies: are Companies Really Green, or are they Pretending to be? *Bus. Soc. Rev.* **2005,** *110*(2), 117–157.

Sarkar, A. N. Green Branding and Eco-innovations for Evolving a Sustainable Green Marketing Strategy. *Asia-Pac. J. Manage. Res. Innovation* **2012,** *8*(1), 39–58. DOI: 10.1177/2319510X1200800106.

Taghian, C. D. M. Green Advertising Effects on Attitude and Choice of Advertising Themes. *Asia Pac. J. Mark. Logistics* **2005,** *17*(3), 51–66.

Tan, J. J.; Litschert, R. Environment – Strategy Relationship and Its Performance Implications: An Empirical Study of Chinese Electronics. *Strategic Manage. J.* **1994,** *15,* 1–20.

Van Dam, Y. K.; Apeldoorn, P. A. C. Sustainable Marketing. *J. Macromarketing* **1996,** *16*(2), 45–56.

Venkatraman, N.; Ramanujam, V. Measurement of Business Performance in Strategy Research: A Comparison of Approaches. *Acad. Manage. Rev.* **1986,** *11*(4), 801–814.

Walker, O. C., Jr.; Ruekert, R. W. Marketing Role in the Implementation of Business Strategies: A Critical Review and Conceptual Framework. *J. Mark.* **1987,** *51,* 15–33.

Zhu, Q.; Sarkis, J.; Cordeiro, J. J.; Lai, K.-H. Firm-Level Correlates of Emergent Green Supply Chain Management Practices in the Chinese Context. *Omega* **2008,** *36,* 577–591.

CHAPTER 10

ANALYZING LONG-TERM BENEFITS IN THE FACE OF HIGHER UPFRONT COSTS FOR GREEN AFFORDABLE HOUSING: A STUDY OF GHAZIABAD, UP (INDIA)

SIDDHARTH JAIN[1], PRATEEK GUPTA[2,*], and DEEPA[2,3]

[1]*Department of Civil Engineering, KIET Group of Institutions, Ghaziabad—Meerut Highway, NH-58, Ghaziabad, UP 201206, India, Ph.: 08126270776, E-Mail: siddharth.jain@kiet.edu*

[2]*Department of Management Studies KIET Group of Institutions, Ghaziabad—Meerut Highway, NH-58, Ghaziabad, UP 201206, India, Ph.: 09634067469, 9997661845*

[3]*deepa@kiet.edu*

[*]*Corresponding author. E-mail: prateek.gupta@kiet.edu*

10.1 INTRODUCTION

A green building is one which uses less water, optimizes energy efficiency, conserves natural resources, generates less waste, and provides healthier spaces for occupants as compared to a conventional building. Considering the tremendous benefits that it offers, green building concept is gaining major importance in India. A common man wants an affordable house to live in which may fit in his pocket whereas the demand of the society is to go green as the pollution is increasing day by day. Affordable housing refers to housing units that are affordable by that section of society whose income is below the median household income. Though different countries have different definitions for affordable housing, it is largely the same, that is, affordable housing should address the housing needs of the lower or middle-income households. Affordable housing becomes a key issue especially in developing nations where a majority of the population is not able to buy houses at the market price.

To go green with green building is a bit costlier for common man somehow in the short run but studies say that a green building uses less water, optimizes energy efficiency, conserves natural resources, generates less waste, and provides healthier spaces for occupants, as compared to a conventional building which becomes quite cheaper in long run. Hence, there are four basic elements of the green building that is smart design, energy efficiency, eco-materials, water conservation (Cleveland, 2012).

According to urban Economist Mark Smith and architect Deborah Weintraub, green building "includes three important components: resource conservation during design and construction, resource conservation during operations, and protection of occupants' health, well-being, and productivity." Others emphasize the protection of more ephemeral things like "community and cultural sensitivity," saying that green projects "blend in with the natural environment and protect open space, increase a sense of community, and address cultural issues."

The Government of India has already taken too much initiative toward Go Green. The green buildings are identified by the environment ministry on the basis of the use of water\energy conservation methods, recycled materials, solar power, natural lighting, and energy self-sufficiency under "Go Green" initiatives. Apart from this, The Indian Green Building Council (IGBC), part of the Confederation of Indian Industry (CII) offers a wide array of services which include developing new green building rating programs, certification services, and green building training programs. In spite of having so many facilities and services, the real estate world, as well as the residents, are still not so much concerned about the urgent need for green housing. Keeping this in view, the present study is focused on the following:

1. This chapter aims to highlight the said three issues related to green housing and to identify why affordability of such house becomes unreachable due to higher upfront cost. An analysis is proposed for its long-term benefits against the inclusive higher cost of green housing.

2. A systematic review of green housing is discussed with a review of related literature. Primary data is analyzed which was collected through personal interview with 300 respondents from Ghaziabad, Raj Nagar Extension, Vasundhara, Sahibabad of Uttar Pradesh (India).

3. This study reveals that India is still lagging behind in the effective implementation of green housing concepts due to the high cost

involved in construction. This may be justified that "if the intention is to construct a new home to live in, it is advisable to go for a green home rather than the ordinary conventional home. Because the percentage increase of 12.94% in the total cost is not a negligible amount when the intention is just to renovate or retrofit an old home."

4. The research is based on original thought and views collected from respondents and reliable resources.

5. This research will benefit the researchers, academicians, innovators, real estate business, residents, and so forth.

10.2 ENVIRONMENTAL ISSUES

The environment is not only a huge area in which we keep on living but also a huge ecosystem in which millions of creatures live. The environment provides raw materials for the economy. This raw material gained through the production process turns out to be a consumer product. Then, these raw materials and the energy used in production return to the environment as pollution. Therefore, sustainability of environment is reasonably important with regard to economic development (Kolukisa, 2012).

In relation to the environment, the first comprehensive arrangements concerning international cooperation have been considered in the early 1970s. In 1972 in Stockholm "United Nations Human Environment Declaration" (Stockholm declaration), the first global evaluation of many countries whose socioeconomic structures and development levels differ from one another was accepted. Sustainable development concept was first defined in 1987 with the Brundtland Report, prepared by World Commission on Environment and Development and since then began to be widely used.

One of the important developments which support the sustainable development vision has come into effect with the 1992 Rio Earth Summit. Rio Earth Summit, the largest meeting including 172 countries as well as India, was an important meeting in which the said countries have approved of the economic and global issues to be achieved in accordance with the environmental values and sustainable development principles.

At this summit conclusion regarding sustainable development were declared in a declaration called "Agenda 21." In spite of not being legally bounding, with this declaration, the countries' political responsibilities about the environment are defined. In Agenda 21, some basic principles of sustainable development which focus on the environmental dimensions are:

- Humankind is the center of sustainable development. She/he has the right to have a healthy and efficient life in accordance with nature.
- On condition that they are not disturbing other countries' environment, all countries have the right to use their natural resources according to their own politics.
- In view of "right to development for current and forthcoming generations," it is required to acknowledge the environment and mankind.
- Environmental protection should be seen as a complementary part of the development process.
- In terms of ecosystem protection and development, countries shall cooperate in global association spirit.
- In environmental protection, countries have common but different levels of responsibility.
- National authorities, in terms of "the one who pollutes pays it" principle, should make the environmental costs international and the use of economic devices improved.

As to activities, which have effects beyond borders, the activist country should inform the related countries in time and provide them with necessary information.

Briefly, the living environment provided by sustainable development means using of water resources so carefully that current and forthcoming generations can benefit adequately, keeping the weather clean in order to breath fresh air, making use of the benefaction of nature, and living in good health today and in the future (Taylor, 2012).

Providing sustainable development is not only limited to management of natural resources, supplying equipment, or controlling them but also to their proper utilization and allocation. Individuals should fulfill their duties as responsible citizens in order to achieve sustainable development with regard to environmental protection. Therefore, it is compulsory that individuals be taught in the subject of sustainable development education beginning from primary school, as well as making them aware of the environment.

UNESCO (2003) has declared the vision of sustainable development education as "societies should work in order to improve their level of civilization and economical power while being aware of the value of natural resources." Sustainable development education improves the ethical attitudes, which are necessary for lifetime learning, and encourages us to use the natural resources necessary for our planet according to our needs. The aims of sustainable development education in terms of the environment are:

- Improving the perception in accordance with economy and environment, and resulting in social peace.
- Encouraging the studies for the protection and improvement of the values of people living in the ecologic and social environment.
- Altering the daily behavior of individuals for supporting sustainable development.
- Preparing individuals as active and participant citizens working for their society and environment.

There are additional urgent needs associated with the investments in infrastructure. The first is the global challenge to the Earth's environment including the long-term availability of nonrenewable resources, concentrations of pollution and waste from human activity, and global climate change. For example, the recent report from the Intergovernmental Panel on climate change could not have been clearer regarding the impact of human activity on global climate change and its potential impact on the planet. People may disagree on the causes of climate change and the exact nature and severity of the impact, but there is a near-universal acknowledgment that the planet is warming, the climate is changing, and resources are being depleted. The impact of human activity upon the environment is not limited to climate change alone. For example:

- Poor land management and the overuse of fertilizer are causing land degradation, soil erosion, and desertification on a massive scale in agricultural areas from the Amazon to the Yangtze.
- One-third of the world's population does not have access to adequate sanitation.
- Armed conflict affects more than 20 of the world's 34 poorest countries, mainly in Africa.
- Almost half the world's population will be living in areas of high water stress by 2030.
- Irrigation accounts for 70% of the world's water demand. More than half the water distributed by irrigation systems is lost due to leaks and wasteful practices.

Clearly, the action is required if we intend to sustain a planet that can support a human society in perpetuity that provides the opportunity for all people to realize the quality of life enjoyed in the developed world. The choice is ours that we go for a global society in regard to infrastructure investments that will directly affect the level of the quality of human life and the long-term health of the planet.

10.3 ISSUES ON ENERGY, ENVIRONMENT, SUSTAINABLE CONSTRUCTION, AND BUILDINGS

Indian construction industry is one of the largest in terms of economic expenditure, the volume of raw materials/natural resources consumed, the volume of materials and products manufactured, employment generated, environmental impacts, and so forth. A large variety of materials are manufactured and consumed in the construction industry (UNESCO, 2003). Production levels and energy expenditure of some of the building materials consumed in bulk quantities are given in Table 10.1. Total energy expenditure on bricks, cement, aluminum, and structural steel consumed in bulk quantities is 1684×10^6 GJ per annum. It has been estimated that 22% of greenhouse gas (GHG) emissions are contributed by the construction sector in India. There is an ever-increasing demand for building materials. For example, demand for houses has doubled in about two decades from 1980. Projected demand for the building materials like bricks, steel, and cement consumed in bulk quantities is given in Table 10.2. Compounded growth rates of 2.5, 5, and 5% have been assumed for bricks, steel, and cement, respectively, to compute the projected demand. In case of brick making activity, at present topsoil equivalent of 300 mm from 100,000 ha (1000 sq km) of fertile land, 22×106 t of coal and 10×10^6 t of biomass are consumed annually. We have an arable land area of 1.62×10^6 sq km comprising alluvial, black, red, laterite, and desert soils. Alluvial, laterite, and red soils are suitable for brick making. The area under the soils suitable for brick making may not exceed 50% of the arable land. Brick-making activity to meet the present and future demand can result in consuming the 300 mm depth fertile topsoil of arable land in about 90 years (assuming 2.5% compounded growth rate).

TABLE 10.1 Volume and Energy Consumption of Building Materials in India.

Material	Volume of materials manufactured per annum	Thermal energy (MJ per kg)	Total energy (GJ)
Bricks	160×10^9 nos.	1.5	650×10^6
Cement	98×10^6 t	4.25	410×10^6
Structural steel	14×10^6 t	45.0	476×10^6

Similarly, the pressure on raw materials like limestone to manufacture cement and energy requirements to produce these materials has to be addressed. Production of building materials has slowly and steadily moved from highly decentralized and labor-intensive methods and processes to centralized, machine-dependent industry mode. Centralized mode of production necessitates hauling of raw materials and distribution of finished materials over great distances. These activities again require the expenditure of fossil fuels for transportation. Transportation of raw and finished building materials is another key issue that can contribute to the cost of materials, increased energy requirements, and environmental issues. Energy (fossil fuel energy) spent in the transportation of some of these building materials using trucks is given in Table 10.3. Sustainability of the present mode of production, consumption, and distribution of building materials and currently adopted construction practices are questionable.

TABLE 10.2   Projected Demand for Building Material (the Years 2000–2020).

Material	2000	2020
Bricks (Nos.)	160×10^9	256×10^9
Structural steel (t)	12×10^6	40×10^6
Cement (t)	98×10^6	265×10^6

TABLE 10.3   Energy in Transportation of Building Materials.

Building material	Unit	Energy in transportation for 100 km (MJ)
Bricks	m^3	210
Sand	m^3	180
Cement	t	110
Steel	t	110

10.4   NEED FOR SUSTAINABLE ALTERNATIVES

Steel, cement, glass, aluminum, plastics, bricks, etc. are energy-intensive materials, commonly used for building construction. Generally, these materials are transported over great distances. Extensive use of these materials can drain the energy resources and adversely affect the environment. On the other hand, it is difficult to meet the ever-growing demand for buildings by adopting only energy efficient traditional materials (like mud, thatch, timber, etc.) and construction methods (TEDDY, 1990). Hence, there is a need for optimum utilization of available energy resources and raw materials to produce simple, energy efficient, environment-friendly, and sustainable building alternatives and techniques to satisfy the increasing demand for buildings. Some of the guiding principles in developing the sustainable alternative building technologies can be summarized as energy conservation, minimize the use of high energy materials, concern for environment, environment-friendly technologies, minimize transportation and maximize the use of local materials and resources, decentralized production and maximum use of local skills, utilization of industrial and mine wastes for the production of building materials, recycling of building wastes, and use of renewable energy sources. Building technologies manufactured by meeting these principles could become sustainable and facilitate sharing the resources, especially energy resources, more efficiently, causing minimum damage to the environment.

10.5   APPLICATION OF SCIENCE AND TECHNOLOGY FOR RURAL AREAS (ASTRA)'S INITIATIVES AND DEVELOPMENTS IN SUSTAINABLE BUILDING TECHNOLOGIES

Centre for ASTRA (Application of Science and Technology for Rural Areas) was formed in 1974 at Indian Institute of Science (IISc), Bangalore, to cater to developing technologies for sustainable development. Recently, this center has been renamed as "Centre for Sustainable Technologies." Developing environment-friendly, energy efficient, simple and sustainable building technologies utilizing maximum local resources and skills is one of the thrust areas of ASTRA's activities. R&D and dissemination of building technologies became an interdisciplinary work, where the Department of Civil Engineering actively pursued this work since over 2.5 decades of time.

Large number of building technologies were developed and successfully disseminated. ASTRA's approach to develop sustainable building technologies was not confined to laboratory work. Field trials and laboratory work went hand in hand to develop viable technologies. ASTRA's made its sincere efforts in alternative building technologies since 1976 on issues of R&D, dissemination, training and establishing mechanisms for spreading the technologies, and the recent developments. The table indicates that considerable amount of time has been spent initially at the Ungra Extension Centre (UEC) in field experimentation of building technologies initially. There was a need for some buildings at UEC for carrying out other activities of ASTRA and this need had thrown up an open ground for the buildings research group to experiment. This opportunity gave scope for experimenting and monitoring long-term performance over a period of several years. Important lessons on building technologies were learned during the initial period. Some of these building technologies are: stabilized mud blocks, steam cured blocks, fine concrete blocks, rammed earth blocks, mud concrete blocks, lime-pozzolana cements, soil-lime plaster, composite mortars for masonry, composite beam and panel roofs, Reinforced brickwork/tile-work roof, ferrocement and ferro-concrete roofing systems, unreinforced masonry vaults and domes, ribbed slab construction, filler slab roofs, rammed earth foundations, reinforced block-work lintels and precast *chajjas*, solar passive cooling techniques and containment reinforcement for earthquake-resistant masonry. A large number of buildings (>12,000) have been built using these alternative building technologies.

10.6   GREEN BUILDING MOVEMENT IN INDIA

The green building movement in India was triggered off when CII-Sohrabji Godrej Green Business Centre building in Hyderabad was awarded the first and the prestigious platinum-rated green building rating in India. Since then, green building movement in India has gained tremendous impetus over the years. With a modest beginning of 20,000 sq. ft green built-up area in the country in the year 2003, today (as on 31 December, 2016) more than 3921 green buildings projects coming up with a footprint of over 4.48 Billion sq. ft are registered with the IGBC, out of which 942 projects are certi-fied and fully functional in India. This growth has been possible with the participation of all stakeholders in the green building movement. Today, all types of buildings are going the green way—government, IT parks, offices, residential, banks, airports, convention center, institutions, hospitals, hotels, factories, SEZs, townships, schools, metros, and so forth.

10.7   THE INDIAN GREEN BUILDING COUNCIL (IGBC)

IGBC, part of the CII was formed in the year 2001. The vision of the council is "to enable a sustainable built environment for all and facilitate India to be one of the global leaders in the sustainable built environment by 2025." The council offers a wide array of services which include developing new green building rating programs, certification services, and green building training programs. The council also organizes green building congress, its annual flagship event on green buildings. The council is committee-based, member-driven, and consensus-focused. All the stakeholders of construction industry comprising architects, developers, product manufacturers, corporate, government, academia, and nodal agencies participate in the council activities through local chapters. The council also closely works with several state governments, central government, world green building council, bilateral multilateral agencies in promoting green building concepts in the country.

IGBC has about 1700 members, 1200 accredited professionals, and 15 vibrant chapters in all major metros. Now, all types of buildings are going the IGBC "greenway" including airports, banks, colleges, convention centers, factories, government buildings, hospitals, hotels, institutions, IT parks, malls, metros, offices, residential buildings, schools, SEZs, townships, etc. varying from 1200 sq. ft to 120 million sq.ft. Any IGBC-rated green building mirrors India's rich architectural heritage blending with modern technological innovations. IGBC-rated green buildings ensure that energy is saved to the tune of 40–50% and water is saved by 20–30%, and intangible benefits like enhanced ventilation, daylighting, and good design with eco-friendly materials improve the productivity of the occupants (https://igbc.in/igbc/redirectHtml.htm?redVal=showAboutusnosign&id=about-content).

10.8   GREEN BUILDING AS PER IGBC

Features and basic requirements of a green building are given below:

Basic requirements of green buildings are to reduce or eliminate negative environmental impacts of development, conserve energy, conserve water, reduce usage of natural resources and construction materials, and improve workplace environmental quality.

1. Challenge for architects, a green building: It is, indeed, a challenge for both architects and developers. However, considering the rapid global warming taking place and depletion of natural resources, we cannot sustain ourselves too long. Therefore, it only requires concern for the environment, commitment to social responsibility, and application of mind and passion to do something to conserve resources to help future generations.

2. The cost involved in green compared to a nongreen structure: It really does not cost extra to develop a green building. It is a simple application of conventional wisdom, the orientation of the building, concern for our neighborhood, and application of mind to minimize the use of materials, best described by reduce, reuse, recycle. Even

the platinum green buildings of IGBC that used to cost 15% more about 8 years ago, now cost just around 9–12% more than nongreen buildings.

3. Top green buildings in India: Some of the best green buildings in India are the buildings designed to house the new Tamil Nadu Legislative Assembly at Chennai (now converted into a hospital), IGP Office at Gulbarga, Suzlon "One Earth" at Pune, ITC Hotel–The Royal Gardenia at Bangalore, Godrej Plant II-IT Park at Mumbai, Infosys Pocharam at Hyderabad and Bearys Global Research Triangle at Bangalore. Energy efficiency, water use reduction, construction waste management, and use of local materials are some of the salient features of these projects. Bangalore is in the forefront of the green building movement and the construction community represented by CREDAI has taken it very seriously and soon it could be in their DNA. Bangalore has some of the highest rated buildings of India. NCR, Chennai, Hyderabad, Pune, and Mumbai are doing well in popularizing the green building movement and registering buildings for green certification. With the concept of green buildings gaining prominence in India, developers are now focusing on developing structures that are eco-friendly and use energy efficient techniques. Some other top green buildings in India are as below:

a) ITC Green Centre, Gurgaon
b) Patni Knowledge Centre, Sahibabad
c) Olympia Tech Park, Chennai
d) Infinity Benchmark, Kolkata
e) CRISIL House, Mumbai
f) Indira Paryavaran Bhawan, New Delhi
g) ITC Maurya Hotel, New Delhi
h) Infosys Hyderabad
i) CISCO Building, Bangalore
j) CII building, Hyderabad

10.9 COMPARISON OF MATERIALS USED IN CONVENTIONAL BUILDING AND GREEN BUILDING

The table below shows the materials used in the conventional and green home, respectively for different items:

TABLE 10.4 Comparison of Materials used in Conventional Building and Green Building

S. No.	Materials	In conventional building	In green building
1	Windows and openings	Aluminum paneled plain glasses	Insulated glass (IG units)
2	Lighting fixtures	Tube lights and CFLs	Low Watt LED tube lights and bulbs
3	Plumbing fixtures	Conventional fixtures	Special green fixtures
4	Flooring	Vitrified and glazed tiles and China mosaic	PVC flooring, glazed tiles, and China mosaic
5	Doors	Pinewood	Engineering Wood
6	Paints	Plastic VOC	Plastic non-VOC
7	Bricks	Clay bricks	Fly ash bricks
8	Cement	OPC	PPC
9	Installation of rainwater harvesting system	Not provided	Provided

LED: light emitting diode; PVC: polyvinyl chloride; VOC: volatile organic compounds; PPC: Portland pozzolana cement; ODC: ordinary Portland cement

10.9.1   THE PRIME MATERIALS USED IN GREEN BUILDINGS (VENKATARAMA REDDY AND LOKRAS, 1998)

1. LED fixtures:

- In green buildings, the only type of the lighting fixtures used is LED (light emitting diode) fixtures.
- These types of lighting fixtures are somewhat costly, approximately 4 times higher price than the normal ones.
- These are solid lights which are extremely energy-efficient.
- A significant feature of LEDs is that the light is directional, as opposed to conventional bulbs which spread the light more spherically.

2. Plumbing:

- In the green plumbing, the plumbing fixtures discharge ranges from 5 to 15 l p m.
- Green plumbing includes the fixtures as below:
  - Faucets discharge ranges between 5 and 8 l p m.
  - Showers discharge ranges between 10 and 15 l p m.
  - Water closet discharge ranges between 7 and 12 l p m.

3. Polyvinyl chloride (PVC) flooring:

- PVC flooring is a type of synthetic flooring.
- Floors such as wood and marble are made from natural materials, but PVC flooring is made from a synthetic plastic called polyvinyl chloride.
- It is easy to install.
- PVC flooring is affordable, water resistant, and very durable. Due to its synthetic nature, PVC flooring is also more customizable than natural flooring.
- To keep PVC flooring in optimal condition, regularly vacuum and polish the floor. Never use any rough-surfaced cleaning tools on the floor as they often result in scratches.

4. Green paints:

- Paints with reduced levels of volatile organic compounds (VOCs) are more eco-friendly than conventional paints, some house paints have an even lower environmental impact.
- Homeowners can select premium-grade zero-VOC paints that also use VOC-free colorants, are free of vinyl and other plasticizers and include no toxic biocides.

5. Bricks:

- Fly ash bricks are considered as the green material of construction.
- So, in this case, it is used in the green construction.

6. Cement:

- If PPC (Portland pozzolana cement) is used in construction, it will be green material because PPC contains the fly ash as the main ingredients.

10.10   CHALLENGES TO BUILDING GREEN AFFORDABLE HOUSING

Notwithstanding the motivations for a community-based nonprofit in green housing development, barriers remain that hinder the ability of community-based organizations (CBOs) to successfully build green housing development projects. Such barriers include the following:

1. Perceived risk: CBOs have little room for risk or project failure. There is a reluctance to use new materials and methods for publicly funded projects. Anything new is considered risky, innovative, or untested green features can reduce confidence. Many developers, funding sources, and contractors fear that following a green agenda will delay project schedules and raise costs. This has led to the widespread perception in the nonprofit affordable housing community that it is difficult to retain the full developer's fee if a project is going to concern itself with environmental issues. Developer's fees are crucial to CBOs' abilities to finance future projects.

2. Multiple funding sources: Affordable housing projects often have many funding sources, making it difficult for all parties to agree to and negotiate the inclusion of new and innovative ideas. In addition, funding sources are becoming more and more difficult to find even for conventional projects, and it follows that innovative green development projects face stiff competition.

3. Many players: There are often more players in an affordable housing project than in a conventional market-rate private development (e.g., underwriters, development consultants, builders, maintenance staff, residents, and the surrounding community), all requiring buy-in. Moreover, new affordable housing projects often face local opposition.

4. Regulatory burdens: Affordable housing projects which include public financing having even a harder time building green than privately financed projects. In addition to per unit cost caps, they are often subject to local design requirements that limit the opportunity for green design.

5. Lack of documented success: With the exception of energy efficiency, green building principles have not been widely applied to affordable housing, and actual experience in terms of incremental costs and benefits has not been well documented.

6. Contracting constraints: CBO construction contracts must often be granted to the lowest bidder, making it difficult to select a contractor with specialized training and knowledge in green building.

7. Limited institutional capacity: Low salaries, high turnover, lack of experience drafting green specifications, limited construction supervision expertise, and limited resources to adequately document innovative projects are common problems at CBOs.

8. Learning curve: There is a significant learning curve required of leaders in any field, and that is especially true with green development. Many CBOs that would like to develop projects that are more

environmentally responsible lack readily available information on green contractors and consultants, materials, systems, techniques, and technologies.

9. Short-Term Cost Focus: Developers and funders often think front-loaded planning and design for green projects will cost more and delay project schedules.

10.11 PROBLEMS IDENTIFIED

In the present study, the following problems have been identified:

1. People having low income even dare to dream for a self-owned house.
2. Urban poor people are not able to get the good environment neither for themselves nor for their children.
3. There is a lack of awareness regarding greenhouses.
4. Greenhouses are much costlier than conventional houses.
5. It is a general belief that a green home will cost much more than a conventional home, but some middle way is required to be found by analyzing the real situations and conditions in the market.
6. In these days, when everyone is talking about the green construction, there is a need for a way by which a common person can afford a green home.

10.12 RESEARCH METHODOLOGY

The present research primarily focuses on residents of West UP, India. The judgmental sampling method has been used to select the sample. Primary data has been analyzed through a personal interview conducted with 300 respondents from Ghaziabad, Raj Nagar Extension, Vasundhara, Sahibabad of West Uttar Pradesh, India with the help of the scheduled questionnaire. Various statistical tools are applied to analyze and conclude the study. The secondary data has been collected from various research journals, periodicals, research reports. The research methodology used in the present study is divided into the following subparts:

- Sample size:300 (Persons having own houses or wish to buy own house)
- Sampling method: Judgmental sampling

- Data collection:
  - Primary: Through a structured questionnaire
  - Secondary: Books, journals, magazines, and internet
- Mode of the survey: Personal interview
- Tools: Tabulation, graphical presentation as pie chart, bar graph, line graph, and so forth
- Places of the survey: West Uttar Pradesh, India (Ghaziabad, Raj Nagar Extension, Vasundhara, Sahibabad)

10.13 HYPOTHESIS

This research is based on the following hypotheses:

H_0: Greenhouses are much costlier than conventional houses.
H_1: Greenhouses are not much costlier than conventional houses.

10.14 ANALYSIS OF CONVENTIONAL BUILDING AND GREEN BUILDING THROUGH COST COMPARISON

The table below shows the cost comparison of each item:

TABLE 10.5 Cost Comparison of Conventional Building and Green Building.

Sr. No.	Item Name	Cost in conventional home	Cost in green home	Difference
1	Windows and openings	140,800	217,350	76,550
2	Lighting fixtures	15,800	46,150	30,350
3	Plumbing fixtures	45,885	108,300	62,415
4	Flooring	230,540	295,295	64,755
5	Doors	79,830	165,510	85,680
6	Paints	166,380	168,880	2,500
7	Bricks	60,175	40,105	(20,070)
8	Cement	976,000	995,250	19,250
9	Rainwater harvesting system	0	90,700	90,700
	Total	1,715,410	2,127,540	412,130

Total development costs for the green buildings reviewed in this study ranged from 9 to 18% above the costs for comparable conventional affordable housing.

10.15 ATA ANALYSIS

1. Question: Do you have your own house or are you interested in buying a new house?

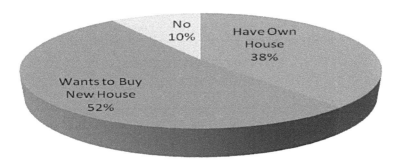

FIGURE 10.1 Response toward owning or buying house.

Interpretation:

52% persons showed interest in buying a new house, 38% already have their own house, whereas 10% were not interested to buy.

2. Question: Are you aware of green housing?

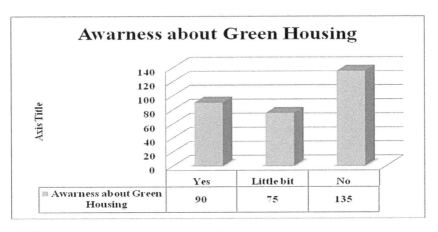

FIGURE 10.2 Response toward awareness about the greenhouse.

Interpretation:

The above graph presents that out of surveyed 300 persons, 90 are aware of the concept of green housing, 75 had little bit information whereas the rest of 135 respondents did not have any idea about this. The researcher made them aware of this concept.

3. Question: Is the environment good in the society due to building construction?

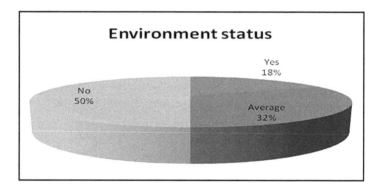

FIGURE 10.3   Response toward environment status.

Interpretation:

We can see in the above graph that the from surveyed 300 persons 50% feel that the environment of the society is not good to live, 32% said that they are just compromising, whereas 18% did not have any complaint.

4. Question: Why you don't want to go for a greenhouse?

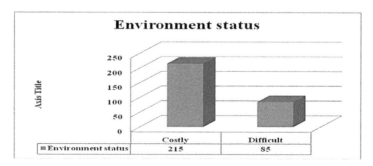

FIGURE 10.4   Response toward greenhouse.

Interpretation:

The above graph presents the feedback of 300 persons about the factors which stop them to purchase a greenhouse. It showed that 215 persons feel that greenhouses are very much costlier then general conventional houses, the rest, 85 respondents feel that it is difficult to construct.

10.16 FINDINGS

The present study has the following findings:

1. If the home is constructed as identified in this study, the total cost increases by 402,130 in addition to the total construction cost of the conventional home which is found to be 3,107,415.
2. The percentage increase in total construction cost is 12.94% (say, 12–15%).
3. The saving in money, which is about 10,000, will be the clear saving after the payback period of LED fixtures is completed.
4. Water saving is about 444 (say, 440) l/day.
5. Payback period, considering savings only in the electricity bills, is found to be 35 years.

Now, this may be justified that "If the intention is to construct a new home to live in, it is advisable to go for a green home rather than the ordinary conventional home. Because the increase of 12.94% in the total cost is not a negligible amount when the intention is just to renovate or retrofit an old home."

10.17 CONCLUSION

1. The following points may be concluded as:
   Creating an infrastructure that sustains human society: As we have outlined here, infrastructure is absolutely vital to realize the global sustainability objectives for society at large, specifically as it relates to meeting the basic human needs of all people on our planet. The basic needs of all people will not be met without investments in infrastructure.
2. Creating an infrastructure that sustains the global environment:
   Infrastructure assets mediate the impact of human activities on the environment. Infrastructure assets themselves impact the

environment. The impact in both cases can be either sustaining or destructive. Infrastructure can remediate the destructive impact of past activities and even increase the regenerative capacity of the planet. Therefore, the nature of our investments in infrastructure will have a direct impact on sustaining our planet

3. Sustaining the world's infrastructure so that it can continue to provide critical services to society and the environment:
    Infrastructure assets are long-lived, but they are subject to entropy and degradation as well. In order to continue providing the services to sustain society and the environment, these assets must themselves be sustained.

4. Sustaining the environment (World Development Report, 2010):
    The second sustainability challenge is to become good stewards of our planet including its environment and its resources. One element of sustaining the environment is increasing the sustainability factor above 1.0 and then keeping it there, maintaining sufficient bio-capacity to continue renewal while accommodating current and future activities by humans. Approaches include increasing bio-capacity, reducing ecological footprint, and more efficient use of nonrenewable resources.

5. Increase bio-capacity:
    This includes not only increasing the bio-capacity of the earth per se but taking advantage of the existing, untapped bio-capacity of the earth. Finally, this would include initiatives to increase the bio-capacity of the infrastructure itself such as buildings that are net producers of power or factories that emit clean water as a by-product.

6. Take advantage of natural energy potential:
    The earth provides many natural energy potentials such as solar, wind, geothermal, ocean temperature gradients, hydroelectric, nuclear, and so on. The technology for taking advantage of these potentials for commercial energy production has long been a reality.

7. Expandability to absorb waste:
    This refers primarily to increasing the natural ability of the earth to absorb GHG by increasing green space through the recovery of waste sites, plankton growth in the oceans, reforestation, recovery of arid land, increased urban green space, green building roofs, and so on.

8. Remediate impact of human activity:
    The bio-capacity of the earth can be increased by recovering bio-capacity lost to prior human activity. This includes solutions such as

hazardous waste recovery and disposal, treatment of polluted water, cleaning polluted air (with urban forests, for example), recovering green spaces lost to activities such as mining and landfills, mining landfills for resource recovery, and so on.

9.   More efficient use of nonrenewable resources:

Given the current global dependence on many nonrenewable resources, consumption of nonrenewable resources continues for the foreseeable future. More efficiently extracting, processing, and consuming these resources will extend the window for replacing them with renewable sources. Infrastructure projects that result in the more efficient use of resources have the potential of improving both the numerator and denominator of the sustainability factor.

10. More efficient consumption:

The obvious initiative in regard to nonrenewable resources is to consume them more efficiently. This involves, for example, greater energy efficiency, alternate modes of transportation (such as mass transit), elimination or replacement of unnecessary activities (by telecommuting, video conferencing, and so on), and use of new materials that require less energy to produce and transport. These are all examples of more efficient consumption. Likewise, designs which result in products that are more readily and efficiently recycled, including "Cradle to Cradle" certified products, are further examples of more efficient consumption.

11. Expand global resource supplies:

Given the ultimate limits on nonrenewable resources, expanding the global resource supplies is also an approach to extending the availability of nonrenewable resources, and ensuring their continuing availability even as they are being replaced for some uses by renewable sources. These approaches could include more complete and efficient extraction methods, more effective and accurate exploration, more eco-friendly exploration and extraction, and so on.

The environmental issue is not a new issue. Even it is not limited to global warming, climate change, pollution, chronic hunger, unsafe bridges, public health, or contaminated water—it is all of these and more. It is not a problem that we will solve and then move on—it will require constant, continuing, and unrelenting attention. The issues and challenges surrounding environment are broad, complex, and interrelated.

To completely satisfy our objectives will mean more investment in infrastructure, not less. It will mean more economic development, not less. If our goal is for all people to be concerned with global sustainability, then the prerequisite is to enable all people to enjoy a quality of life which affords them that luxury. This is a significant challenge for society, a significant challenge for the world's infrastructure, and a challenge for all of us as members of the infrastructure professions. Currently, R&D efforts in developing green building technologies are limited in the Indian context. There is a large scope for R&D efforts in developing alternative building technologies, addressing the following issues.

- A clear understanding of the sector-wise demand and growth of the Indian construction scenario.
- Estimating current building stock and the contribution of unorganized sector in manufacturing and supply of energy-intensive building materials.
- Assessing the availability (region wise) of local resources, raw materials/traditional materials for developing and manufacture of building products.
- Developing alternative building technologies to meet the region-specific needs/demands for buildings.

The null hypothesis is rejected and it is concluded that greenhouses are costlier than conventional houses but only in short run. As per the need of society and livelihood, greenhouses are very much required and will surely be much cheaper than conventional houses in long run.

KEYWORDS

- **green housing**
- **affordability**
- **real estate companies**
- **green housing policies**
- **technology**
- **cost**

REFERENCES

Cleveland, A. B. A Bentley White Paper, 2012.

Kolukisa. Environment Education Projects for Sustainable Development. **2012.**

Taylor, J. Sustainable Development: A Dubious Solution in Search of a Problem. *Policy Analysis*, **2002,** *449.*

UNESCO, Education for Sustainable Development Information Brief, Section for Education for Sustainable Development (ED/PEQ/ESD) Division for the Promotion of Quality Education, 2003.

TEDDY, Tata Energy Research Institute, New Delhi. 1990–91.

https://igbc.in/igbc/redirectHtml.htm?redVal=showAboutusnosign&id=about-content.

Venkatarama Reddy, B. V.; Lokras, S. S. Steamed-Cured Stabilized Soil Block for Masonry Construction. *Energy Build.* **1998.**

World Development Report. Development and Climate Change. The World Bank: Washington DC, 2010.

CHAPTER 11

INNOVATION IN GREEN PRACTICES: A TOOL FOR ENVIRONMENT SUSTAINABILITY AND COMPETITIVE ADVANTAGE

NOMITA SHARMA*

Department of Management Studies, Keshav Mahavidyalaya, University of Delhi, Pitampura, New Delhi, India, Mob.: +91 9818077706

Corresponding author. E-mail: nomitasharma@gmail.com

11.1 INTRODUCTION

The main objective of this chapter is to highlight green innovation practices as a tool of environment sustainability and competitive advantage. It also explains their influence on environment and how they contribute to competitive advantage of an organization and help in maintaining environmental sustainability. Sustainability and environmental issues are rapidly emerging as one of the most important topics for strategic business, management, manufacturing, and product development decisions. This heightened awareness of the natural environment has been reflected in the innovative and environmentally conscious products offered to consumers in recent years. Firms develop sustainable programs with the purpose of greening their own products and process while reducing the impact of their activities. In order to eliminate the problems of environmental pollution, the concepts of environmental management such as green management, green marketing, green production, green innovation, etc. are now being pursued. Apart from innovation, the concerns of enterprises regarding environment have become main factor for achieving competitive advantage in the competitive business environment. Organizations tend to rely on developing innovative products to stay competitive. According to Porter (1989), organizations create competitive advantage through better and new ways to compete in an industry and bring them to the market. Quality innovation gives the company an opportunity to grow faster and better. Through this quality innovation, organizations develop green products that differentiate them from other competitors. The use of green innovation products and processes result in a green economy. This can be thought of one which is characterized by less use of carbon, better in efficiency, and something that makes financial, technical, and social position of society better. In a green economy, there are investments in processes and procedures that limit pollution and carbon emissions. They enhance overall efficiency of resources and control damages to the environment.

In the present scenario, innovation in green products is attracting a lot of attention among organizations that aim for higher growth. Green innovative products allow lower energy consumption and result in minimum waste as a by-product. Organizational performance is influenced by adoption of green innovation products (Rave et al., 2011). The adoption of innovative green products results in reduction in pollution, better productivity reduction of waste, and better substitution with sustainable resources (Kemp and Arundel, 1998). As per Ramus (2002), green innovation can be either product or process that mitigates the environmental effects of an enterprise, or solves the environmental problems or develops environment-friendly and efficient products. Green innovative products are either in the form of improvements to existing products or changes in raw material or other characteristics which enhance the performance of products (Tübitak, n.d.). These products also have positive impact on environment (Durif et al., 2010). On the other hand, green process innovations are those that use different kind technologies or production facilities to reduce negative impact on an environment. There are many green innovation products that are manufactured by enterprises worldwide. There is an example of electric cloth dryer that dries cloths in just 20 min and also uses 70% less energy. Then, there is eco-friendly and cost effective ice battery method of controlling indoor temperature. The system uses a simple concept that cold ice can make air cold, too. There is also a thermostat that is smartphone-connected device which allows regulation of temperature of home. Another green innovative product is a smokeless wood-burning camp stove with a USB-rechargeable airflow system that controls flame size. All these green innovation products have changed the way people live their life. They are all sustainable products that reduce the negative impact on environment. Looking at these innovations, it can be said that innovation is the key factor to sustainability for firms and countries.

Considering all these reasons, it becomes imperative to study innovation in green products and to explore their effects on organization adopting green innovation products worldwide.

Enterprises that adopt strategies of environmental management result in meeting the targets of environmental protections and improvements by utilizing the innovative product or process and green practices (GP) (Greeno and Robinson, 1992). In order to achieve targets of sustainable development, enterprises redesign products and adapt new technology for processes, (Nidumolu et al., 2009). Shrivastava (1995a) suggested that companies can differentiate their products, improve product quality, and lower the cost of production through product and process innovations. Sharma (2016) suggested the use of innovative techniques in different sectors, such as information technology and pharmaceutical, that can differentiate small and medium enterprises from large enterprises. Innovation in GP has propelled the position and status of enterprises. Sometimes, it is the barriers, for example, high cost of technology, nonavailability of funds, or negative impact on already existing resources that motivates enterprises to think out of box and create something that is sustainable and competitive (Sharma, 2014b).

The chapter highlights some of the green product innovations by enterprises. It further suggests that green innovations act as a tool for sustainability and competitive advantage of enterprises. Enterprises that adopt green innovative products tend to use less production material during production process. They tend to recycle waste resources and develop them into useful products. A green innovation practice is a relatively new concept that can be viewed as a product of the 1990s. These can be defined as an economically-driven, system-wide, and integrated approach to the reduction and elimination of all waste streams associated with the design, manufacture, use, and/or disposal of products and materials (Handfield et al., 1997). Green innovation practices can be reflected in production plan or technology program that uses less resources and energy consumption. It also results in less environmental pollution. These practices aim at zero potential safety problems, zero health threats on the operators and product users, and zero environmental pollution, waste recycling, and waste disposal during the production process as much as possible (Gao et al., 2009). Few examples of green innovative products and their influence on society have been discussed in the chapter. These products provide sustainability and competitive advantage to the enterprises. Wong (2013) suggested that enterprises are able to meet eco-targets and integrate environment benefits through green innovation.

11.2   CONCEPT—INNOVATION IN GREEN PRACTICES (GP)

Innovation has been described as the "Industrial Religion of the 21st century." One can create new and differentiating market opportunities by being innovative. Firms need to develop new products, at least on occasion to gain competitive advantage. In this chapter, we define innovation in Indian context, as a process where small and medium-sized enterprises (SMEs) adopt a group of activities that help in creating new wealth from the existing resources (Sharma, 2014a). The concept of green innovation revolves around exploitation, production, or assimilation of all activities including service, production process, management of product and different methods that are new to the organization that adopts and develops them (Kemp and Pearson, 2008).

Innovation has been recognized by the top executives due to its strategic importance. In the recent times, focus has shifted from innovation to green innovation where enterprises are focusing on adopting green manufacturing practices in order to protect the environment. "Adopt Green" is the buzz going around in the contemporary business environment. Heavy reliance on natural resources has raised concerns regarding environment among corporate business houses (Qi et al., 2010; Panwar et al., 2011). Due to this, many countries have focused on environmental regulations in order to manage environmental damages. One of the main reasons of increase in awareness among enterprises about adopting green innovative products is the rise of international environmental regulations such as Montreal Convention, Kyoto Protocol, Restriction of the Use of Certain Hazardous Substances in EEE (RoHS), and Waste Electronics and Electrical Equipment (WEEE), restrictions on chlorofluorocarbons, output of Johannesburg world summit on sustainable development, restrictions on the use of certain hazardous substances, Restriction of Hazardous Substances Directive by European Union (Zhu and Sarkis, 2004; Claver et al., 2007). In order to meet these standards for environmental regulations, enterprises have had to adopt environment-friendly practices to improve their environmental images and branding (Chen, 2008; Hillestad et al., 2010). Green innovation has been categorized into three types green product innovation, green process innovation, and green management or system innovation (Chen et al., 2006; Chen 2008; Madrid-Guijarro et al., 2009). Green product innovation is introducing improved or new products in a significant way because of the environmental concerns [e.g., energy savings, use of natural resources, reduction of waste and its recycling (Chen et al., 2006)]. On the other hand, green process innovation is the process of producing environment-friendly products that help in preserving environment through changes in manufacturing processes and systems. There are very less studies done on green manufacturing and concept. One group study dealt with the overall concept of green manufacturing and the other one focused on various analytical tools and models to implement the concept of green manufacturing at different levels (Deif, 2011). Examples of the first group study is the work done by Mohnty and Deshmukh (1998) which highlights the importance of green productivity in gaining competitive edge. Jovane et al. (2003) also projected use of nano/bio/material technologies for futuristic green manufacturing practices. Another framework is presented by Burk and Goughran (2007) for understanding sustainability in green manufacturing. Their study is based on small and medium enterprises with ISO 14001 certification. Examples for the second group include the work of Fiksel (1996). This study explained different analytical tools for green manufacturing with respect to product/process design research. Few examples are life cycle analysis (LCA), screening methods, design for the environment (DFE), and risk analysis. Innovation was differentiated in environmental innovation by Klemmer et al. (1999). This results in improvement of environment. Another definition of green innovation as given by Chen et al. (2006), which defined green innovation technologies that save energy, prevents pollution, recycles waste or based on green product designs. There is also a concept of eco-innovation that reflects on concept of GP. Eco-innovation is also referred to as green innovation. It can also be differentiated into three main categories: eco-product innovation, eco-process innovation, and green managerial innovation. As per Halila and Rundquist (2011), eco-innovations are those innovations that contribute to a sustainable environment through ecological improvements. Beise and Renninsg (2003) have defined eco-innovation as applications consisting of either new or modified processes, techniques, practices, systems, and products in order to protect environmental. Since late 1990s, researchers have given different viewpoints of eco-innovation. There are different studies that define factors affecting eco-innovation and corresponding performance output (Kammerer, 2009; Dangelico and Pujari, 2010). Another study has explored different dimensions of eco-innovation (Carrillo-Hermosilla et al., 2010). Other researchers have tried to quantify eco-innovation (Arundel and Kemp, 2009; Cheng and Shiu; 2012). Kemp and Pearson (2008) have defined eco-innovation as the process of utilization of production techniques that result in a reduction of environmental damage. Further, it has been explained as a process of introduction of new/improved good or service having less negative impact on environment (Jin et al., 2008).

11.3   APPLICATION OF GP FOR ENVIRONMENT SUSTAINABILITY

Enterprises are promoting environment-friendly manufacturing methods and practices for competitive advantage and sustainable development. This results in significant reduction of pollution levels, other environmental damages. Recently, Delhi Metro has applied strategy of recycling debris from its construction sites. The construction site has around 20–25 t of waste. This waste has been used to make around 12 sculptures at new Eco Park. This park has been built on 42000 m², designated as green zone by Delhi Development Authority (DDA). Waste materials, for example, iron scrap, waste sheets, pipes, rolling stock waste, and so forth have been used to make sculptures and the other structures in the park. Footpath was constructed with the help of broken or chipped tiles, Kota stone, granite, and so forth. Another interesting feature of this park is that "green measures" such as solar energy generation; medicinal plants as well as herbs and rare plants that decorate the park. In order to save water, special care is taken while selecting plants for planta-tion, for example, species that require less water have been planted. For adding colors to the park, seasonal flower plants have been planted. Apart from adding colors and planting plants that use less water, landscaping has been done with the help of native species of shrubs and trees. The park will only utilize the treated water from sewage plants. It is also done using green waste, leaves, and grass for making manure. The solar panels have been installed for energy generation for lightening purpose. There is also artificial lake that connects the rainwater harvesting pit. All these uses highlight the conversion of waste into GP that help in making environment a livable place. Another example of the application of GP for environment sustainability is the use of solar energy to power mode of transports. Recently, in Noida, 29.7 km Noida metro corridor has become India's greenest metro. After getting operational, it will have capacity to generate enough solar power to run. This energy produced will be sufficient not only for all 21 stations but also for running offices and train depot as well. Though this facility is being run in Delhi Metro, Phase 3, but Noida Metro will become India's most environment-friendly metro. This is because the entire corridor including head office, parking lot, and footbridge will homogenously use solar power. It also has a target of generating 12 MW solar power daily. In order to meet this target, solar panels have been installed on the rooftops of all stations, footbridges, main office building, depot, and parking lot boundary walls. They further plan to reduce the total power consumption by adopting better engineering practices, sleek design, recycle, and reuse. Solar power will be used for running lights, fans, elevators, escalators, and air-conditioners at stations and offices. Though they do not plan to cut the conventional power supply altogether, they will use it as a supplementary source if required or as a backup as per need. The corridor's total power consumption will be less than 12 MW. In case of surplus, they plan to claim rebate on its power usage and cut operational costs. This metro is also recycling construction waste. Deport in Greater Noida will have zero discharge, which will recycle its whole waste. Fly ash used in the construction will be used to preserve top soil, curb dust. They also plan to make kerbstones and tiles from waste concrete at stations and use wasted iron for grills and railings of stations.

Another application of solar energy has been seen in Bombay Suburban Electric Supply (BSES) DISCOM. They have reportedly energized 206 net metering connections with a sanction load of over 7 MW (over 7000 kW). Around 50 connections with a standard load of over 1.5 MW are under various stages of commissioning at the consumer's end. Some other organizations that have implemented green energy projects are Delhi Cheshire Home, Vasant Valley School, East Point School, Dayal Singh College, DMRC, Church in Mayur Vihar, Delhi, Kohli Imports and Exports Spartan Management Services, and the Delhi Secretariat. This makes BSES the first power distribution utility in the country to have such a large number of net metering connections. Net metering (or net energy metering, NEM) is a process that allows consumers to generate some or all of their own electricity anytime, instead of when it is generated. The concept of net metering has been accepted quiet well by educational institutes, commercial establishments, and domestic users. Consumers have begun to see the benefits of rooftop solar net metering and its impact on electricity bill. The consumers are able to save around Rs.1000–10,00,000/month. Another application of GP for environment sustainability is in the tourism industry. One such study shows that in the tourism industry, focusing on green innovation offers both environmental benefits and opportunities. Use of energy efficient refrigerators, microwave oven, dishwashers, and heating in hotels offers ample scope of application of green innovation products. There can also be use of advanced technology with low carbon emissions. Green innovation practice can also be used in water management. One such prominent case is in Aurangabad, Maharashtra, India. It is called as *Panchakki,* that is, water mill. In this, mill is used to grind grain for the pilgrims who visit this monument. There is a scientific process that was used in medieval Indian architecture. In this water mill, energy is generated through water that is brought down from a spring on a mountain. The water-mill gets water by an underground channel that starts from a well just above junction of river, called Harsul with a tributary stream 8 km away. The water pipe proceeds to the *Panchakki*. The water falls on the cistern of *Panchakki* from quiet a height. This generates the required power to drive the mill. The bottom of cistern forms the roof of spacious hall that remains cool during scorching summers. The extra water is let in the Kham River. Several countries such as Denmark, Israel, South Korea, Sweden, USA, Finland, Germany, Canada, the UK, and Ireland have explored the potential of green technologies. South Africa has developed an environment-friendly method to clean highly toxic water and convert it into drinkable water. The techniques require freezing of the acidic water to generate not only drinking water but also useful salts. The technique is environment-friendly and cost effective. The most important difference is that it does not produce any toxic waste like the other techniques.

Apart from electricity and water management, GP have their application in waste management as well. Agricultural by-products such as straw, sawdust, and corncobs can be used to create environment-friendly biofuel to power cars. Recent study has identified five strains of yeast capable of turning agricultural by-products into bioethanol—a well-known alcohol-based biofuel. It is estimated that more than 400 billion liters of bioethanol could be produced each year from crop wastage. The new study has found five strains of naturally occurring yeast which could be used successfully in the fermentation process. Bioethanol is a very attractive biofuel to the automotive industry as it mixes well with petrol and can be used in lower concentration blends in vehicles with no modification. These five strains can produce highest ethanol yield. Moreover, they are resistant to the toxic compound furfural. Whisky waste could be used as a fuel. There is a possibility of use of whisky by-products as a next generation biofuel, which is being explored by a Scottish start-up. The company is working on tons of waste produced by one of Scotland's most valued industries. This will turn the dregs of whisky-making into fuel. Celtic Renewables has refined its process based on the century old fermentation technique and is now taking a step toward a commercial plant. The process of making whisky requires three ingredients water, yeast, and a grain, primarily barley. But only 10% of the output is whisky. Each year, industry produces 500,000 metric t of residual solid called draff and 1.6 billion liters of yeasty liquid known as pot ale. The firm has taken the old industrial process to turn molasses and other sugar into chemicals and fine tune it into chemicals and to convert draff and pot ale into acetone I, 1-butanol and ethanol. The latter can be used as a fuel.

Another innovation is possible through tannery waste where scientists have made gelatin from tannery waste. This innovative process will yield a high quality product and also help reduce the waste generated at tanneries. Council of Scientific & Industrial Research-Central Leather Research

Institute (CLRI) at Chennai has developed a technology to make a high quality gelatin from animal skin, which is usually disposed of as solid waste tanneries. High grade gelatin is usually made from collagen that is predominantly extracted from animal bone and pig skins. Now, gelatin has been produced from collagen protein which is extracted from the remains of the animal skin or hide that is converted to leather. These remains called trimming wastes are those edges that are usually from the neck, flank, and tail area of the whole animal skin. The collagen is chemically processed (hydrolysis) to make gelatin.

There is also an innovative use of leather waste in building body of bike, car, or aircraft. While leather is used to make shoes and other accessories, leather waste can be used to make an aircraft. A new way has been found out by CLRI where leather solid waste can be used to make a nanocomposite material that is tough enough to make a body of a car, bike, or aircraft, besides light weight construction material, electrical switches, and computer cabinets. Dust generated from buffing leather, a process to get a smooth on the leather, is combined with a polymer and certain nanoparticles to make the material that is almost closed to metal in terms of strength. The polymer could be epoxy or synthetic rubber while nanoparticles, which act as a reinforcement could be titanium dioxide or silicon dioxide. Once they are combined, heat is passed to harden the material. The proportion of the three components used in developing the composite material varies with the thickness and toughness; this technology is patented by CLRI. Buffing dust is a microfine solid particulate that has chromium, synthetic fat, oil, tanning agents, and dye chemicals. Nearly 745 kg of solid waste is generated during the processing of 1000 kg raw hide into leather. Approximately 1% of it will be buffing dust. About 2–6 kg of buffing dust is formed per ton hide processed. Unused chemicals present in leather solid waste, which includes buffing dust, cause pollution. But when this buffing dust is combined with a polymer and nanoparticles, it adds porosity and resistance to the composite due to the presence of chemicals such as chromium. Further winery waste could be used to produce cheap biofuel. The solid grape waste leftover from wine-making could be used to produce low-cost biofuel. Up to 400 l of bioethanol could be produced by fermentation of a ton of grape marc. Global wine production leaves about 13 million t of waste each year. Majority of carbohydrates could be converted to ethanol through fermentation with a yield of up to 270 l/t of grape marc. All these examples emphasize that there is a need for having continuous quality initiatives and green technology to improve competitive advantage. Organizations need to apply GP such as green balance score card, green supply chain management practices, and Green Lean Six Sigma practices. Various industries like automotive and brick industry have achieved better competitive advantage by practicing green innovative practices. As per Comoglio and Botta (2012), the implementation of GP has a positive influence on employee commitment and focus on greater investments towards improvements in environment.

11.4   CHALLENGES IN ADOPTING GP

There are many bottle necks in adoption of GP. There is a need for enterprises to focus on important drivers and antecedents in their businesses, (Routroy, 2009). They must look into the needs of the customers (Thøgersen and Zhou, 2012). Government regulations play an equal role in deciding what practices can be adopted and how they have to be implemented (Kammerer, 2009; Qi et al., 2010). It is necessary to consider the effects of GP on each member of the society. Apart from considering customer needs and government regulations, feasibility study of GP should be done. Organizations should evaluate cost and benefits of GP before implementing. There must be a system to monitor suitability of such practices. There are internal and external barriers that affect adoption of GP. There is a lack of research on barriers to green innovation. The innovative activities in enterprises are affected by financial and technical barriers. One such study shows that most of the enterprises face barriers in innovation such as shortage of technical know-how, shortage of skilled manpower, and high cost of innovation (Sharma, 2014a). Main challenges faced by enterprises are resource availability and competition. These enterprises consider high competition in manufacturing, quality, human resources, and marketing. In general, main barriers faced by small and medium enterprises are complex funding procedure, shortage of technical training, high cost of technology, difficulty in procuring technology, shortage of technical manpower, shortage of sources of funds, and making innovative products, (Sharma, 2014b). Different empirical studies have confirmed the presence of different barriers like innovation value-added chain such as deficient technical support from vendors (e.g., Baldwin and Lin, 2002), limited supply (Carlsen and Edwards, 2008), competitive pressure (Ozgen and Olcer, 2007), and lacking customer responsiveness (Galia and Legros, 2004; Tourigny and Le, 2004).

These barriers can arise either due to strategic issues or structural issues (Galia and Legros, 2004; Kim et al., 2005; Ren, 2009; Tourigny and Le, 2004). Other studies have tried to focus on barriers related to organizational culture (Anumba et al., 2006; Storey, 2000; Vermeulen, 2004; Zerjav and Javernick-Will, 2009). Barriers to green innovation practices adopted by an organization are lack of consensus at the top level, long-term costs and economic conditions, lack of standards and mismatch of short-term and long-term strategic goals, according to Giunipero et al. (2012). Another study done by Wooi and Zailani (2010) in Malaysia suggests the lack of resources as the key barriers that impede GP as resources. It further suggests that technical barriers affect more in case of manufacturing sectors. Muduli et al. (2013) explored barriers like lack of information or less awareness about green technologies among management, lack of knowledge regarding health hazards, absence of society pressure, reluctance of employees for working with green technologies, shortage of green consumerism, absence of motivation among employees, weak laws regarding green innovation practices, absence of proper supervision, corruption, and last, limits of an organization with respect to technology, financial, and production constraints, absence of proper infrastructure to manage or disposal of waste.

Another study done by Jayant and Azhar (2014) explained barriers such as cost, less knowledge of information technology, the overall culture within the organization, less commitment of top management, reluctance for using advance technology, less knowledge of green practice, more competition, reluctance on part of consumer due to less awareness, fear of rejection by society, lack of policies that address reuse and recycling concept, less knowledge about ISO certifications. Similar barriers were identified by Pellegrino and Savona (2013) as cost, knowledge, market factors, and regulation factors. Similarly, in case of agricultural waste which can solve biggest problem of world by providing an alternative fuel but it has its own set of challenges. The processes to generate bioethanol from straw and other by-products are currently complex and inefficient. This is because high temperatures and acid conditions are necessary in the glucose-release process. But this treatment process causes the waste to breakdown into compounds which are toxic to yeast, making fermentation difficult. These are some of the challenges that have reduced the speed with which green innovation should be adopted in society.

11.5 ROLE OF GOVERNMENT IN GP

In order to develop a holistic approach for development and implementation of GP, there is a need for strategic approach for required sustainable environment. It should include horizontal and vertical policy coordination and close integration of different policies for sectors that are interrelated or dependent on each other. For example, in case of sectors such as transportation and agriculture, there can be the use of GP in transportation of farm produce. GP can also be helpful in promoting tourism sector. These days, enterprises are promoting eco-tourism. Entrepreneurs have developed their personal lands into eco-tourism sites. There is a need for policies to develop supply and demand-side policies, for example, regulations to promote and develop GP, use of alternatives of hazardous materials. Such policies should also be subjected to continued evaluation and monitoring. This will improve their effectiveness and efficiency over a period of time. There is also need for educating people about the use of green innovation practices in day-to-day life. Government can conduct educational programs for such practices from time to time. For example, they can make it mandatory for government employees or government organizations to adopt GP. The policy mix is essential for supporting and encouraging innovation, especially for green innovation. It is dependent on various factors, keeping in mind that innovation output and nature of firms differ within industries and across countries. Therefore, one policy for all sectors cannot be feasible and is not applicable. The OECD/Nordic Innovation project on green business model innovation (BMI) (2013) in tourism indicates that active implementation of green innovative activities results in increase in partnership innovations. Government can take measures to develop competencies for innovative partnerships for implementing green innovative practices.

Action from central authority is essential to give proper shape to activities on green innovation initiatives. They become more important under the light of market failures. Like in India, there is a need for promoting green alternatives to be used during festivals. India is a land of festivals. During *Ganesh Visargan*, environment-friendly idols of Ganesha can be used. During Diwali, green crackers can be developed that do not produce any emissions. For Holi festival, there can be environment-friendly options. But often due to less profit or awareness about alternatives, these products often fail in recurring good return on investments. With government interventions, they can deliver optimal level of return on investments. In the absence of such interventions, these products can fail to deliver optimal results. Due to cost-benefit imbalance, enterprises are not motivated to try green innovative practices. As per OECD study, strengthening local innovative capacity is the best way to benefit from global innovation networks. This should be supported by developing international talent and attracting more foreign direct investment in research and development. Similarly, there are several barriers such as gap in available information, reluctance of consumers, capacity of industry, overall cost of investment, financial constraints, and lack of education, thus, highlighting potential areas of focus for government policy responses.

Following specific suggestions can be adopted by government for promotion of GP in Indian business environment:

1. Instead of plastic bags, compostable bags can be used for shopping. This has been implemented by Big Bazar in their retail operations. Compostable bags get converted to invaluable compost/manure under composting conditions which helps to improve the quality of soil.
2. Biodegradable waste can be used as land filler. Roads can be built on them after conducting feasibility study.
3. Make policies to utilize solar energy in maximum areas.
4. Grant benefits to organizations that switch to greener innovative practices such as tax rebates, etc.
5. Develop policies that enforce use of filters at the industrial level for reducing air and water pollution.
6. Switch to zero budget farming practices.

11.6   GP AS A TOOL FOR COMPETITIVE ADVANTAGE

Enterprises try to develop resources into their competitive advantage in order to manage tough competition. They build competitive advantage around quality, speed, variety, safety, design, reliability, use of natural resources, low cost, and low price. Mostly, competitive advantage is the combination of these variables (Kotler, 2003). For example, in case of Noida Metro, they have combined the use of natural resources and cost saving techniques. Organizations have realized that protection of environment is a fundamental factor for achieving competitive advantage. Worldwide, many countries adopt car-free days. This helps in reducing the pollution levels, traffic jams, saves time and, the most important thing, and helps in maintaining good health of people. The green process innovation reflects on changes in environmental performance and competitive advantage. Innovation in GP within an enterprise propels the company to a better position among its competitors and also reduces the negative effects of an enterprise on environment (Küçükoğlu and Pınar, 2015). Another study reveals that there is a positive relation between green product innovation, firm performance, and competitive capability. At the same time, the relation between green product innovation and firm performance is affected by managerial concern for environment (Murat, 2012). Enterprises must focus on performance derived from environment sensitive areas. Though this is difficult to measure but they need to direct their initiatives toward environment-friendly issues. Several studies have been that relate environmental performance and competitive advantage.

As per these studies, both green product and process innovation performance of an enterprise have a positive impact on its competitive advantage (Chen et al., 2006; Chang, 2011). Further, there is a positive relation between green innovation, green performance and environmental performance, (Conding and Habidin, 2012). Another study investigates green innovation with respect to green supply chain management (Chiou et al., 2011). The main aim of implementing GP is to minimize the effects of environmental damages while enhancing the production in the facilities (Cheng et al., 2014). They target reducing pollution levels, manage waste, and improve efficiency (Shrivastava, 1995b). Enterprises mainly focus on two approaches controlling and preventing (Fernandez et al., 2003; Azzone and Noci, 1998). Many enterprises consider protecting environment as an opportunity. These opportunities can be used to improve overall image of an enterprise. Also, worldwide issues related to environment have become the most relevant strategic areas that affect enterprises (Guziana, 2011). Green innovation practices also help in enhancing corporate image and competitiveness. Green product innovation applications improve corporate images, develop new markets, and obtain competitive capability (Porter and Van der Linde, 1995; Shrivastava, 1995b). There is a study that describes relation between international competitiveness and adoption of environment innovation (Brunnermeier and Cohen, 2003). Competition among enterprises in managing environmental practices defines the competitive advantage of enterprises (Porter and Van der Linde, 1995). This competitive advantage can be made stronger if enterprises focus on product innovations. Enterprises also gain competitive advantage by applying environmental techniques in different products (Reinhardt, 1998). Further, it was studied by Chen et al. (2006) that there is a positive relation between competitive advantage and green product innovations. Yalabik and Fairchild (2011) suggested an economic analysis technique. This technique explored relation between competitive pressure and investments by firms. And, the result showed that enterprises can improve the impact of environmental innovations if they compete on the basis of customers that are inclined toward environmental innovations. Guziana's (2011) study identified competitive advantage as one of these motivations to engage in GP. Therefore, engaging in green product innovation actively has positive influence upon corporate competitive capability.

11.7   CONCLUSION

The chapter highlights green innovation practices that can be used as a tool of environment sustainability for competitive advantage. Enterprises are developing strategic plans focusing on sustainable development. They aim at increasing competitive advantage through this strategy. There is increased awareness about sustainability and creating green environment. They adopt innovative environment-friendly products and processes that reduce burden on the environment. This also helps them in maintaining competitive edge over others. This strategy results on green economy which is using less carbon, and is more efficient in utilizing resources. Green innovation concept is adoption of group of activities that help in making new value from the existing resources. These innovations generate better value from existing resources. There is also a growing international concern about environment. Countries have made regulations such as Montreal Convention, Kyoto Protocol, and Restriction of the Use of Certain Hazardous Substances in EEE (RoHS) to manage environmental damages. Green innovations prevent pollution and recycle waste material. There are many examples of environment-friendly manufacturing methods and practices adopted by enterprises. Delhi Metro has created value addition through recycling techniques. BSES has implemented concept of solar energy. Even in the tourism industry, green innovation has been applied. Leather is now being used in building body of bike, car, or aircraft. The implementation of GP increases employee commitment and improves environment. The execution of GP is affected by many barriers. These problems are related to feasibility, target market, acceptability of consumers, and high cost. The enterprises consider high competition in manufacturing, quality, human resources, and marketing of green products as the main barriers. There are also organizational barriers that slowdown the progress of greener products. Less information about greener technologies is also major barrier. Government can play a major role in developing green economy. There is a need for holistic approach for promoting greener practices. Awareness programs can be conducted for general public. Measures can be taken to develop competencies for innovative partnerships. Greener products and practices can be promoted. Policies to utilize solar energy in maximum areas can be made. Strategies should aim at competitive advantage through combined use of natural resources and cost-saving techniques. Studies have shown positive relation between competitive advantage and green product innovations.

KEYWORDS

- **green process innovations**
- **eco-innovations**
- **life cycle analysis**
- **green practices**

REFERENCES

Anumba, C. E. H.; Dainty, A.; Ison, S.; Sergeant, A. Understanding Structural and Cultural Impediments to ICT System Integration: a GIS-Based Case Study. *Eng. Constr. Archit. Manage.* **2006,** *13*(6), 616–633.

Arundel, A.; Kemp, R. *Measuring Eco-Innovation, UNI-MERIT Research Memorandum.* United Nations University: Maastricht, The Netherlands, 2009.

Azzone, G.; Noci, G. Identifying Effectiveness for the Deployment of Green Manufacturing Strategies. *Int. J. Oper. Prod. Manage.* **1998,** *18*(4), 308–335.

Baldwin, J.; Lin, Z. Impediments to Advanced Technology Adoption for Canadian Manufacturers. *Res. Pol.* **2002,** *31*(1), 1–18.

Beise, M.; Rennings, K. *Lead Markets of Environmental Innovations: A Framework for Innovation and Environmental Economics;* Centre for European Economic Research (ZEW): Mannheim, 2003.

Brunnermeier, S. B.; Cohen, M. A. Determinants of Environmental Innovation in US Manufacturing Industries. *J. Environ. Econ. Manage.* **2003,** *45,* 278–293.

Burk, S.; Goughran, W. Developing a Framework for Sustainability Management in Engineering SMEs. *Rob. Comp. Integr. Manuf.* **2007,** *23,* 696–703.

Carlsen, J.; Edwards, D. BEST EN Case Studies: Innovation for Sustainable Tourism. *J. Tourism Hospitality Res.* **2008,** *8*(1), 44–55.

Carrillo-Hermosilla, J.; del Rio, P.; Könnölä, T. Diversity of Eco-Innovations: Reflections from Selected Case Studies. *J. Cleaner Prod.* **2010,** *18*(10–11), 1073–1083.

Chang, C.-H. The Influence of Corporate Environmental Ethics on Competitive Advantage: the Mediating Role of Green Innovation. *J. Bus. Ethics* **2011,** *104*(3), 361–370.

Chen, Y.-S. The Driver of Green Innovation and Green Image—Green Core Competence. *J. Bus. Ethics* **2008,** *81,* 531–543.

Chen, Y.-S.; Lai, S. B.; Wen, C. T. The Influence of Green Innovation Performance on Corporate Advantage in Taiwan. *J. Bus. Ethics* **2006,** *67*(4), 331–339. http://dx.doi.org/10.1007/s10551-006-9025-5.

Cheng, C. C.; Shiu, E. C. Validation of a Proposed Instrument for Measuring Eco-Innovation: an Implementation Perspective. *Technovation* **2012,** *32*(329), 344.

Cheng, C. J.; Yang, C.; Sheu, C. The Link between Eco-Innovation and Business Performance: a Taiwanese Industry Context. *J. Cleaner Prod.* **2014,** *64,* 81–90.

Chiou, T.-Y.; Chan, H. K.; Lettice, F.; Chung, S. H. The Influence of Greening the Suppliers and Green Innovation on Environmental Performance and Competitive Advantage in Taiwan. *Transp. Res. Part E: Logistics Transp. Rev.* **2011**, *47*(6), 822–836.

Claver, E.; López, M. D.; Molina, J. F.; Tarí, J. J. Environmental Management and Firm Performance: a Case Study. *J. Environ. Manage.* **2007**, *84*(4), 606–619.

Comoglio, C.; Botta, S. The Use of Indicators and the Role of Environmental, Management Systems for Environmental Performances Improvement: a Survey on ISO 14001 Certified Companies in the Automotive Sector. *J. Cleaner Prod.* **2012**, *20*, 92–102. http://dx.doi.org/10.1016/j.jclepro.2011.08.022 (accessed Oct 5, 2016).

Conding, J.; Habidin, N. F. The Structural Analysis of Green Innovation and Green Performance in Malaysian Automotive Industry. *Res. J. Finance Accounting* **2012**, *3*(6), 172–178.

Dangelico, R. M.; Pujari, D. Mainstreaming Green Product Innovation: Why and How Companies Integrate Environmental Sustainability. *J. Bus. Ethics* **2010**, *95*(3), 471–486.

Deif, A. M. A System Model for Green Manufacturing. *J. Cleaner Prod.* **2011**, *19*(14), 1553–1559.

Durif, F.; Bolvin, C.; Julien, C. In Search of a Green Product Definition. *Innovative Mark.* **2010**, *6*(1), 25–33.

Fernandez, E.; Junquera, B.; Ordiz, M. Organizational Culture and Human Resources in the Environmental Issue: A Review of the Literature. *Int. J. Human Res. Manage.* **2003**, *14*(4), 634–656.

Fiksel, J. *Design for Environment: Creating Eco-efficient Products and Processes.* McGraw-Hill: New York, 1996.

Galia, F.; Legros, D. Complementarities between Obstacles to Innovation: Evidence from France. *Res. Pol.* **2004**, *33*(8), 1185–1199.

Gao, Y.; Li, J.; Song, Y. Performance Evaluation of Green Supply Chain Management Based on Membership Conversion Algorithm 2009. *ISECS International Colloquium on Computing, Communication, Control and Management,* **2009**, 237–240.

Giunipero, L. C.; Hooker, R. E.; Denslow, D. Purchasing and Supply Management Sustainability: Drivers and Barriers. *J. Purch. Supply Manage.* **2012**, *18*(4), 258–269.

Greeno, L. J.; Robinson, S. N. Rethinking Corporate Environmental Management. *Columbia J. World Bus.* **1992**, *27*(3/4), 222–232.

Guziana, B. Is the Swedish Environmental Technology Sector 'Green'? *J. Cleaner Prod.* **2011**, *19*(8), 827–835.

Halila, F.; Rundquist, J. The Development and Market Success of Eco-Innovations: A Comparative Study of Eco-Innovations in Sweden. *Eur. J. Innovation Manage.* **2011**, *14*(3), 278–302.

Handfield, R.; Walton, S.; Seegers, L.; Melnyk, S. Green Value Chain Practices in the Furniture Industry. *J. Oper. Manage.* **1997**, *12*(5), 38–53.

Hillestad, T.; Chunyan, X.; Haugland, S. A. Innovative Corporate Social Responsibility: the Founder's Role in Creating a Trustworthy Corporate Brand through "Green Innovation". *J. Prod. Brand Manage.* **2010**, *19*(6), 440–451.

Jayant, A.; Azhar, M. Analysis of the Barriers for Implementing Green Supply Chain Management (GSCM) Practices: an Interpretive Structural Modeling (ISM) Approach. *Procedia Eng.* **2014**, *97*, 2157–2166.

Jin, J.; Chen, H.; Chen, J. Development of Product Eco-Innovation: Cases from China. Paper Presented at the XXXI R&D Management (RADMA) Conference "Emerging and New Approaches to R&D Management", Ottawa, 2008.

Jovane, F.; Koren, Y.; Boer, N. Present and Future of Flexible Automation: Towards New Paradigms. *CIRP Ann.* **2003,** *52*(2), 543–547.

Kammerer, D. The Effects of Customer Benefit and Regulation on Environmental Product Innovation: Empirical Evidence from Appliance Manufacturers in Germany. *Ecolog. Econ.* **2009,** *68*(8–9), 2285–2295.

Kemp, R.; Arundel, A. *Survey Indicators for Environmental Innovation.* IDEA (Indicators and Data for European Analysis) Paper Series #8 1998.

Kemp, R.; Pearson, P. *Final Report MEI Project about Measuring Eco-innovation;* Maastricht, 2008. http://www.merit.unu.edu/MEI/papers/Final%20report%20MEI%20project%20 DRAFT%20version%20March%2026%202008.pdf (accessed Oct 12, 2016).

Kim, Y.; Lee, Z.; Gosain, S. Impediments to Successful ERP Implementation Process. *Bus. Process Manage. J.* **2005,** *11*(2), 158–170.

Klemmer, P.; Lehr, U.; Lobbe, K. *Environmental Innovation: Incentives and Barriers*, 1st ed.; Analytica: Berlin, 1999.

Kotler, P. *Marketing de A a Z: 80 Conceitos Que Todo Profissional Precisa Saber;* Elsevier: Rio de Janeiro, 2003.

Küçükoğlu, M. T.; Pınar R. İ. Positive Influences of Green Innovation on Company Performance. *World Conference on Technology, Innovation and Entrepreneurship,* 3 July **2015,** *195,* 1232–1237. DOI:10.1016/j.sbspro.2015.06.261.

Madrid-Guijarro, A.; Garcia, D.; Van Auken, H. Barriers to Innovation among Spanish Manufacturing SMEs. *J. Small Bus. Manage.* **2009,** *47*(4), 465–488.

Mohnty, R. P.; Deshmukh, S. D. Managing Green Productivity, Some Strategic Directions. *Prod. Plann. Control* **1998,** *9*(7), 624–633.

Muduli, K.; Govindan, K.; Barve, A.; Kannan, D.; Geng, Y. Role of Behavioural Factors in Green Supply Chain Management Implementation in Indian Mining Industries. *J. Res. Conserv. Recycl.* **2013,** *76,* 50–60.

Murat, I. A. In *Impact of Green Product Innovation on Firm Performance and Competitive Advantage and Competitive Capability: the Moderating Role of Managerial Environmental Concern,* World Conference on Business, Economics and Management (BEM-2012), Antalya, Turkey, May 4–6, 2012. DOI:10.1016/j.sbspro.2012.09.144, 2012.

Nidumolu, R.; Prahalad, C. K.; Rangaswami, M. R. *Why Sustainability is Now the Key Driver of Innovation.* Harward Business Review, 87/9, September, 2009, 57–64.

OECD. *Green Innovation in Tourism Services;* OECD Tourism Papers, 2013/01, OECD Publishing, **2013**. http://dx.doi.org/10.1787/5k4bxkt1cjd2-en (accessed April 15, 2016).

Ozgen, H.; Olcer, F. An Evaluative Study of Innovation Management Practices in Turkish Firms. *Int. J. Bus. Res.* **2007,** *7*(2), 53–63.

Panwar, N.; Kaushik, S.; Kothari, S. Role of Renewable Energy Sources in Environmental Protection: A Review. *Renewable Sustainable Energy Rev.* **2011,** *15*(3), 1513–1524.

Pellegrino, G.; Savona, M. *Is Money All? Financing Versus Knowledge and Demand Constraints to Innovation.* UNI-MERIT Working Paper Series. 2013.

Porter, M. E. *A Vanategem Competitiva das Nacoes;* Elsevier: Rio de Janeiro, 1989.

Porter, M. E.; Van der Linde, C. Toward a New Conception of the Environment Competitiveness Relationship. *J. Econ. Perspect.* **1995,** *9*(4), 97–118.

Qi, G. Y.; Shen, L. Y.; Zeng, S. X.; Jorge, O. J. The Drivers for Contractors' Green Innovation: An Industry Perspective. *J. Clean. Prod.* **2010,** *18*(14), 1358–1365.

Ramus, C. A. Encouraging Innovative Environmental Actions: What Companies and Managers Must Do. *J. World Bus.* **2002,** *37*(2), 151–164.

Rave, T.; Goetzke, F.; Larch, M. *The Determinants of Environmental Innovations and Patenting;* Germany, Reconsidered. IFO Working Paper#97, 2011.

Reinhardt, F. L. Environmental Product Differentiation: Implications for Corporate Strategy. *Calif. Manage. Rev.* **1998,** *40*(4), 43–73.

Ren, T. Barriers and Drivers for Process Innovation in the Petrochemical Industry: a Case Study. *J. Eng. Technol. Manage.* **2009,** *26*(4), 285–304.

Routroy, S. Antecedents and Drivers for Green Supply Chain Management Implementation in Manufacturing Environment. *ICFAI J. Supply Chain Manage.* **2009,** *6*(1), 20–35.

Sharma, N. Determinants of Innovation: a Study of Small and Medium Enterprises in India, *World SME News,* (Publication of WASME), ISSN-0973-1261, Issue no-05, 2014a, pp 5–14.

Sharma, N. Management of Innovation in SMEs in India: a Barrier Approach, *World SME News,* (Publication of WASME), ISSN-0973-1261, Issue no-086, 2014b, pp 8–14.

Sharma, N. Management of Innovation in Small and Medium Enterprises in India: Comparison of I.T. and Pharmaceutical Firms. *IUP J. Knowl. Manage.* **2016,** *XIV*(2), ISBN: 9788131427958.

Shrivastava, P. The Role of Corporations in Achieving Ecological Sustainability. *Academy Manage. Rev.* **1995a,** *20*(4), 936–961.

Shrivastava, P. Environmental Technologies and Competitive Advantage. *J. Strategic Manage.* **1995b,** *16*(S1), 183–200.

Storey, J. The Management of Innovation Problem. *Int. J. Innov. Manage.* **2000,** *4*(3), 347–369.

The Nordic Innovation. Report of Findings on, Green Business Model Innovation in the Tourism and Experience Economy. 2012. http://www.nordicinnovation.org/Publications.

Thøgersen, J; Zhou, Y. Chinese Consumers' Adoption of a 'Green' Innovation—The Case of Organic Food. *J. Mark. Manage.* **2012,** *28*(3–4), 313–333.

Tourigny, D.; Le, C. Impediments to Innovation Faced by Canadian Manufacturing Firms. *Econ Innov. New Tech.* **2004,** *13*(3), 217–250.

Tübitak. Vizyon 2023 Sürdürülebilir Kalkinma Paradigmasi Üzerine ön Notlar, (n.d.), http://www.tubitak.gov.tr/tubitak_content_files/vizyon2023/csk/EK-16.pdf (accessed March 9, 2014).

Vermeulen, P. A. M. Managing Product Innovation in Financial Services Firms. *Eur. Manage. J.* **2004,** *22*(1), 43–50.

Wong, S. K. S. Environmental Requirements, Knowledge Sharing and Green Innovation: Empirical Evidence from the Electronics Industry in China. *Bus. Strategic Environ.* **2013,** *22*(5), 321–338.

Wooi, G. C.; Zailani, S. Green Supply Chain Initiatives: Investigation on the Barriers in the Context of SMEs in Malaysia. *Int. Bus. Manage.* **2010,** *4*(1), 20–27.

Yalabik, B.; Fairchild, R. J. Customer, Regulatory, and Competitive Pressure as Drivers of Environmental Innovation. *Int. J. Prod. Econ.* **2011,** *131*(2), 519–527.

Zerjav, V.; Javernick-Will, A. Motivators and Critical Factors for Worksharing in Design and Engineering Networks. LEAD 2009-Global Governance in Project Organizations, Lake Tahoe, California, 2009.

Zhu, Q.; Sarkis, J. Relationships between Operational Practices and Performance among Early Adopters of Green Supply Chain Management Practices in Chinese Manufacturing Enterprises. *J. Oper. Manage.* **2004,** *22*(3), 265–289.

CHAPTER 12

COMMUNICATING SUSTAINABILITY AND GREEN MARKETING: AN EMOTIONAL APPEAL

MOTURU VENKATA RAJASEKHAR[1],
KRISHNAVEER ABHISHEK CHALLA[2],
DHARMAVARAM VIJAYLAKSHMI[3], and
NITTALA RAJYALAKSHMI[4]

[1]*Department of Commerce and Management Studies, Andhra University, Vishakhapatnam, Andhra Pradesh 530003, India, Ph.: 9885279425*

[2]*Department of Foreign Languages, Andhra University, Vishakhapatnam, Andhra Pradesh 530003, India, Ph.: 9908742869, E-mail: com2mass@gmail.com*
[3]*Gayatri Vidya Parishad, Gayatri Valley, Rushikonda, Visakhapatnam, Andhra Pradesh 530045, India, Ph.: 9866777148, E-mail: vijayalucky28@gmail.com*
[4]*Department of Commerce and Management Studies, Andhra University, Vishakhapatnam, Andhra Pradesh 530003, India, Ph.: 8500362509, E-mail: nittalarl@yahoo.co.in*
Corresponding author. E-mail: raj3727@gmail.com

12.1 INTRODUCTION

12.1.1 GREEN MARKETING AND THE CHALLENGE OF SUSTAINABILITY

Seen from the vastness of outer space, the Earth is a very small place. So small that Carl Sagan (1997) compared the Earth to a speck of dust suspended in a sunbeam.

The universe was born after the great explosion of the Big Bang 13,700 million years ago[1]; the Sun and the Earth came into existence about 4500 million years ago; the first populations of the genus *Homo* lived on Earth 2.5 million years ago and it was not until 200,000 years ago when human beings (*Homo sapiens*) began to exist. This is how society was formed: it is a social system and physical infrastructure that humans have created to satisfy our collective and individual needs (Robèrt, 2010).

There are various elements that allow you and me to live on Earth, the biosphere is one of them. The biosphere is one of the systems that support life on Earth: it ranges from the atmosphere's top layer to the soil's bottom layers, both on the continents and in the ocean. Society and all earthlings depend on the biosphere for food, fresh water, a habitable climate and to maintain nature's integrity and diversity.

Over the years, mother earth is deteriorating because of overexploitation of human needs, wants, and desires. Balbus et al. (2013) have pointed that the biosphere is constantly becoming worse, and stated that it should be of vital concern for all *Homo sapiens* existing on mother earth. Human consumption has proven to be one of the major sources of deteriorating environment (De Bettignies and Lepineux, 2009).

12.1.1.1   CURRENT SOCIETY

In the history of humankind, society represented a small part of the biosphere and its impact was relatively small. In 1750, with the beginning of the Industrial Revolution, society grew in size and technological advances. Consequently, agriculture, goods production and transportation were redefined and cures for different diseases that caused death, epidemics and pandemics were found. Nowadays, despite differences between the rich and poor countries, life expectancy has considerably increased on a global level.

12.1.2   THE CHALLENGE OF SUSTAINABILITY

As a result of technological advances that led human beings to live in great prosperity, society has had a growing impact on the biosphere (Fig. 12.1).

---

[1]There is a hypothesis that suggests the existence of more universes: the theory of parallel universes.

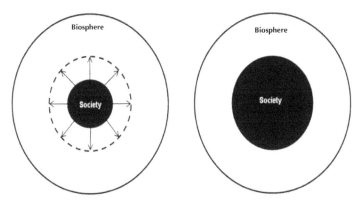

FIGURE 12.1   Growth of Human Society. Historically, human society has been small compared with the biosphere (left). However, since the Industrial Revolution, it has had major negative and continuous effects on the biosphere (right).

Source: Adapted from: "Crowded World" by Herman Daly (2004).

Before the modern industrial society developed, global sustainability was not an issue; therefore, it is right to argue that the cause of the challenge of sustainability is due to dominant trends in modern society.

1. *Environmental Effects:* For the first time in the history of the planet, between 1970 and 1980, human demand exceeded nature's ability to regenerate. Society impacts the planet by using 50% more of the natural resources that Earth can provide. In other words, if this trend continues, by 2030, we will need two planets in order to maintain our current lifestyle (WWF, 2012).
Greenhouse gas emissions from burning fossil fuels and deforestation have caused climate change on a global level (IPCC, 2007).
Society is also using a large percentage of fertile land to build colossal infrastructure projects, such as cities and large-scale agro-industrial plantations and pasture fields (Ellis, 2008). Furthermore, thousands of new synthetic chemicals are produced during industrial processes external to nature. All of this has caused environmental effects of immense proportions that the planet cannot cope with.
2. *Social Effects:* Society faces continuous and increasingly difficult problems in inequality, disease, and malnutrition. Violation of human rights, inequality, corruption, labor abuse, discrimination, lack of access to education, and other problems, are still a reality in many countries.

The progressive degradation of the biosphere is limiting the ability to live beings to have prosperous lives (Robèrt, 2010).

The systematic degradation of the biosphere implies continuous degradation: its growth is more accelerated and intense. Therefore, from this perspective, society is in a process of decay that threatens its welfare.

How did we get to the point of threatening the lives of humans and other species on the planet? The consumption of energy produced from fossil fuels, the use of pesticides in food and the increase in consumption of animal products are some of the examples that, on first glance, seem isolated; they come from different economic activities but generate, as a whole, different greenhouse gasses. This has contributed to global warming. Let us not lose sight, though, global warming is merely one aspect of the challenge of sustainability.

The challenge is to address the combination of mistakes that are systematically increasing to change the paradigm and lifestyle of a society that is encouraging unsustainable human effects on the socio-ecological system[2].

12.1.3   ACTIONS TAKEN

After scientists from diverse fields concluded that society is on an unsustainable course that cannot continue indefinitely, several society sectors called for a new global goal: a socially and ecologically sustainable society (Steffen et al. 2004; Springer-Verlag, 2005). The goal is to be able to continue to develop without further deteriorating the systems that sustain life, creating human well-being within ecological boundaries.

The United Nations Brundtland Commission (1987) defined sustainable development; nowadays, it is the best known and globally accepted definition:

Sustainable development meets the needs of current generations without compromising the ability of future generations to meet their own needs.

Since then, many movements have emerged all over the world; an example is the *Earth Summit* in Rio de Janeiro, in 1992, which created global awareness regarding the seriousness of the planet's environmental crisis. In addition to this movement, many organizations that support this cause have appeared around the world.

---

[2]Socio-ecological system: a system made up of the biosphere, human society, and their complex interactions.

Likewise, multinational companies launched initiatives to take action; nongovernment environmental organizations (NGOs) emerged, environmental laws were passed, and financial support was allocated to environmental programs, among other examples.

However, despite these global movements, global environmental and social statistics have not shown progress; on the contrary, in some cases, statistics show major setbacks.

According to the Living Planet Report 2012 published by the World Wide Fund NGOs for Nature (WWF, 2012), from 1970–2008, the world's biodiversity dropped by 30%; moreover, society doubled the demand for natural resources between 1961 and 2007 due to the increase of urban consumption.

If we have known about this problem for over 25 years[3], why have statistics not improved, what stops us from heading toward a sustainable society?

There are many variables to consider; next, I will introduce some factors that can help us reflect on it.

a) Communication with the Public:

Considering that a significant part of the public's knowledge regarding sustainable development is acquired through newspapers, magazines, the radio and, above all, the television (Hansen, 1995), one of the variables to consider is mass media communications.

People's participation in democratic decision-making regarding sustainable development largely depends on the extent to which the arguments, analysis, tests, and evaluations are available. Media coverage of sustainable development topics is not only a matter of political debate forums but also a matter of managing and presenting the type of information that allows citizens to make "appropriate" decisions (Hansen, 1995).

At the end of the Earth Summit in Rio de Janeiro in 1992, the topic of the environment remained on the agendas of some media members. Even though specialized media was born, the topic has been dropped (Voisey and Church, 1999).

Although currently, it is not a priority, the media has increasingly paid more attention to topics like the green economy, clean energies, global warming, and biodiversity.

Journalists face a number of limitations while practicing their profession (Kunst and Witlox, 1993). Some of them believe that committing to cover

---

[3]In 1962, the first report on the environment's deterioration was released in the book "Silent Spring" by Rachel L. Carson. Thanks to this report, the use of the dichlorodiphenyltrichloroethane (DDT) pesticide was banned and the Environmental Protection Agency was created.

long-term sustainable development, they risk their careers by committing "professional suicide." Many journalists and editors believe their work serves only to inform and not to educate; they see themselves as guardians following investigative reporting.

In some countries, the media is questioned on various aspects: the media has its own personal interest and they provide information that does not always reflect the truth (Bacchetta, 2012) rather than looking after the interests of the community. The educational aspect is questioned and transparency in its management is considered poor (Vieira, 2012b).

The media has the challenge to convey the complexity of sustainable development in a way accessible to all citizens. The media demands speed and sometimes, the reality is oversimplified (Guillen, 2009).

These topics have commonly been spread via environmental catastrophes (Guillen, 2009; Larena, 2006). In the end, it only manages to cause fear, frustration and, eventually, disinterest among the population.

On the other hand, many organizations have focused on urging people to take action by recycling, stopping the use of plastic bags or changing from light bulbs to the energy-saving versions (Leonard, 2012a).

b) Mental models:

Another possible factor that has prevented us from improving is our mental models. Mental models are the thoughts through which we interpret reality and the world; these have an effect on our behavior. Some mental models can be identified very easily while others are in our minds, silent but present.

Below are some examples of mental models [4]:

- Things are as they are and there is nothing that I or we can do about it.
- If no one is doing something, why do I have to do something? I am alright this way.
- Natural resources can be replaced one way or another.
- If others start doing something about it, I probably will, too.
- What is important is economic development, we need jobs.
- My responsibility is with me and my family, I will look after them. You take care of yourself and your own family.
- Either technology, innovation or someone smart and powerful will solve it.

---

[4]Inspired by the presented models at The Cloud Institute for Sustainability Education http://www.cloudinstitute.org/

- My company is small, we cannot afford to care for the environment in the same way that large companies can.
- I want to stay as I am and I do not care about the consequences.
- Our politicians do not agree, we do not move forward because of them.
- Yes, I care about the situation but do not know what to do; I will wait and follow others.
- I do what the government and NGOs recommend, but I do not think what I am doing will be of much use because not everybody does it.

The most dominant mental model is that social problems such as poverty, corruption, and malnutrition are not related to environmental degradation. These facts are apparently isolated but are interconnected; we have been trained to not think critically (Buján, 2010).

A second dominant mental model is that economy is the goal to solve the world's problems, when in fact, the economy is a means. Social and ecological sustainability are the goals and the economy is one of many ways to achieve it (Robèrt, 2010).

c) Unclear definition of sustainable development and sustainability:

A third factor that has stopped us from progressing is confusion about what sustainable development and sustainability mean: they are usually connected only to the environment. An already popular phrase attributed to Australian architect Andrew Maynard is "Sustainability is like teenage sex. Everybody says they're doing it; very few people actually are doing it. Those that are doing it are doing it badly."

In the first few pages, I gave the most recognized definition of sustainable development given by the United Nations Brundtland Commission. I consider it to be an inspiring and even philosophical definition.

There are many other ways to define these concepts, but how do we present this definition to make it practical that is, how can we understand it in a way that will teach us to satisfy the needs of current generations without compromising the future?

Quoting Albert Einstein, he said that if he had only 1 h to save the world, he would spend 55 min defining the problem and 5 min finding the solution. With this, I just want to emphasize the importance of knowing what is the root of the problem we are facing because, once the problem is understood, the solution will be there.

If the challenge consists of realizing that our society is badly designed, which are the design faults, what is the root of the problem?

12.2   HOW TO DEFINE A SUSTAINABLE SOCIETY?

Let us start by exploring the topic of basic human needs. If we talk about meeting the needs of current generations without compromising the ability of future generations to meet their own needs, what are those needs?

Basic human needs are innate requirements to be met so that people stay physically, mentally and socially healthy. There are various theories and classifications; a popular one is the definition of the nine fundamental human needs, defined by the Chilean economist Manfred Max-Neef.[5]

Based on several studies, Max-Neef classified the needs into nine categories which are universal in terms of culture and history; they cannot replace each other, cannot be replaced by others and must be met continuously. These nine needs are essential to all people and are intrinsically related:

Let us differentiate the basic needs of satisfiers:

Satisfiers are ways in which humans try to satisfy their needs and desires. Satisfiers change throughout time and cultures, that is, what is culturally determined are not fundamental human needs, but the satisfiers of those needs; cultural change is the consequence of abandoning traditional satisfiers and replacing them with others.

Subsistence	**Idleness**	**Understanding**
Protection	**Creativity**	**Identity**
Participation	**Affection**	**Freedom**

FIGURE 12.2   Basic needs of human beings (Neef, 1991).

Two key points here are: to deprive others of the opportunity to meet their individual basic needs because they carry consequences for everybody on a global level (Neef, 1991) and it is important to reflect on how we meet our needs today.

Furthermore, let us assume that any system, including the Earth, has parts, interconnections, a role or purpose and within the structure of these interconnected parts, there is a behavior (Meadows, 2008). What this tells us is that the system can only be understood if we observe its behavior in its

---

[5]Max-Neer says there is no correlation between the degree of economic development (industrial) and people's relative happiness; on the contrary, economic development seems to increase solitude and alienation in developed societies.

entirety, not only one aspect (Senge, 2006). In other words, to understand the behavior of the Earth (the system where we live), we need to know its parts, its interconnections, their roles, structure, behavior, and so forth.

Let us examine briefly some basic concepts of the system we live in: even though society has different ways to satisfy its basic needs, everyone as well as all ecosystems, depending on the Sun, a biosphere and all the natural process that occur among them. Earth exchanges energy with the universe and the exchanged matter is scarce; matter stays on Earth due to gravity, but energy enters Earth in the form of sun radiation and leaves as heat radiation. The Sun is our main source of energy and plants make the most of this energy to conduct photosynthesis, which produces the food that animal species depend on (Robèrt, 2010).

Let us continue with a very simple definition of sustainable development: the transition from unsustainable to a sustainable society (Fig. 12.3).

How would we define an unsustainable society, what are the society's problems that make it unsustainable? It is clear that we have deteriorated the socio-ecological system that sustains our lives and those of future generations but in what way have we deteriorated it?

In 1992, Dr. Karl-Henrik Robèrt launched a scientific exploration to understand what root causes have brought about the degradation of the socio-ecological system.

FIGURE 12.3 Sustainable Development. The transition from the current unsustainable society to a sustainable society is known as sustainable development. The term can also refer to a continuous development of society because society has become sustainable (Robèrt, 2010).

Dr. Robèrt began writing a paper about possible root causes of the socio-ecological system's degradation and sent it to over 50 ecologists, chemists, physicists, and doctors in Sweden and asked for their opinion; after rewriting the paper 21 times, they finally reached consensus on the basic elements that sustain the socio-ecological system on the planet (Robèrt, 1991).

The consensus resolved that we are only doing four things that result in the degradation of the Earth's natural processes that sustain us and future generations.

There are only four key mechanisms to achieve a sustainable society. The first three refer to mechanisms that cause the degradation of the biosphere and its ability to support life, the fourth refers to social sustainability.

The question is if we eliminate these four mechanisms from our daily agendas, can we achieve a socially and ecologically sustainable society?* According to Dr. Robèrt's proposal, yes (Robèrt, 2010). To create a sustainable society, we must:

1. Eliminate our contribution to the systematic increase of substances extracted from the Earth's crust to the biosphere.

    For example, increased concentrations of fossil fuels, such as CO_2, in the atmosphere or the increase in the concentrations of cadmium in the Earth's crust in agricultural soil. The concentrations can increase from activities throughout the products' life cycle, from mining to production, to end use and disposal.

    Complying with this principle requires replacing the use of certain minerals that are scarce in nature with others that are more abundant, using all mined materials efficiently and reducing dependence on fossil fuels. This does not mean that mineral extraction is not possible, but that society's collective actions should not lead to a systematic increase in the concentrations of these substances in nature.

2. Eliminate our contribution to the systematic increase of chemicals and compounds produced by society within the biosphere.

    For example, toxins such as dichlorodiphenyltrichloroethane (DDT)[6] and polychlorinated biphenyls (PCBs)[7] and substances which deteriorate the ozone layer, such as chlorofluorocarbons (CFCs).[8]

    Complying with this principle requires replacing certain persistent and artificial compounds with others that normally abound or break down more easily in nature. It also requires reducing the flow of

---

*Note: It is unrealistic to think that in a sustainable society, mining or forest degradation or corruption will not exist; these negative effects may occur in a sustainable society as long as these do not increase systematically.

[6]Dichlorodiphenyltrichloroethane: insecticides that kill the species they are meant to and affect other species in the ecosystem.

[7]Polychlorinated biphenyl: dielectric and coolant fluids used in electric systems such as transformers or rectifier stations.

[8]Chlorofluorocarbon: deteriorate the ozone layer that protects the earth from ultraviolet radiation.

substances that decompose in nature due to great influxes in society. Finally, it requires efficiently using all substances that the society produces.

3. Eliminate our contribution to systematic physical degradation as well as the destruction of nature and the biosphere's natural processes.

    For example, overexploitation of forests, overfishing, loss of soil due to overuse of land, urbanization of wild habitats of critical importance to the survival of ecosystems.

    Complying with this principle requires only using natural resources from well-managed ecosystems, seeking greater efficiency and productivity. One needs to be cautious with changes done to nature, including the harvest of resources and species introductions.

4. Eliminate our contribution to conditions that undermine people's capacity to meet their own basic needs.

    For example, unsafe working conditions or wages too low to allow a decent life, unequal conditions, and so forth, generate less social trust and inequality, therefore causing weakening of the social fabric.

    Complying with this principle requires adjusting any behavior that restricts the people's opportunities, today or in the future, to meet their own basic needs.

Once unsustainability mechanisms were categorized, the principles of sustainability were defined.

Why principles? A principle is a condition that must be met so that the system (in this case, the biosphere) remains in a specific state. A strategic way to define a vision or a goal is through principles that have the following characteristics:

a) Necessary to achieve the goal
b) Sufficient to achieve the goal
c) General enough to be used in different contexts
d) Concrete enough to guide actions and
e) Non-overlapping, or mutually exclusive.

The four basic principles of sustainability[9] state: "In a sustainable society, nature is not subject to a progressive increase in:

---

[9]These four conditions to a sustainable society in the biosphere are called "sustainability principles." They are also widely known in the scientific and business community as the "TNS System Conditions," referring to the NGO, The Natural Step. http://www.thenaturalstep.org/

1. Concentrations of substances extracted from the Earth's crust,
2. Concentrations of substances produced by society,
3. Degradation by physical means, and

In that society, people are not subjected to conditions that systematically undermine[10] their capacity to meet their basic needs.

12.3   THE PRACTICE OF COMMUNICATING FOR ENVIRONMENTAL AND SOCIAL WELLBEING:

12.3.1   BACKCASTING

Returning to the example of the game of chess, the list of considerations (Section 1.4.2) was planned to follow the rules of the game (principles) and achieve checkmate; it is not possible to know in advance where the pieces will be placed, however, by following the rules of the game from the beginning, we can reach the goal. The game follows planning in backcasting, which means that one has to start playing while thinking about success (the four sustainability principles), then, one must return to the present (by asking oneself, what can I do today to move forward, following the rules) and finally, one must head toward the vision of success, step-by-step (Fig. 12.3) (Ny, et al. 2006).

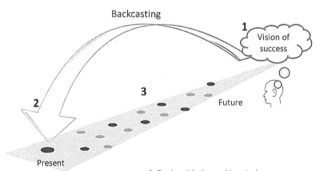

FIGURE 12.4   Steps for planning in backcasting (Ny et al., 2006).

Source: Adapted from The Natural Step (Natural Step, 2011).

---

[10]Definition by the Royal Spanish Academy: to weaken something or someone, especially in a moral aspect.

One of the great advantages of using backcasting from the beginning is that it maintains a constant focus on the "rules of the game" in relation to success while allowing flexibility in the ways to reach a vision of success.

The vision of success in this context would be the practice of journalism in the service of a sustainable society (defined with the four principles of sustainability). There are many ways in which communicators and journalists can achieve it; the list of considerations shows some tools that will head in that direction.

Notes:

- It is assumed that communicators and journalists reading this chapter know and apply the basic principles of journalism and communication such as accuracy, contrast, and honesty.
- The list of considerations presented here is not an absolute truth or a magic formula. These do not intend to show how to do journalistic or, even less, give moral lessons, it merely seeks to collect and present key points so communicators and journalists can consider applying when integrating strategic sustainable development in their daily work.
- The list of considerations does not aim to limit creativity; on the contrary, it aims to reveal relevant aspects that can stimulate creativity.

12.3.1 LIST OF CONSIDERATIONS

The objective of this list is to offer communicators, a series of considerations to integrate strategic sustainable development into their daily agendas to generate learning in their audience and concrete actions that will steer the society to sustainable development.

1. Keep in mind that everything is connected:

There is nothing, action, thing or person, which is independent. Everything is connected, nothing can change by itself.

A Sufi proverb says: "You think that because you understand one you must understand two, since one and one make two, but you must also understand why."

When we follow this perspective, we enter a new panorama of connections and phenomena that cannot be reduced only to causes and effects but must also include the constant flow of dynamic processes.

Throughout history, different cultures have had and still have the belief that we are all one single entity. The Mayans, for example, greeted each other saying, *In LaK'ech* which means "I am another you."

2. Devote time to know yourself:

Gabriel García Márquez said, "I think, with absolute seriousness, that to always do what one likes, and only that, is the master formula to have a long and happy life."

Find activities that make you feel satisfied and happy; it has a greater effect on the economy than any other factor.

Why and for what do you do what you do? What motivates you? What do you enjoy doing the most? What gives meaning to your life? What is your purpose? What are your goals and dreams? Are you being the person that you really are? If you had to make a wish, what would it be?

These are just some questions that might be worth exploring. Albert Einstein used to say that we are all geniuses if you judge a fish by its ability to climb a tree; it will live its whole life believing that it is stupid.

"If, after a serious and rigorous evaluation, you are willing to be a communicator, I welcome you because I am sure that those who test their own calling, will make important social contributions that will improve the conditions of human coexistence, which ultimately is what our profession must do," Professor Miguel Ángel Granados Chapa.

3. Furnish your brain:

Reading is one of the secrets to becoming a better communicator and journalist. Furnish your brain with all that is available, from training, online forums, and the press, to classic literature, and even junk content. You have to know the bad to appreciate the good.

4. Begin with the end in mind:

Any human being can contribute to the design of a society that can meet the needs of current and future generations in a way that does not deteriorate the social and ecological systems; every one of us can contribute with what they are good at and enjoy doing.

A practical way to contribute to sustainable development, which can be performed along with any endeavor, profession or activity, is using the four principles of sustainability.

The principles guide you, provide a long-term vision and will also help, in a very practical way, your personal and professional lives. In order to assess the sustainability of an action, an event or an organization, you can ask the following questions:

- *Sustainability Principle 1*: Does it depend on fossil fuels? Does it use extracted materials that are scarce in nature? If so, are they kept in safe places, are they subject to strict technical control or is there a significant leak?
- *Sustainability Principle 2*: Does it depend on substances that are lasting and alien to nature? Are these chemicals kept in safe places and under strict technical control during their entire life cycle or is it probable that the substances will systematically increase in the biosphere? Does the emission of natural substances increase systematically due to the large emissions caused by society?
- *Sustainability Principle 3*: Does it depend on activities that are administered with no control? Does it depend on unnecessary ecological degradation? Does it depend on massive landfills or extreme, open-air mining, without restoration?
- *Sustainability Principle 4*: Are social costs covered along the entire value chain, including suppliers in other countries? Is there any abuse of power, economical, political or any other, along with the value chain?

5. Question the 6 W's + 3:

Who, what, how, where, when, and why are the six basic journalism questions. To work from the sustainable development approach, you can incorporate three questions and find support in the four principles of sustainability:

a) *Where? Source:* For example, what is the root cause? What is the history of the environmental and social problem? Why is it happening?
b) *Toward where? Implications:* Ask critical questions and analyze the event critically. For example, you can ask questions that at first glance may seem inappropriate or absurd, such as: Is development equal to high sales in the automotive industry? What are the possible consequences of this event? Are the environmental and social standards sustainable?
c) *Where are the connections? Systematic perspective:* For example, how is this case with others? What is the money circuit? Where and when does the case go back to? Build a memory that is able to refer to previous events.

6. Research beyond official versions:

"The pleasure of writing an article is to investigate," said Gabriel García Márquez. Documentation is essential, one must delve into issues.

The information coming from governments, international agreements, the United Nations or NGOs is important, nevertheless, go beyond these official versions. Obtain information from multiple sources that allow you to provide a context that includes economic, political and other perspectives (as well as that of citizens).

If you cannot always do it, tell your audience that the information provided does not include other viewpoints. Remember that things are not black and white in life, avoid falling into demonizing. Furthermore, remember that it is wise to rectify. If you make a mistake, it is preferable to correct it as clearly and as soon as possible.

7. Include and listen to everyone:

Look for diversity in your work team and your information sources. Diversity is enriching. Include everyone, people from any religion, culture, gender, sexual preference, political party, and so forth.

But who fits into our idea of "everyone"? It is important for journalism to guarantee that the meaning of the word "everyone" belongs to "everyone" and not everyone except indigenous groups, small farmers, analphabets, and children.

There are so many ways in which we exclude people; we find differences instead of finding what we have in common. How can we be so different and yet feel so much alike?

8. Encourage questioning and critical thinking:

We need to regain critical thinking and give feedback so that society can think on its own.

T. S. Eliot wrote in his poem, The Rock: "Where is the knowledge we have lost in information?" To explain, to present topics from different angles, to ask the audience questions and encourage it to share its opinion, and involve it from the root of problems are some options that you can consider. According to Winston Churchill, "People like to change, not be changed."

9. Encourage to reinvent the world:

"These times demand a new viewpoint: sustainable development and journalism challenged to tell the story of the future." Investigate and provoke possible solutions:

a) *Investigate:* There are countless stories related to progress or solutions around the world. Solutions often have a specific context, that is, they have been developed for very specific circumstances; however, you can provide new and interesting information to local citizens. Keep in mind people's problems and what is perceived as an obstacle to take action. Think globally and inform locally.

b) *Promote:* Antoine de Saint-Exupéry once said that if you want to build a ship, do not drum up people together to collect wood and do not assign them tasks and work, but rather teach them to long for the endless immensity of the sea.

"Nature has benefited from a period of 3700 million years of research and development. And given that level of investment, it makes sense to learn from it" says Michael Pawlyn, who firmly believes that studying how nature solves problems will provide many solutions.

An example is that in ecosystems, an organism's waste is nutrient for another; so if one has waste resources that are apparently not used, one should think what to add to the system to increase its value and not how to get rid of it.

It is considered that we receive 10,000 times more solar energy, in all its forms, than we actually need each year. Energy problems are not insurmountable; they are a challenge for our ingenuity.

This could be the most exciting period of innovation we have seen. "It is a privilege to be alive, to reinvent the world and to add ideas to the public agendas," Professor Vieira said, also sharing Gabriel García Márquez's views about journalism, "the best job in the world."

10. Tell stories:

Beyond sending messages of "be good or you will go to hell," what is needed is to promote actions and to find ingenious ways to communicate.

Put catastrophes aside. A large part of environmental and social narrative has used fear- and guilt-provoking languages and that has stalled us.

The only things that human beings really understand, the only things that they keep in their memories are stories. "Reality does not pass in front of our eyes like still nature, but like a story with dialogues, diseases, love, statistics, and speeches." Journalism was born to tell stories, however, a journalist or communicator is not a novelist; good reporting is not a branch of literature, although it could have the same intense language and the same ability of seduction that great literary texts have. Despite this, if you can, make your reports so interesting that the readers will burn their breakfast.

Some advice from the experts:

- Recognize which is the core idea, the essence or subconscious dominant purpose mobilizes the characters. Whenever possible, humanize or analyze the information. Convey the soul and face of the story.
- "Where I get bored of writing, I know that the reader will be getting bored."
- People's appearance must be described as well as their gestures and expressions.
- The beginning of the story must pledge that the rest of the story will be worthwhile. Good first sentences instantly catch the reader's attention and show them that the narrator knows something they ignore. "At the end of each line, there must be an element of suspense that will take readers to the next line."
- Help the audience draw its own conclusions; do not say "4," instead say "2+2." Audiences want to earn their bread but do not want to know they are doing it. We human beings want to fill in gaps, finish phrases, solve and deduce. Moreover, people will not take action because they have to, but because they want to.
- A story has a beginning, an end, a hero, a plot, a conflict, and solution. The hero is the audience.

11. Empower citizens:

Sometimes we do not remember, as a society, that we can redirect the world. If you have the opportunity, remind your audience.

For a long time, we held on to the notion that governments and experts were the ones who solved global issues; however, in the last few years, we have witnessed the great social changes that society can bring about.

i. On the other hand, to try to live in and build a sustainable society as individuals is similar to swimming against the current. To take individual actions is a great place to begin but terrible to stop it. It is necessary to take action as a society, empowering society to generate collective changes. The world's greatest social changes (like those initiated by Gandhi,) have three things in common:

ii. A great idea that benefits everybody is shared; that idea comes from the heart of the problem, even if it means changing the systems that do not want to be changed. "Common good allows us to engage in higher purposes while being responsible for each other."

    iii. The millions of people who have achieved extraordinary changes did not only say, "I will be more responsible," but also said, "let's work together until we solve the problem."

    iv. For example, "If we want alternative energy, we should not ask an oil-dependent administration to produce it, we should produce it ourselves.

    v. To take action and make real changes, including all kinds of people and not just protesters.

12. Make it personal:

People are more interested in matters closely related to their lives. "That could be me" is a powerful element to spark awareness.

There are times when people are more open to receiving messages: when marrying, moving houses, changing jobs, having children or retiring. People are more open to change in moments of transition because their habits and costumes are changing. Encourage people to reflect when they are already going through changes.

Also, take into account that you do not necessarily have to use the words "sustainable development" or "sustainability." To be honest, those words tend to draw people away instead of drawing them closer.

13. Provide context:

Give context to your audience, avoid assuming that your audience knows the background information or is knowledgeable about the topic you are discussing.

In order to achieve a clearer understanding, it is advisable to relate new or abstract ideas with something that is familiar to the audience. Offer information that connects the spectator with his surroundings.

Using analogies is highly recommended. For example, to come up with a city's water consumption in million liters, it might be a good idea to determine how many Olympic pools that number equals to.

14. Look for simplicity without reducing complexity:

To talk about sustainable development is complex, yet we can find ways to convey its complexity without being superficial; superficiality does not indicate better communication.

One of the factors that lead to superficiality is considering facts in an isolated manner. You can try to step away from facts, the more we widen our perspective, the better our understanding and the more complete our information will be (see consideration no 5).

Two examples of simplified communication that is not less complex are:

I.   Seen by more than 12 million people, the Story of Stuff video by Leonard (2012b) didactically explains the production and consumption system that we live in. The video was made in a simple way that did not reduce its complexity.

II. Using metaphors can be useful when explaining complex topics. For example, to talk about the principles of sustainability, one can think of a tree. Its trunk and branches represent the system's basic principles, such as technology. The basic principles are fixed and unchangeable; similarly, a tree's trunk and branches are relatively solid and constant. However, details constantly change depending on the situation, just as tree leaves grow, move and change colors in autumn.

Without forcing it, try to inject some humor into your work.

15. Go beyond text:

Why limit ourselves to texts when we can incorporate audio, video, images or hyperlinks? You can tell stories in multi-format by using videos, texts, photos, multimedia support, and on the web, Facebook, Twitter, or YouTube. Your involvement will be more active. Beware not to forget the people who do not have access to the web.

Research shows that communication that has a visual component can be much more effective than communication that lacks it. Studies show that people remember 10% of what they hear, 20% of what they read and 80% of what they see and do.

16. Stay close to nature:

Beyond knowing about all the studies that prove the benefits of being close to nature, explore (if you have not already done it) the way that closeness benefits you. You can have plants at home, grow fruits, and walk in a park or gaze at the sky, a tree, insects, and other animals. Observe nature, without thinking about it, just be aware of it; observe how each plant and animal is completely itself. Explore what this means to you and your work.

12.4 ROLE OF EMOTIONS

12.4.1 INTRODUCTION

One of the most debated topics in communication ethics these days revolves around the roles of reason and emotion in decision-making. More precisely, is it thoughtful reasoning that leads to meaningful attitude change or is it automatic intuition? And by extension, which has a greater impact on behavior? This debate may seem academic, but it is of vital interest to environmental communicators. Should green messages focus on strong rational arguments or is it more effective to emphasize under the radar emotional appeals?

While social scientists and philosophers must painstakingly build empirical and logical support for their theories and arguments, advertisers just go with what works. Emotion works. While peer-reviewed journal articles and books take months or years to come to fruition, marketers roll out and cancel campaigns on a dime. For these practical reasons, advertising professionals have been out in front using new methods and approaches for persuasion; researchers follow up with empirical evidence and theories that explain and predict why the new techniques are so effective.

While brands have been remarkably successful at feeding universal human drives, such as the desire for adventure, power or status, sustainability has not been seeing the same success in its messaging. What sustainability needs to create the same impact is a similar level of insight into the best way to embrace the full range of human emotions. Because it is a human emotion that is at the heart of what motivates us.

Effective neuroscience identifies four basic human drives aligned to a handful of neurochemicals in the brain. These four drives each have their own useful role to play in promoting human survival and out of these have emerged the full range of human emotions:

i.   **Contentment**: to minimize harm and probability of bodily destruction
ii. **Nurturance**: to facilitate familial and social bonding
iii. **Seeking**: to reward curiosity, survival abilities, achievement and excitement about achieving the desired goals, for example, food and stimulation
iv. **Assertiveness**: to overcome restrictions on freedom of action.

This insight presents a more balanced view of consumerism and how brands meet our basic needs. For example, a new mother is not buying extra baby products because of an outer-directed motivation to demonstrate her status in the world; she is buying them because of her innate drive toward the nurturance of her child. Rather than trying to change us, brands have worked out what makes us tick. They successfully appeal to our full range of drives, our sense of freedom, home, playfulness, power or sensuality.

12.4.2 FACTORS

Much of the psychological research examining the relationship between environmental attitudes and behavior uncover a value-action gap. Research suggests that while a majority of people might endorse pro-environmental beliefs, few would be willing to forego price, convenience, and ease in favor of a product's "greenness". However, support is also found for rationalist approaches, showing a consistency between environmental attitudes, identity, and some types of green purchases. For instance, in some studies, an aggregate measure of "environmental consciousness" is more predictive of green purchasing intentions than demographic or personality variables. In others, only particular kinds of pro-environmental beliefs (e.g., those about product packaging or labeling) appear to predict green consumerism but not engagement in recycling or other environment-friendly actions. This finding seems to suggest that green purchases may be a distinct type of pro-environmental behavior, one that may possess a separate set of antecedents from behaviors such as recycling, using public transportation, or participating in environmental activism.

Other cognitive variables, such as consumers' belief in the efficacy of their individual behavior on the environment, have been shown to reliably impact subsequent green consumerism. Interestingly, this appears to be the result of a more generalized internal locus of control (i.e., not particular to the environmental domain); for example, respondents who were less likely to believe in the role of luck or other external influences in their life were more willing to choose a more environment-friendly laundry detergent over a conventional one. Finally, researchers have found that the cognitive construct and hope impacts green consumerism. The rhetoric of global climate change can often consist, by necessity, of dire and ominous predictions. Unfortunately, this may have the (unintended) consequence of instilling a sense of fatalistic helplessness in some, leading to a decrease in environmental engagement. The antidote appears to be a sense of constructive optimism—believing that the future is positive and believing that one has a potential path to that future. This construction of hope combines cognitive (e.g., agency) and emotional (e.g., positive feeling) perspectives and motivates pro-environmental behavior, particularly among young adults—even when controlling for the types of values described in the previous section.

In addition to personal norms that people embody, social norms play a large part in encouraging green consumerism. Field experiments set in hotels used a variety of appeals to persuade guests to reuse their towels. Appeals that invoked social norms (e.g., "Join your fellow citizens in helping to save the environment.") were more successful than direct appeals (e.g., "Help the hotel save energy.") or ones centering on cooperation (e.g., "Partner with us to help save the environment."). Peer group behavior may also be an effective route to increasing the salience of green norms and hence, encouraging green consumerism. Social psychological research into conformity has shown that people often change their own behavior to adapt to normative standards set by one's social group. Similar processes appear to affect green consumerism. For instance, the adoption of solar photovoltaic cells is dependent on how pervasive this technology is in one's neighborhood; consumers who see their neighbors choosing to buy green products may be more likely to do the same.

Green consumerism appears to be influenced by social norms that push for conformity but recent evidence suggests that it can also serve as a signaling device for attaining social status or a pro-social reputation. "Conspicuous conservation" behaviors indicate to others that an actor is able and willing to incur personal costs (because green products sometimes entail a pecuniary premium—at least at the onset) for the betterment of society. This perspective suggests that green consumerism should be more likely in public rather than private settings, and it indeed appears that when the behavior is public (versus private), individuals are willing to pay more to uphold a common environmental resource and show a preference for green products relative to conventional ones. However, note that the need to demonstrate one's commitment to conservation action is only active if the consumer herself or the group she cares about believes that the environment is in need of saving. Sexton and Sexton recently demonstrated the interaction between environmental values and social signaling: the value of a green signal (e.g., the purchase of a distinctive hybrid vehicle) was several times greater in a city with demonstrable green values than in a comparable "brown" city.

On the one hand, environmental behaviors have the potential to create positive spillover effects, such as when the initial adoption of smaller green behaviors (e.g., the purchase of green consumer products) increases support for larger green projects further down the road (such as the adoption of wind energy). On the other hand, environmental behaviors can allow individuals to feel morally licensed and subsequently behave less pro-socially. For example, households that managed to conserve water subsequently increased their energy consumption, and recycling decreased people's likelihood of using reusable grocery bags. In another study by Mazar and Zhong, participants who were randomly assigned to purchase products from a green rather than a conventional store, subsequently, not only acted less altruistically but also more unethically to earn more money. However, hope is not lost as recent research has begun to identify the cases in which negative versus positive spillovers occur. In particular, focusing on a long-term commitment, highlighting identity or societal obligations, and providing psychological closure might be effective solutions to fight negative spillover effects and promoting continued pro-environmental behavior. For instance, the extent to which one focuses on self-motivated reasons to engage in pro-environmental behaviors may predict whether positive or negative spillover may occur. In addition, pro-environmental behaviors that are performed for self-transcending reasons rather than self-interested ones appear to increase other green behavior.

12.5 CONCLUSION

Since its origins, the Earth has gone through extraordinary changes, from changes of mass destruction to unimaginable evolutionary changes and the planet continues to transform. The human race will cease to exist or will become a more evolved species, yet today, here we are.

The 21st century began with a challenge that has resulted in the extinction of many species and continually threatens our own existence. This is a great opportunity to transcend.

A society's prosperity has been called sustainable development because it meets the needs of its current generations without compromising the needs of future generations. How to achieve this?

There are various theories to understand sustainable development; however, there are a few that provide a practical and generic enough definition that can be used by any person or organization; there are few that give a definition based on systematic thinking, resulting in a greater and effective impact on everything (the Earth). This chapter introduced the proposal of strategic sustainable development, which has the aforementioned characteristics.

Apart from the importance of understanding what sustainable development is, that is, which would be the minimum requirements for a sustainable society, we need tools, techniques, and practices that help us move toward a sustainable society. Undoubtedly, technological advances and innovations that arise daily may help us forward. In addition to this, if what we want is a change right from the root, we need people, we need society. For over 250 years, we have been building our society in a way that has harmed us, that was the direction we took as a society. We can now take a new direction, we can learn from what we have had to face and we can head in the direction of prosperity and social as well as environmental well-being. In these situations, shifting toward green consumerism is mandated.

A green behavior that is viewed as a conventional norm and/or becomes habitual rather than an effortful, personal sacrifice (i.e., a morally motivated action) may not earn moral credits or credentials and hence, reduces subsequent licensing. In other words, consumers might be more likely to engage in green consumerism if they think of it as a "conventional" behavior or if it becomes automatic. Massive population growth combined with an increasing demand for consumer goods suggests that developing economies are where researchers should focus future investigations in. Substantial environmental benefits could be reaped by modestly shifting consumption patterns away from conventional products to sustainably produced ones. Green consumerism is a culturally learned, context-dependent behavior and further research outside of traditional research populations could help develop strategies for matching green consumerism (and accompanying marketing communication) with specific contexts and cultures. A more inclusive approach to green consumer behaviors would also involve shifting the focus from an individual consumer as the decision-maker to a family, a community or other unit of actors. Belatedly, the conceptualization of green consumerism would need to be broadened in order to be inclusive of diverse social, economic, and ecological constraints.

12.6 FURTHER READINGS

Quote by Paul Hawken: environmentalist, entrepreneur, journalist, and writer. Since he was 20 years old, he has dedicated his life to sustainability and the changes in the relationship between companies and the environment. His practice has included creating and managing green businesses and he has written seven books about the impact of trade in nature. http://www.paulhawken.com/

Quote by Donella Meadows in Wheatley, Margaret J. 1992. Leadership and the New Science: Organization viewed from the XXI century. Ediciones Granica.

Quote from Claudia Vernés cited by Vieira, Geraldinho. 2012a. Video interview with Vieira Geraldinho. The future of Journalism: solutions beyond allegations. Fundación Nuevo Periodismo Iberoamericano. http://www.fnpi. org/actividades/2012/storify-periodismo-de-futuro/(Consultation: July 5th, 2012). Tolle, Eckhart. 2006. A new earth.

Research carried out by the psychologist Jerome Bruner from New York University, Paul Martin Lester, "Syntactic Theory of Visual Communication," California State University at Fullerton, 1994–1996. Original not consulted.

KEYWORDS

- **sustainable communication**
- **green marketing**
- **emotional appeal**
- **mass communication**
- **marketing sustainability**

REFERENCES

Antonetti, P.; Maklan, S. Feelings That Make a Difference: How Guilt and Pride Convince Consumers of the Effectiveness of Sustainable Consumption Choices. *J. Bus. Ethics* **2014**, *124*, 117–134.

Bacchetta, M. Relaxing Export Constraints: the Role of Governments. In *Trade Infrastructure and Economic Development*; Ajakaiye, O., T. A. Oyejide, Eds.; Routledge: London, 2012.

Balbus, J. M.; Boxall, A. B.; Fenske, R. A.; McKone, T. E.; Zeise, L. Implications of Global Climate Change for the Assessment and Management of Human Health Risks of Chemicals in the Natural Environment. *Environ. Toxicol. Chem.* **2013**, *32*, 62–78. http://dx.doi. org/10.1002/ etc.2046

Brundtland Commission. *Framing Sustainable Development;* The Brundtland Report (1987) – 20 Years On, 1987.

Buján, S. What Should be the Priorities of Latin America? *Environ. Responsib. semin. Journalistic Qual.* **2010**.

Byggeth, S. H. Integration of Sustainability Aspects in Product Development. Thesis for the degree of Licentiate of Engineering. Blekinge Technology Institute and Chalmers Technology University, Gothenburg: Sweden, 2001.

Calderon and Wendy. Electronic survey "Communicating for Environmental and Social Well-Being, 2012.

Cannon, Jannel and Stellaluna. Editorial Juventud: Spain, 1993.

Carreon and Areli. Electronic Survey "Communicating for Environmental and Social Well-Being. **2012**.

Cialdini, R. B. Crafting Normative Messages to Protect the Environment. *Curr. Dir. Psychol.* **2003**, *12*, 105–109.

Daly, H. E.; Farley, J. C. *Ecological Economics: Principles and Applications*, 2004.

Davis, J. J. The Effects of Message Framing on Response to Environmental Communications. *J. Mass. Commun. Q.* **1995**, *72*, 285–299.

De Marco, G. Environmental Journalism-Module 3, Task 1. Remote Master in Environmental Journalism. Environmental Training Institute: Madrid, Spain, 2011.

Devinney, T. Survey Applied to 10,000 Individuals. Make It Personal: How to Get People to Care About Sustainability, **2012**.

Ellis, R. *Task-Based Language Learning and Teaching*. Oxford: Oxford University Press, 2008.

Evans, L.; Maio, G. R.; Corner, A.; Hodgetts, C. J.; Ahmed, S.; Hahn, U. Self-Interest and Pro-Environmental Behaviour. *Nat. Clim. Change* **2013**, *3*, 122–125.

Fernandez-Checa, J. L. Electronic Survey "Communicating for Environmental and Social Well-Being, 2012.

Futerra Sustainability Communications. Sizzle, the New Climate Message: United Kingdom, 2010a.

Futerra Sustainability Communications. Branding Biodiversity: United Kingdom, 2010b.

Garcia Marquez, G. Manual for a Child (Extract from Volume 2 of the series "Mission Papers, Science, Education and Development: Education for Development", 1995.

Garrone, V., Jaime, M. Workshop on RSE/Sustainability–Fundamentals. Journalists Network for Sustainable Development. Externado University of Colombia and Colombian Business Council for Sustainable Development: Bogota, Colombia, 2012.

Granados, Miguel A. Miguel Angel Granados Chapa Interview. Journalism workshop 1. Communication Sciences Faculty, Universidad Vasco de Quiroga. Michoacan, Morelia, Mexico, 2010.

Goldman, P.; Rossana, R. Inside, on the Edge or Outside: Chronicles of Latin American Youth Diversity. Anfibio Reguillo Workshop. Workshop organized by FNPI-New Iberoamerican Journalism Foundation and UNSAM-National University of San Martín: Buenos Aires, Argentina. 2012.

Guillen, M. F. *The Limits of Convergence: Globalization and Organizational Change in Argentina, South Korea, and Spain*. Princeton University Press: Princeton, NJ, 2009.

de Bettignies, H. C.; Lepineux, F. *Finance for a Better World: the Shift Toward Sustainability*. Palgrave Macmillan: New York, 2009.

Hansen, J.; Rossow, W.; Carlson, B.; Lacis, A.; Travis, L.; Del Genio, A.; Fung, I.; Cairns, B.; Mishchenko, M.; Sato, M. Low-Cost Long-Term Monitoring of Global Climate Forcings and Feedbacks. *Clim. Change* **1995**, *31*, 247–271. DOI:10.1007/BF01095149.

IPCC Report. Climate Change: Impact, Adaptations and Vulnerability, Cambridge University Press, Cambridge, UK, 2007.

Jimenez, R. Electronic Survey Communicating for Environmental and Social Well-Being, 2012.

Kollmuss, A.; Agyeman, J. Mind the Gap: Why Do People Act Environmentally and What are the Barriers to Proenvironmental Behavior? *Environ. Educ. Res.* **2002**, *8*, 239–260.

Kunst, M.; Witlox, N. Communication and the Environment. *Commun. Res. Trends* **1993**, *13*(1) 1–31.

Larena, A. Environmental Journalism, Think Globally and Report Locally. Collection of Essays: Environmental Journalism-Risks and Opportunities in news coverage. Smashwords Inter American Press, 2012.

Larena, Arturo.; La Naturaleza del Periodismo Ambiental. Taller de Cobertura de Temas Ambienales. Fundación Nuevo Periodismo Iberoamericano. Caracas, Venezuela, 2006.

Lazarus, D. S. *A Green Battle for the Truth. In Environmental Journalism for the1990s: Held at Ranche House College, 6th-8th March 1990.* National Seminar Series Report, 1991.

Lee, A. Why Emotion Beats Reason in Green Marketing? 2013.

Leonard, A. The Story of Change–Why Citizens (not Shoppers) Hold the Key to a Better World. Berkeley, California, USA, 2012a.

Leonard, A. The Good Stuff—Episode 5: How You Show Up in the World- Podcast. The Story of Stuff Project, 2012b.

Marquez, B. G. Gabriel. Journalism is a dangerous profession. Three days in Barranquilla with Gabriel Garcia Marquez. La Nacion: Colombia, 1997.

Martinez, T. E. Presented at Journalism and Narrative: Challenges for the XXI century. Lecture delivered before the assembly of the SIP-Inter American Press: Guadalajara, México, 1997.

Max-Neef, M. A. *Human Scale Development: Conception, Application and Further Reflections.* The Apex Press: New York, 1991.

McKenzie-Mohr, D.; William, S. Fostering Sustainable Behavior. New Society Publishers: Gabriola Island, BC, Canada, 2008.

Meadows, D. H. *Thinking in Systems: a Primer.* Earthscan: London/Sterling, VA, 2009.

Meisel, A. More Deep Coverage. Workshop on Media and Sustainable Development in the Colombian Caribbean. Conference carried out by Promigas and FNPI-New Iberoamerican Journalism Foundation: Barranquilla, Colombia, 2012.

Nelkin, D. Selling Science: How the Press Covers Science and Technology (Rev.ed.). W. H. Freeman: New York, 1995.

Ny, H.; MacDonald, J. P.; Broman, G.; Yamamoto, R.; Robert, K.-H. Sustainability Constraints as System Boundaries: an Approach to Making Life-Cycle Management Strategic. *J. Ind. Ecol.* 2006, *10*(1–2), 61.

Pawlyn and Michael. Ted Ideas worth Spreading. Using nature's genius: Ted Salon London, 2010.

Peattie, K. Green Consumption: Behavior and Norms. *Annu. Rev. Environ. Resour.* 2010, *35*, 195–228.

Peri, J. L. Electronic Survey "Communicating for Environmental and Social Well-Being, 2012.

Rademakers and Lisa. Examining the Handbooks on Environmental Journalism: A Qualitative Document Analysis and response to the Literature. University of South Florida: Graduate School Theses and Dissertations, Paper 1207, 2004.

Robèrt J. B. Economic Growth in a Cross Section of Countries. *Q. J. Econ.* 1991, *106*(2), 407–443.

Robèrt, K.-H. Strategic Leardership towards Sustainability. Blekinge Tekniska Högskola: Karlskrona, Sweden, 2010.

Sachdeva, S.; Jordan, J.; Mazar, N. Green Consumerism: Moral Motivations to a Sustainable Future. *Sci. Dir. Curr. Opin. Psychol.* 2015, *6*, 60–65.

Sagan, C. *Pale Blue Dot: A Vision of the Human Future in Space.* Random House: New York, 1994. ISBN 0-679-43841-6.

Sahtouris, Elisabeth. Evolutionary Biologist. Thrive: What On Earth Will it Take? 2011.

Salcedo, A. The Narrative of Journalism, a Different Focus from Reality. *Conference carried out by Promigas and FNPI-Fundacion Nuevo Periodismo Iberoamericano*. Riohacha: Colombia, 2012.

Senge, P. M. *The Fifth Discipline: the Art and Practice of the Learning Organization*. Doubleday/Currency: New York, 2006.

Steffen, W.; Sanderson, A.; Tyson, P. D.; Jager, D.; Matson, P. M.; Moore, B., III; Oldfield, F.; Richardson, K.; Schnellnhuber, H. J.; Turner, B. L., II; Wasson, R. J. *Global Change and the Earth System: a Planet Under Pressure*. Springer: Berlin, New York, 2004.

Thøgersen, J.; Noblet, C. Does Green Consumerism Increase the Acceptance of Wind Power? *Energy Policy* **2012**, *51*, 854–862.

Uribe. Aliria. Electronic Survey "Communicating for Environmental and Social Well-Being, 2012."

Valenti, J.; Wilkins, L. An Ethical Risk Communication Protocol for Science and Mass Communication. *Public Understanding Sci.* **1995**, *4*, 177–194.

Vieira, Geraldihno. The Journalism of the Future, Seeking Solutions Beyond the Allegations. Workshop on Media and Sustainable Development in the Colombian Caribbean, 2012a.

Vieira, Geraldihno. Expert Sources and Visual Elements: Keys to Better Narratives. Workshop on Media and Sustainable Development in the Colombian Caribbean. *Conference carried out by Promigas and FNPI-Fundacion Nuevo Periodismo Iberoamericano.* Barranquilla, Colombia, 2012b.

Voisey, H; Church, C. *Media Coverage of Sustainable Development and Local Agenda 21*, CSERGE Working Paper, University of East Anglia: Norwich, 1999.

Wheatley, M. J.; Leadership and the New Science: Organization of the XXI Century Viewed from the Borders. Doubleday Publishing, 1992.

Wilkins, L Communicating and Environmental Journalism. In *Mixed news: The public/civic/ communitarian debate;* Black, J. Ed.; Lawrence Erlbaum Associates, Publishers: Mahwah, New Jersey, **1992**, pp 200–214.

WWF. Living Planet Report: Biodiversity, Biocapacity and Better Choices, 2012.

Young, W.; Hwang, K.; McDonald, S.; Oates, C. J. Sustainable Consumption: Green Consumer Behaviour When Purchasing Products. *Sustain Dev.* **2010**, *18*, 20–31.

PART III

Ecological Dimensions of Green Consumer Behavior

CHAPTER 13

ECO-AWARENESS: IMBIBING ENVIRONMENTAL VALUES IN CONSUMERS

ANJALI KAROL[1,*] and C. MASHOOD[2]

[1]*Institute for Financial Management and Research (IFMR), 24, Kothari Road, Nungambakkam, Chennai 600034, India,
Mob.: 9177170570*

[2]*SPI Global, 6th Floor, Block-9B, DLF-IT Park, Chennai 600089, India,
Mob.: 9447436386, E-mail: a.mashood@spi-global.com*

*Corresponding author. E-mail: anjali.k@ifmr.ac.in

13.1 INTRODUCTION TO ECO-AWARENESS

A classic moral dilemma is the conflict between biocentrism and anthropocentrism. Popularized in 2010 by Doctor Robert Lanza, biocentrism (through a book by the same name) (Lanza and Berman, 2010) puts nature at the center and treats every element of it with equality. It predisposes that humanity has direct and moral responsibilities toward its conservation. Anthropocentrism, on the other hand, perceives humans as the center and believes that environmental responsibilities unfold as reactions to safeguard human interests. It tries to evaluate everything in terms of human values. While the reception to biocentrism is mixed, anthropocentrism has been cited as a major reason for human exploitation of nature due to its inbuilt human supremacy. Supporters of anthropocentrism are often accused of placing sheer priority to human needs and neglecting the needs of nonhuman nature. Anthropocentrism views all other elements of nature as being dominated by humans and as mere facilitators for human living. For this reason, biocentrism treats anthropocentrism as anti-environment.

Taking an ideological digression, we neither need biocentrism nor anthropocentrism nor for that matter, any centrism. We need a system where every element of nature coexists without the need to dominate over one another. We do not need a system where there are exploiters and exploited, but need a system of cohabitation. Only then can we voice for the plausibility of conserving nature and its ecosystem at its best. What we need is thoughtful changes and responsiveness that will impart us awareness to regard nonhuman life-forms. This is where we need eco-awareness to play a role. Eco-awareness should create an understanding among people that all life forms have an equal stake in the resources of the earth and have equal rights to live.

Eco-awareness (also known as green awareness) tries to strike a balance between the two centrist views by continuous and conscious efforts to create a change in the mind-set of the people to adopt eco-friendly practices and to deviate from anthropocentrism and move toward sustainable development. Another fascinating perspective that explicates the need for awareness is that we cannot altogether avoid polluting activities as they create value (Kolstad, 2000). We should, therefore, ascertain the optimal amount of pollution and conservation. Only sustainable practices can ensure longevity and quality of human lives in particular and the environment in general.

To start with a simple understanding of eco-awareness, it can be defined as the awareness of ecology. The word ecology is derived from two Greek root words, *Oikos*, which means household or habitation, and *logia*, which means the study of. As we began to realize and accept the fact that the environment is one supreme structure of which all life forms are part of, the implications of the word ecology broadened. Ecology now connotes the branch of knowledge that studies organisms and their interaction with the environment. Awareness signifies the knowledge, realization, consciousness, and the ability for eclectic thinking. Eco-awareness can then be defined as a process of breeding a responsible and eco-friendly society with an adequate understanding of environmental issues to address the indispensable need for safeguarding the environment. In short, eco-awareness encourages humanity to protect, preserve, and respect the natural world despite their anthropogenic afflictions. Eco-awareness designed with the right perspective and knowledge support can ensure that the balance of the environment is maintained. Only then can our present and future generations endure on this planet contented, while preserving the fauna and flora in their natural habitats.

The chapter proceeds with a discussion on the history of eco-awareness, various perspectives of eco-awareness and then the need for eco-awareness. It goes on to explain how eco-literacy and eco-awareness are two sides of the same coin. The various ways and means to promote eco-awareness are discussed next. It is found that eco-awareness of a society depends on their geography, demography, socioeconomic culture and the levels of eco-literacy. Eco-awareness is aimed at inculcating environmental values and to creating eco-friendly mind-sets. We also present some success stories of companies and countries that used eco-awareness toward improving consumerism. The chapter ends with a discussion on the future prospects of eco-awareness.

13.1.1   HISTORY OF ECO-AWARENESS

Until the 19th century, the availability of natural resources was considered to be in plenty and people did not bother much about the environment and its importance on the sustenance of mankind. Exploitation of environment started with felling trees. Mankind benefitted from trees in several ways. It was used for the construction of dwellings, making animal-drawn vehicles, furniture, utensils and so forth. The ever-mounting human needs enhanced the demand for felled trees and such demand created a market value for trees than trees in a forest. This market value exacerbated the speed of felling trees. With the advent of the industrial revolution, feller benchers were invented to motorize the process of tree felling. The newly established and mushrooming factories required huge supplies of raw materials which were met by intensive mining, quarrying, and deforestation. This exploitation of natural resources gained meteoric speed as we started exploiting natural resources with the help of the ever-evolving advanced technologies.

On one hand, there was rampant exploitation to get the inputs for factories and on the other hand, the factories expelled industrial effluents polluting the water, air, and land severely damaging their quality (Eliasson, 2004). Anthropocentric behavior of mankind and market value of natural resources crimped environmental aspects, as a result of which natural resources became normal products like any other product in the market. Due to swift changes in the environment over the period of time, the mind-set of people witnessed a paradigm shift from threat to threatened. At the beginning of the 20th century, the attitude of people was that "nature poses threats to man" which underwent an antithesis view to become "nature is threatened by man" at the beginning of the 21st century.

In the later half of the 20th century, the concept of protection and conservation became a significant research area for historians (Eliasson, 2004). There began the importance of eco-awareness for the protection and preservation of natural resources and to cultivate respect toward nature. Vigorous attempts were made by organizations and countries across the world to impart eco-awareness, which pioneered in the 1970s with the observation of the World Environment Day, the observation of which was started on the first day of the United Nations Conference on the Human Environment in Stockholm in 1972 (United Nations, 1972).

These days, eco-awareness is the buzz word all over the world. It is a reiterating ideology that stimulates the necessity to safeguard the nature and natural resources in the minds of each human being. It is progressively used by individuals, organizations, corporate, and the government through different channels such as public notices, social media, media advertisement, classroom lecture, social drama and filming, observing environment weeks, campaigns, and so on. By disseminating awareness, effort is taken to alleviate the obstinate mind-set of human beings and to reinstate the indispensable need for safeguarding natural resources. Thereby, eco-awareness is the best way to a balance between biocentrism and anthropocentrism, which will bring a sustainable development for both man and nature.

13.1.2   NORMATIVE AND POSITIVE PERSPECTIVES

Eco-awareness can be looked at from both a positive as well as a normative perspective. In the positive view, eco-awareness attempts to bring about appreciable changes in the way people perceive the environment. It captures the present level of awareness and the cost-effectiveness of each program to create awareness. It tries to explain and understand how people have incorporated environment in their day-to-day living by doing daily chores in a more eco-friendly way and how they make decisions after considering the environmental externalities of their decisions. Eco-awareness works as an instrument to alleviate redundant exploitation of natural resources.

Eco-awareness also has an inbuilt normative aspect, as it prescribes eco-friendly and nature-oriented living as ideal. It encourages the public to engage in eco-friendly practices, purchasing green products, developing a green culture and to erase the gap between man and nature as clashing binary opposites.

13.1.3   ISSUE SPECIFIC VERSUS GENERAL AWARENESS

Environmental awareness can be imparted in two broad ways. Traditionally, environmental activists select a specific issue that needs attention like oil drilling, soil erosion, landfill pollution, encroachment and construction in eco-sensitive zones, effect of industrial pollution on water and air quality, decongestion of cities to improve air quality, protection of heritage sites, rivers and forests to protect flora and fauna in its original habitat, conservation of endangered species, protection of rights of tribes and their eco-system, and so on. Groups of people (usually headed by a leader) organize to guard a single issue that bothers them and start a campaign to garner people's attention to the cause and to make them aware of all plausible environmental hazards that result if the issue remains unattended. The main aim of such awareness programs is to emotionally attack the conscience of various stakeholders to generate changes in the way they perceive the problem. After knowledge is created about the issue, campaigners proceed to both evince and concomitantly extract from people pragmatic, compassionate, feasible and innovative ideas to solve problems. Various remedial actions and projects are undertaken to either force stop the issue or to mitigate the effects of the problem. The mediums used to impart issue-specific awareness are issue-based and, hence, vary from issue to issue.

In contrast to issue-specific awareness, general environmental awareness is akin to environmental education where environmental issues of global and everyday significance are studied, compiled and awareness on them is imparted to family, friends, colleagues, and the society. The mediums used to impart general eco-awareness are films, posters, observation of environment day and the likes, environmental education in school curriculum, environmental compliance classes and certifications for employees, and so forth. Though we cover both issue-specific awareness and general awareness under the umbrella of eco-awareness, usually eco-awareness is issue specific and campaign-based. General awareness is achieved through environmental education.

13.2   NEED FOR ECO-AWARENESS

The need for eco-awareness arises primarily from the realization that natural resources are being depleted at a faster rate than the rate at which they are being or can be replenished. There exists a pressing need to address environmental problems such as resource depletion, pollution, wastage, unsustainable development, and extinction of vulnerable species. Eco-awareness creates understanding about environmental problems and heightens the scope of environmental education as a separate discipline to study the various facets of man-environment interdependence and conflicts. It tries to bridge the gap between biocentrism and anthropocentrism and aims for sustainable development to protect the interests of both humanity and nature. Moreover, it is essential to preserve the environment for cultural, aesthetic, biological, geographic and economic significance. Environmentally aware citizens possess basic knowledge and skills to understand and solve environmental issues at the grassroot level.

Eco-awareness facilitates the voluntary shift of mindset toward eco-friendly measures. It gives people an opportunity to understand the man-nature conflict and to better take up environmental protection voluntarily. Since the decision is purely within the self, it gives them more utility even if they have to give up some activities and products, which were very much, part of their lives or to pay a premium to use green products and services. Moreover, it presents them an opportunity to strengthen the relationship between people, business, institutions, local environmental groups, and nature by helping to realize their interdependencies. Eco-awareness helps in moving from conflict toward a partnership with the environment. Awareness works as a qualitative instrument to help normalize the maddening overuse of natural resources.

13.2.1 SUSTAINABLE DEVELOPMENT

Sustainable development is an integrated approach for the socioeconomic modernization and advancement of a society by both efficient utilization and minimal depletion of natural resources. It is a development that causes minimum negative externalities to the environment. The word sustainable development holds two key features. One, it takes an effort to sustain natural resources. Two, it tries to augment socioeconomic development by making more resources available by way of improving efficiency. By sustaining natural resources, we mean to protect them from undesirable exploitation. It calls for a judicious use of resources so that human activities become less destructive to the balance of the natural system.

Sustainable development principally influences three stakeholders—nature, consumers, and the government. From nature's stance, sustainable development is a propitious practice that protects the environment by attenuating exploitation and restituting renewable natural resources. Consumers are a complex group that includes different socio-politico-economic classes. While some groups are prosperous, some face complex challenges like illiteracy, extreme poverty, chronic hunger, and abysmal nutrition either individually or collectively as a society. As a result, sustainable development is a very complicated objective for consumers. These challenges act as obstacles in front of consumers as not all classes are aware and capable to expend their money and efforts toward the attainment of sustainable development. In addition, a lack of perceivable, immediate, and direct benefits for the personal contributions stultify the motives of the people willing to contribute.

The role of government emanates at this juncture as an institution to balance the needs of people as consumers of natural resources and nature as the provider. The government has to ensure that its development policies and practices cater to the needs of individual consumers, the society at large and the nature by economizing the usage of available resources, preventing unchecked extraction of resources, and minimalizing pollution. The government must also try and ensure that any new policy that calls for environmental protection does not provide much hardship to the people. In addition, policies must be designed considering all probable environmental impacts. With climate change, global warming, changing lifestyle, and land use patterns, governments worldwide have realized that the existence and development of mankind are only possible if the environmental balance is maintained. As a result of this, several green pricing and green tax initiatives have been imposed by the government to preserve natural resources. The income flow generated from these schemes is used for the development of eco-friendly technology like reusable and recyclable materials, afforestation drives, and other conservation programs, thus aiming for sustainability in the long run.

13.3 ECO-AWARENESS AND ECO-LITERACY AS TWO SIDES OF THE SAME COIN

The word eco-literacy was initially used by American educator David W. Orr and physicist Fritjof Capra in the 1990s to mean an ability of an individual or a group of people to understand how the nature functions to make life possible on the earth (Stone and Barlow, 2009). Eco-literacy is considered as a third intelligence apart from social and emotional intelligence as an understanding and empathy toward all forms of life. In terms of their intention, both eco-literacy and eco-awareness are identical and their ultimate focus is to educate and lead mankind toward the conservation of earth. Though they differ in terms of the way both are practiced (McBride et al., 2013), often the procedures overlap. Hence, eco-awareness and eco-literacy are seen as two sides of the same coin complementing each other.

Eco-awareness is a broader and flexible method that has a wider scope. It is more of a cyclical process in the sense that awareness leads to more awareness. For example, green pricing programs of electricity boards, eco-friendly vehicles, and innovation toward existing green products all leave a strand for developing awareness about environmental issues, and their success, in turn, is also determined by the existing level of eco-awareness.

Eco-literacy is a constructive method that operates on the idea that humans should be educated about their impact on the environment if they have to coexist sustainably. Educational institutions typically conduct lectures to stimulate knowledge about how the environment works and the issues to mankind, other life forms, and the nature as a whole. Environmental science is now a compulsory subject the world over for children until high school. Eco-literacy aims to make people respect the environment, reduce harming it, and alter daily actions to minimally exploit nature. Considering the instructional angle, eco-literacy is emphasized in schools whereas eco-awareness is given more attention by non-profit organizations, corporate, and governmental departments for having a wider scope and flexibility. In a nutshell, eco-literacy enhances our knowledge of the environment whereas eco-awareness makes us aware and vigilant about the environmental issues.

13.4   METHODS TO PROMOTE ECO-AWARENESS

Different societies have unique and different belief systems, values, and notions of nature which is often the main reason for conflicting environmental views. Due to this very reason, eco-awareness has to be propagated using diverse techniques. By awareness, we mean gradual and lasting inculcation of environmental values to ensure that people protect and safeguard the environment. Since attitude change can take a long time, favorable results arise only in the long run. Numerous initiatives are carried out by governments, nongovernmental organizations (NGOs), international organizations and naturalists for the promotion of eco-awareness. The popular methods to promote eco-awareness are discussed below.

> Pamphlets/posters: Pamphlets with facts and messages to protect and conserve natural resources can be distributed in malls, supermarkets, streets, schools, parks, and other places where people gather. Low cost and eco-friendly practices that can save the environment, alarming facts about the environment and environmental dos and don'ts can be illustrated with diagrams and presented in colorful posters in public spaces to capture people's attention to eco-friendly means.

Green club: Green club is an emerging and trendy concept in the field of voluntary organization. If we can find some time in our normal living, then it is a progressive method toward the environment by organizing a green club. In many countries, it does not require any legal registrations. All we need is to write down the name and objectives of the club on a plain paper and then distribute copies among the public. At regular intervals, arrange meetings and discuss the environmental issues in a particular locality and take initiatives to solve such issues. Likewise, conducting eco-related programs like painting competitions, green exhibitions, debates (man versus nature) and speech contests on such issues can auspiciously create awareness in our society.

Women forum: Even today, in most countries working women are considered a taboo and men are the sole livelihood earners. Attempts from various angles to empower women to step out of their homes and create livelihood opportunities have led to the formation of women's forums. In some countries, these forums are called as self-help groups (SHGs). In India, SHGs are groups of women from homogenous backgrounds who try to enhance skills, develop capabilities and earn a livelihood. These SHGs can be used to spread awareness among members on social, environmental, health and economic issues, considering their penetration into women from the lower strata of the society. These SHGs have brought in positive environmental results in their localities by way of educating women on the need to preserve water, adopt eco-friendly fuels such as biofuels, rainwater harvesting methods, and so forth. This method of creating awareness is unique because it helps in women empowerment along with solving local environmental issues.

Religious podium: Most countries in the world follow a particular religion and its followers take up the advice and orders of their religious leaders with high revere and importance. The domination of a church, temple or a mosque on its people can be utilized as a good platform to uplift eco-awareness levels in a region by discussing environmental issues through their forums. Most religions have the holy river, the holy city, and the holy tree concepts. By religious mandates, these holy places should not be littered and every effort is taken by the followers to preserve their natural stature. This is one way in which religion contributes to controlling environmental pollution and preservation of the natural habitat.

Green blog and social media: The younger generations are technologically advanced and progressively aware of the use of technology and internet-based communication. Social media and microblogging platforms like Facebook, Twitter, Instagram, Pinterest, and their likes have played a crucial role in the information revolution providing a platform for people to organize opinion and develop public perspective and sentiments. Anyone can easily create a green blog at no cost and can freely express their views on various environmental issues. Online environmental organizations and cyber green communities have mushroomed everywhere connecting individuals in remote locations and bringing up environmental issues in one part of the world known to all around the world. Social media can be used to facilitate communication on environmental topics worldwide through individual and group conversations, sharing videos, live telecasts and live updates on natural disasters, disaster relief programs, green movements, and so forth. Reputed environmental organizations like Greenpeace International uses social media to mobilize support for its environmental battles by e-signing petitions. A recent and very impactful e-petition was one against palm oil production in Malaysia which clears rainforests and natural habitats of many endangered species of orangutan.

Films: Films are audiovisual mediums that can be used to promote eco-awareness. Films have the magic to deliver powerful messages to people, irrespective of their religion, geography, demography, literacy, and culture due to the inbuilt entertainment factor that they have. An influential art, films have proved to promote fast and effective awareness by capturing the reality of environmental tragedies and building imaginations of the possible harmful effects of global warming. The movie "Day after tomorrow" shows how terrifying global warming can be by depicting a new ice age and how the whole world is submerged under water. The movie "Dam 999," a science fiction disaster movie, which depicts the aftereffects of a dam break, created a lot of commotion and a series of dam-protection movements in India for safeguarding Mullaperiyar Dam that was in environmental scanning for issues related to its strength and the safety of people living under its valley. Similarly, several documentaries have been made to bring up issues, such as river pollution, industrial waste management, safe drinking water availability, protection of endangered species, deforestation, ozone layer depletion etc., to the mainstream media.

Mobile networks: Telecommunication companies have easy pathways to reach out to each individual and society through their mobile networks. Mobile companies can use their networks to spread awareness messages. They can periodically send eco-friendly practices as messages to their customers, which if practiced by customers have the potential to create an eco-revolution. They can offer special discounts on environmental days and eco-awareness weeks for bulk recharges to reduce transaction costs. Additionally, NGOs and governments can partner with mobile companies to send awareness messages to the public on environmental issues of present relevance. Ring-back tones can have messages on climate change, pollution as well as conservation of nature and endangered species, and so forth. The frequent reminding and information sharing with the available means of telecommunication networks can stimulate a behavioral change in the society.

Digitalization: Promotion of digitalization of bill payments, bookings, recharges and banking transactions have been possible only by spreading awareness on the need to reduce the use of paper to save trees. Awareness messages to save trees, how much paper is wasted in printing a bill, the harmful effects of ink dye used to print, discounts for going digital are all printed on bills to create awareness among existing customers to go paperless. This way, each time a customer uses paper transactions he is made guilty of harming the environment; awareness is created to make more customers shift to digitalization.

13.5   ECO-AWARENESS AND NATURE OF THE SOCIETY

The world is heterogeneous in terms of geography, religion, language, race, culture, socioeconomic conditions, attitudes, and so on. In the psychological perspective, a child's behavior is primarily influenced by their heredity, microsystem, and mesosystem. Heredity is the coherent behavior of a child that is influenced from the antenatal period. This behavior roots from the behavior of the parents, grandparents, other direct blood relations, and ancestors (genetic behavior). This behavior is very difficult to change. The microsystem is any distinct group of people an individual encounters in his life. Each of these groups creates a society or a community such as family, peer groups, schoolmates, and friends. The mesosystem is a blend of two microsystems, which can mold a new behavior in an individual. For example, the behavior of an individual varies based on the combination of the nature of the school and the friends they mingle with. The mesosystem may influence to create a new style of behavior in the child which may be very different from heredity characteristics. These factors interact and together determine the perspectives that prevail in a particular society. At this note, eco-awareness has no one formula that fits all and should be meticulously planned, keeping in mind the background of the targeted population.

Eco-awareness and geographical differences: The need for eco-awareness varies with geography as geography determines the level of exploitation of natural resources. People from hilly and tribal regions have their culture rooted in conservatism and protection of resources rather than exploitation. The goddesses of forest and tree gods among both African and Indian tribes attach sanctity to the environment and ensure conservation devotedly. Contrary to this, due to the predominance of anthropocentric mind-set, most societies ruthlessly exploit natural resources for the fulfillment of their personal requisites and desires. People began to construct concrete forests instead of rain forests. Harju-Autti and Kokkinen (2014) developed an Environmental Awareness Index (EAI) for 57 countries to enable cross-country comparison of environmental awareness. According to EAI, the best performers are all European countries such as Austria, Sweden, Finland, Germany, and Denmark. Outside Europe, Japan, New Zealand, and Canada fared well. South Asian and African countries performed worst in the list. So, eco-awareness has multiple and varying perspectives that vary with the geographical location which insists the need to have appropriate environmental awareness programs befitting each geographical location.

Eco-awareness and socioeconomic conditions: Socioeconomic factors play a crucial role in the living standards of a region. Lifestyle, culture, and attitude of the society are naturally influenced by their socioeconomic conditions. Criminal experts have pointed out that recidivism is highly visible in regions where the socioeconomic condition is poor. Factors such as income, education, health, and environment are the major determinants of the socioeconomic status. The quality of the socioeconomic culture vastly influences the implementation of eco-awareness programs. Eco-awareness is very simple and can be implemented in numerous ways in developed countries. But, it is very difficult to implement it in regions of poor socioeconomic conditions. Eco-tourism, environmental camps, green taxes, and green pricing are effective methods of eco-awareness in developed countries. Most developed countries have a devoted eco-tourism fund to encourage people to take part in eco-trips and leisure activities that provide direct economic benefits for conservation, livelihood to local people, minimize the impact on the environment, and inculcate environmental values in travelers. However, this method cannot create an ecological mind-set in regions with poor socioeconomic conditions because they lack the adequate education to understand the need for eco-awareness. Thus, each social class requires different methods as per their level of socioeconomic development. Classroom lectures, social drama, films, and regional campaigning are effective among the less developed societies.

Eco-awareness and demography: The key demographic factors are birth rate, age, sex, marital status, income, and occupation. Changes in these demographic factors ultimately lead to changes in the nature of the society. The appropriateness of various eco-awareness methods depends upon the nature of the society. For the attainment of favorable results, demographic differences have to be considered while designing eco-awareness programs. There are huge differences between demographics of countries and hence, eco-awareness practices should be designed in such a way that they are compatible to the demographic factors that exist in each country so that they become digestible to the targeted people. As of 2012, 16.2%

of the total population is old in the United States (65 years plus) whereas the same figure is less than 8% in India (60 years plus). Between 2000 and 2010, the United States witnessed 21% growth in people of this age group while India had a massive 35% growth in the similar age group. Sex ratio is 0.97 in the United States while it is only 1.06 in India. Eco-clubs can be run effectively among old-age communities and university students (Abbas and Singh, 2012) but not for the middle-aged groups because they may not get productive time off their work responsibilities. Children are given eco-awareness with a recreational tint, such as eco-painting, eco-quizzes, outdoor games, trekking trips with small games or by visits to eco-spots. It is to be noted that the demographic factors may change according to the passage of time. It creates several hitches to abolish the present prototype of eco-awareness practices and implement new practices. All these examples denote the significant role of demography in the fruitful implementation of eco-awareness programs.

13.6   ECO-AWARENESS AND CONSUMERS

The world has witnessed a radical change in the perspective of consumers. The consumer is the king of the market and companies are working endlessly to satisfy the burgeoning and evolving tastes and preferences of the consumers. A paradigm shift occurred from needs satisfaction to want satisfaction of consumers, thus instigating marketers to turn to consumerism (which means the ultimate aim of a product or service is to promote the interest of consumers irrespective of all other factors). Companies started using iniquitous means to satisfy consumers by ruthless exploitation of natural resources and manufacturing products that are harmful to nature just to satisfy the wants of the consumers. This dominance of consumerism fueled tormenting the nature. Only if consumers have proper eco-awareness can environmental distress be solved.

Consumer behavior is a key factor that influences product development. Changes in the purchasing power and consumer needs alter consumer behavior and consequently, consumer satisfaction and consecutively impacts the nature. Eco-awareness practices have a vital role to play to make changes in the consumer behavior and attitudes. Most countries wield eco-awareness with the help of green organizations as well as business firms or companies. A study conducted by Martinez et al. (2015) found a positive effect of regulative, normative, and cognitive dimensions of the institutional environment in shaping pro-environmental attitudes such as eco-friendliness of consumers. The promotion of green practices like green purchasing, digitalization, and efforts to reuse and recycle products can develop eco-awareness in consumers. Green purchasing is an emotional practice where consumers make purchase decisions keeping in mind human health and minimal environmental damage. Digitalization initiatives have been very successful in the recent years with more institutions going for paperless offices and online service delivery. Bank of America reduced the usage of paper from 32 to 24% during 2000–2005 as well as recycled 30,000 t of paper every year saving roughly 200,000 trees every year.

Recycling efforts, such as the case of Nokia, have to be lauded for uplifting consumer eco-consciousness. Nokia introduced e-waste management in the year 2008 in India. In the first phase, the company set up drop boxes in different priority Nokia care centers to take back the used mobile phones, accessories, chargers, and so on, irrespective of their brands. Further, they announced and implemented a massive campaign in the year 2009 in four major cities of the country with mass advertisements in all the major newspapers and cited 600 articles in different newspapers. The total collection rose from 3 t in the year 2009 to the 65 t in the year 2012. Since the launch of this campaign, Nokia has collected 160 t of e-wastes. It shows the difference in eco-awareness among the consumers with mass institutional campaign.

13.6.1   ECO-AWARENESS AND PURCHASE DECISIONS OF CONSUMERS

The purchase decisions of consumers are influenced by factors like product quality, price, socioeconomic conditions, brand value, peer recommendations, and customer reviews. Eco-awareness strives to restructure the stereotype purchase habits and decisions of the consumers. When a consumer extends his purchase decisions with the scope of eco-awareness, brand image and price of the product are the key indicators that lead to purchasing decisions. Based on these key factors, government and NGOs have to implement certain policies to promote eco-friendly products that specifically enrich the consumers' brand loyalty.

Recently, numerous eco-friendly products have come to the market with unique features and cost efficiency. Reusable products, energy efficient lightings, and promotion of solar energy are the key examples of the innovation in green products. These types of products predominantly protect the interest of the anthropocentric behavior of consumers by way of cost efficiency and utility. This journey of building eco-friendly consumers starts by creating awareness about the greenness of the products and culminates at ensuring brand loyalty to green products, wherein consumers voluntarily shift from brown products to green alternatives. Ideal purchase decision process has been shown in Figure 13.1.

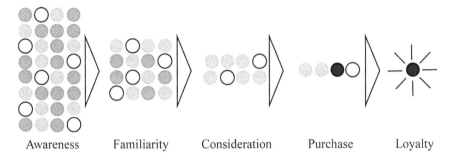

Awareness   Familiarity   Consideration   Purchase   Loyalty

FIGURE 13.1   Ideal purchase decision process.

Step 1—Awareness: In this step, eco-awareness has a tremendous role to attract consumers to green products. This step improves eco-literacy and attitude of each individual in the society to become responsible toward nature by choosing products that have lower eco-footprints and lesser harmful environmental and health effects. For example, the rise in the demand for organic crops is expected to bring about a complimentary fall in the consumption of brown products and, thereby, protect people from the avoidable health issues of harmful pesticides and antibiotics.

Step 2—Familiarity: The awareness on eco-products or green products stimulates the familiarity about such products and stamps its brand in the mind of the consumers and helps them to be aware of the benefits to them as well as the earth. Eco-products are favorable to both our health and nature.

Step 3—Consideration: Both the above steps help consumers to set green brands as preferable brands and choose affordable products among them in their purchase decisions. A consumer should have different alternatives at this stage of consideration.

Step 4—Purchase: At this step, a consumer prefers the best alternative that provides maximum satisfaction and buys that product. Like brown products, consumers will neglect and discard green products unless they are satisfied with the products in terms of their quality and utility. Hence, this stage is very crucial.

Step 5—Loyalty: Loyalty is the cognitive behavior or a mind-set that spurs the consumers to buy a specific brand and reiterate their purchase decisions. The reiteration occurs due to their enhanced consumer satisfaction from the first purchase and stability of satisfaction in the subsequent purchases. Creation of loyalty is the ultimate aim of the eco-awareness toward consumers.

13.7 ECO-AWARENESS AND CHANGE

Change is an inexorable phenomenon. The entry of different factors such as technology, globalization, and industrial revolution pressed the people to change their lifestyle, culture, and perspectives which went against the environment. These dangerous changes or unhealthy tendencies were the specific reasons to raise the slogans of eco-awareness. Eco-awareness gives attention to and implements distinct programs to rethink these changing practices that take people away from the environment and urge them to protect the eco-system. The Figure 13.2 shows the change process of eco-awareness.

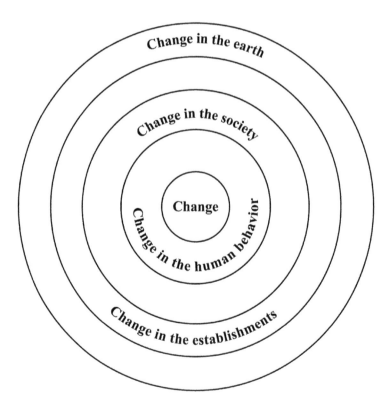

FIGURE 13.2   Eco-awareness change process.

Change in the human behavior: If a society has to achieve environmental compliance and go green, individuals must first transform. From the standpoint of an individual, understanding the threats and changes in the earth are the key elements of eco-awareness. These threats and changes potentially inhibit the options available for eco-awareness. In the first stage, eco-awareness gives attention to change the mind-set of people and stimulates their awareness (Takala, 1991). In the second stage, it favorably influences the people with respect to the level of their awareness and then redirects efforts. For example, power boards provide energy-saving messages to their consumers hoping that consumers would economize on power consumption and prevent wasteful consumption of power. But, many consumers do not heed these messages and keep wasting hard produced power due to lack of proper awareness and irresponsible attitude. Hence, a change of human behavior is the first and foremost objective of eco-awareness.

Change in the society: From changes in individual attitudes and practices come the changes in the society. The change in a society is also highly influenced by the eco-awareness of the social leader. The lack of awareness on the part of the leader may lead to laymen straying from the right paths of eco-awareness. The concepts of plastic-free villages (eco-village, eco-parks), pollution-free townships, tobacco-free states, and voluntary movements against pollution are the best examples of eco-friendly changes visible in the society. Conservation programs like afforestation, water-saving techniques, energy-saving technology, clean city as well as green city programs take place as a part of these eco-awareness practices. In this phase, eco-awareness is an extension of behavioral and cultural changes in the society.

Change in the establishments: Institutions such as government and NGOs have a key role in eco-awareness. These establishments are the pioneers of eco-awareness movements and have positively changed the work environment, work culture, and industrial relations of commercial and governmental establishments (Martinez et al., 2015). Before eco-awareness emerged as an important tool for social responsibility, most business organizations primarily gave attention to profits than any other social objectives. Due to that, overemphasis on profit, unchecked exploitation of resources, and polluting the environment were not considered seriously. After the advent of eco-awareness in mainstream business, drastic changes have taken place in industrial relations and work environment of companies. The stringent industrial laws are also a significant reason for these changes. Gaining inspiration from eco-awareness, international online retailor eBay built a business plan with eco-initiatives through their online retail and auction site by providing an option for easy reuse of the goods instead of disposing them off. As a result, the life span of many goods expanded and reduced wastages and pollution. Hence, awareness performs as a tool in the business and nonbusiness organizations for pollution control, sustainable development, and minimal utilization of natural resources. These eco-practices remind that eco-awareness can make enlightened changes in each of these areas.

Change in the environment: The environment has its own rejuvenation process that can rectify damages but once its threshold is surpassed, the environment expels its repercussions by way of global warming, climate change, ozone depletion, and so forth. To combat this, green reformist practices of individuals, society, and establishments can make a progressive change in the earth. In many countries, environment and forest-related departments and nature lovers take efforts to restitute the tree cover by undertaking afforestation in drought-prone areas, investing in watershed programs, expanding the mantra of the three Rs (reduce, reuse and recycle) to more sectors, and so forth. These efforts have made the earth a lot greener. Likewise, present endeavors of individuals, society, and the establishments will progressively revitalize the environment in the forthcoming days than ever before.

13.8   CASES: SUCCESS STORIES OF ECO-AWARENESS

Radical and alarming changes in the environment augmented the attention of individuals, companies, and governments toward the environment. Green practices such as the development of green products, green innovation, green pricing, green energy, and so on have certain limitations for widespread implementation across all types of organizations due to the lack of adequate coverage and infrastructure. This enhances the scope of eco-awareness to be promoted as a part of the total quality management of organizations. Hence, based on their level of accessibility to people and availability of funds, companies, NGOs, and state-sponsored organizations can focus on suitable eco-awareness practices. Four cases of business and nonbusiness organizations that were successful in their efforts toward eco-awareness are presented here.

13.8.1   ENVIRONMENTAL MANAGEMENT SYSTEM (EMS)-TOYOTA

Toyota's EMS is a comprehensive practice toward eco-awareness and eco-friendly production techniques. Whenever there is a slump in demand, the company suspends production. This is when they enable EMS to find and implement improvements in the current production processes and to find new opportunities to reduce costs. EMS primarily educates employees by providing a framework in place to help employees measure performance targets and identify areas that require improvement. Considering the fact that steel consumption involves large amounts of carbon footprints from its mining phase to the car-manufacturing phase, Toyota cars are built using a process that gets improved constantly to ensure minimum steel usage and wastage. Each of their plants has rainwater-harvesting ponds that cater to a large number of their water needs. Water at its plants gets recycled and 60% gets reused back in production. This framework helped the company improve its eco-footprints and step closer to the environment each time. The skill, creativity, and experience of the employees find pathways to lessen the use of water, energy, and thereby, emissions in the manufacturing units as well as (by their cars) on road.

13.8.2   GREEN COMPANY PROGRAM-ADIDAS

Assuming environmental responsibility, Adidas launched the Green Company Program in the year 2008. Alarmed by the ecological impact of their own operations, Adidas set ambitious targets for continuous improvement at their manufacturing sites, distribution centers, and administration offices to reduce internal environmental footprints. These targets were prescribed by engaging employees in eco-awareness programs and promoting several green practices. The company achieved several positive changes within the targeted year 2015, such as water savings of 28% per employee, reduction of carbon emission by 30% per square meter, and employee household waste volume reduction by 27% to name a few. With the success of the first stage, the management developed the second generation of Green Company Program at the end of the first target year 2015. As per this program, the company extends the scope from environmental stewardship to numerous areas. It includes retail stores of the company, effort to phasing out the use of plastic bags and adopting scientific ways to reduce carbon emissions (annual emission reduction designed to deliver neutrality of the company's own operations). Today, Adidas environmental data management and reporting system covers 80% of its global operations.

13.8.3 ENVIRONMENTAL AWARENESS STRATEGY-CORK COUNTY COUNCIL

Cork county council is the local authority of the county of Cork, Ireland. They introduced an environmental awareness strategy for the target period of 2010–2015. It was an aspiring strategy to promote eco-awareness in all the communities irrespective of the regions of the cork county. This strategy aimed to disseminate environment awareness to the general public. It has under its leeway, the industry, youth, community groups, educators, NGOs, and the media. The council spreads eco-awareness by behavioral change programs, community activities, and environmental discussions. The five-year strategy outlined key areas such as litter pollution, waste prevention, biodiversity and climate change, water conservation and water pollution, and research (Cork County Council, 2010). Special programs catering to these key areas were implemented in coordination with other councils. This collaboration helped to build comprehensive plans and policies such as the environmental directorate's operational plan, waste management, and climate change strategy. The fruitful implementation of these programs forged a new relationship between the society and the government for a collective environment partnership. As a result, the citizens became environmentally educated, eco-aware as well as eco-conscious, and they started protecting their environment. There were notable improvements in the environmental quality in terms of air, land, and water.

13.8.4 WORK WITH MEDIA AND EDUCATING COMMUNITY-UNITED NATION'S ENVIRONMENTAL PROGRAM (UNEP)

Recognizing the role of media to bring about a behavioral change, the UNEP introduced eco-awareness programs making active usage of media support. The media has the potential to educate and persuade the public by sensitizing eco-related issues. In the case of Bulgaria, UNEP worked with the government to devise policies to protect Black Sea marine environment from land pollutants with the active participation of media houses of Bulgaria. Mass media such as newspapers and television took to environmental reporting and spreading environmental dos and don'ts. It was found that educating the media is an effective practice in building capacity to report on eco-awareness matters as they tumble over to the audience as gems of wisdom on the awareness on environmental protection in general and Black Sea marine environment in particular.

Likewise, educating the community and cultural leaders is another strategy of the eco-awareness program that is undertaken by UNEP. In every community, there will be specific iconic personalities who immensely influence individuals of that community. They may be religious, cultural or community leaders as per the nature of the society. Such leaders have a high command, especially in the rural areas of the country. Providing environmental education to such leaders helped to make a radical change in the eco-attitude of the respective societies. UNEP considers and suggests certain measures to work with these leaders. Language is the first measure to express the intention of eco-awareness to these leaders. In most cases, the language of a community may be different from the language of UNEP officials and hence, it poses certain problems to disseminate the intentions of the program. As a result, the program promotes either translated or simple worded scripts for easy understanding of the concepts. Similarly, due to low literacy rates in many regions, the program promotes distinct communication methods such as posters, radio presentation, and social drama to stimulate the level of eco-awareness. UNEP also promotes similar kinds of coherent eco-awareness practices within the United Nations system.

13.9   FUTURE PROSPECTS OF ECO-AWARENESS

Eco-awareness schemes should try to keep up with the pace of techno-logical development and be adapted to the changing needs of the society. Eco-awareness should take place through all modern telecommunication and information-sharing mechanisms so that the future generations do not perceive environmental protection as an outdated concept. What we need is an eco-aware, eco-friendly and eco-responsible society. Eco-awareness should start from home, strengthened at school, and be practiced in life.

Eco-awareness practices can be promoted by both monetary and nonmon-etary terms. Monetary promotion is the method of indirectly providing financial benefits for eco-awareness practices or levying additional charges for undertaking activities that pollute or disturb the balance of nature. Nonmonetary promotion is done by urging people and institutions to spread eco-awareness and by building easily adaptable and flexible models of eco-awareness practices that have enormous effects on the society.

Tax advantages: Tax is an inevitable concern for businesses, governments, and individuals. Companies are expected to spend a part of their profits on any mix of social schemes of which environment is one area. Companies that engage in effective eco-awareness practices can be given special tax rebates to promote more companies to spend on environmental awareness and protection. This will attract more companies to take up environment conservation as a social cause. As more companies mobilize the capital for environmental causes, more people can be covered under the ambit of eco-awareness drives. If this delivers results, then government budgetary allocations to the environment can be reduced significantly and the government can focus more on enhancing civic necessities. This is advantageous to both the government and business concerns. With lower tax rates, compliance will increase generating revenue for the government. Research has shown that companies that fund environmental activities are highly regarded by the consumers. Thus, companies earn customer loyalty by engaging in eco-awareness programs.

Eco-linked educational system: Integrating eco-awareness with education is a widely practiced strategy to assimilate environmental ethos from an early age. Sadly, environmental science proves to be yet another subject that children learn at school, only to forget when they move to higher classes. Rather than educating students only about the various facts and issues about the environment, they should also be taught to respect and protect the environment. The system should develop their scientific temper and creativity so that when they grow up they can find creative solutions for the environmental problems (Bjorkland and Pringle, 2001).

Eco-awareness model building: Eco-awareness practices are heterogeneous around the globe but the nature of environmental changes and issues are homogeneous across regions. This provides the possibility to develop a perpetuity formula for eco-awareness practices irrespective of the geographical location. World eco-organizations can identify the present state of environmental issues and develop the best possible measures that can overcome the present obstacles and issues. A detailed analysis of eco-awareness programs around the globe and adapting successful programs of one country in other countries will help to build a new model of eco-awareness that is expedient for the entire world. This model has to be flexible, scalable, adaptable, and futuristic.

Eco-awareness index: Indices are appraisal tools to understand the past and predict the future performance of a particular asset or information. Environmental awareness is the state of being aware and knowledgeable about the environment which we inhabit. However, not all countries are equally aware of the importance of environmental protection and making normative changes in their resource utilization patterns. The development of an eco-awareness index can ascertain the relative position of each country in terms of their environmental awareness. An environmental awareness index was proposed by Pekka Harju-Autti and Eevi Kokkinen. It ranks countries based on an online survey, conducted globally by asking five questions on the availability of environmental information, personal skill sets, the quality of environmental education, motivation, and possibilities to act toward the betterment of the environment. This self-reporting index ranks European nations as highly aware and African countries as less aware. This is logical as most European countries are developed and literate when compared to the countries of Africa with their poor socioeconomic conditions. Low-faring countries can learn from successful eco-awareness programs of high ranked countries and implement similar programs at home. An eco-awareness index is so far the best available tool to make SWOT (strengths, weaknesses, opportunities and threats) analysis of the prevailing eco-awareness practices. Like an awareness index, there is also a performance index. The environmental performance index published annually by Yale University ranks countries' performance in the protection of human health and conservation of natural ecosystems.

KEYWORDS

- **eco-awareness**
- **eco-literacy**
- **anthropocentrism**
- **biocentrism**

REFERENCES

Abbas, M. Y.; Singh, R. A Survey of Environmental Awareness, Attitude, and Participation Amongst University Students: A Case Study. *Int. J. Sci. Res.* **2014,** *3*(5), 2319–7064.

Bjorkland, R.; Pringle, C. M. Educating our Communities and Ourselves About Conservation of Aquatic Resources Through Environmental Outreach. *BioScience* **2001,** *51*(4), 279–282.

Cork County Council. Environmental Awareness Strategy 2010–2015, Cork, 2010. http://www.corkcoco.ie/co/pdf/828098920.pdf

Eliasson, P. *Learning from Environmental History in the Baltic Countries*, 1st ed.; Liber: Stockholm, 2004.

Harju-Autti, P.; Kokkinen, E. A Novel Environmental Awareness Index Measured Cross-Nationally for Fifty Seven Countries. *Univers. J. Environ. Res. Technol.* **2014,** *4*(4), 178–198.

Kolstad, C. *Environmental Economics*; 2nd ed. Oxford University Press: New York, 2000.

Lanza, R.; Berman, B. *Biocentrism: How Life and Consciousness Are the Keys to Understanding the True Nature of the Universe*. BenBella Books: Dallas, TX, 2010.

Martinez, C. P.; Castaneda, M. G.; Marte, R. B.; Roxas, B. Effects of Institutions on Ecological Attitudes and Behaviour of Consumers in a Developing Asian Country: The Case of the Philippines. *Int. J. Consum. Stud.* **2015,** *39*(6), 575–585.

McBride, B. B.; Brewer, C. A.; Berkowitz, A. R.; Borrie, W. T. Environmental Literacy, Ecological Literacy, Ecoliteracy: What Do We Mean and How Did We Get Here? *Ecosphere* **2013,** *4*(5), 1–20.

Stone, M. K.; Barlow, Z. *Smart by Nature: Schooling for Sustainability.* Watershed Media: Healdsburg, CA, 2009.

Takala, M. Environmental Awareness and Human Activity. *Int. J. Psychol.* **1991,** *26*(5), 585–597.

United Nations. Report of the National Conference on the Human Environment, Stockholm, **1972**.

CHAPTER 14

ENVIRONMENTAL MARKETING AND EDUCATION

KUNAL SINHA[1,*] and S. N. SAHDEO[2]

[1]Department of Management, Birla Institute of Technology, Mesra, Lalpur Campus, Ranchi, Jharkhand 834001, India, Mob.: +91–9386454755

[2]Department of Management, Birla Institute of Technology, Mesra, Lalpur Campus, Ranchi, Jharkhand 834001, India, Mob.: +91–9431596629, E-mail: snsahdeo@bitmesra.ac.in

*Corresponding author. E-mail: phdmb10051.13@bitmesra.ac.in

14.1 INTRODUCTION

The beginning of the modern environmental movement started after the *Rachel Carson's Silent Spring* publication in 1962. This gave a foretaste of the current environmental scenario. The conference on the human environment was held at the United Nations in Stockholm in 1972, which resulted in the declaration "to protect and improve the environment for present and future generations has become an imperative good for mankind." The environmental marketing and education concept, World Environment Day (June 5, every year) emerged from this conference.

14.1.1 SCENARIO OF INDIA

Concern for nature and natural resources is not a new concept for Indians. Admiration for nature in India and the urge to conserve and protect it has been a part of our civilization. India's wealth of scriptures, literature, and folklore are replete with examples which show that our ancestors were environmentally conscious and advocated concepts of sustained usage of available resources through many social customs, myths, traditions, and religion. Thus, in traditional society, environment education was an integral part of the learning, but in new society, environmental education (EE) and marketing both are required. However, with the onset of industrial revolution resulting in alienation of societies from the natural environment, this kind of education has ceased to be a part of the natural learning process. Indiscriminate and unrestricted exploitation of environmental resources is necessitated by population growth, poverty, and illiteracy. Government politics, lack of awareness and values among people in India have created ecological imbalance, resulting different types of environmental problems, pollutions, and other kinds of ecological disorders. This environmental crisis in future may become more worsted in the coming days simply because of lack of concern for the common issues and the absence of basic sense of responsibility for sustaining a balanced ecosystem. Therefore, what is required today is education and environmental marketing for the people and reorientation of the people toward the desirable attitudes and values, to maintain the ecosystem, besides marketing and teaching them how to save environment from further degradation. The emergence of the concept of EE is a new dimension in educational (both in formal and nonformal) system (United Nations Conference, Stockholm, 1972). India also recognized the significance of environmental marketing and EE in the direction of environment protection and took initiative in this regard and marched ahead to put into practice.

14.2 CONCEPT AND SCOPE OF ENVIRONMENTAL EDUCATION (EE)

14.2.1 CONCEPT OF EE

While EE has its roots in nature study, conservation education, and outdoor education, it is distinctly different from these earlier movements. While these areas focus on nature, wise use of natural resources, and the use of the outdoors to teach, EE is fundamentally concerned with the interconnection between humans and the environments that surround them (Disinger and Monroe, 1994).There are a plethora of definitions of EE given by different organizations, authorities, scientists, educationists, and politicians, reflecting each one's own philosophy and perception of the subject. Some of the definitions which are comprehensive and commonly accepted are cited below. In 1969, Dr. William Stapp at the University of Michigan published the first definition of EE: Environmental education is aimed at producing a citizenry that is knowledgeable concerning about the biophysical environment and its associated problems, aware of how to help and solve these problems and motivated to work toward their solution. (Stapp, 1969) The concept of EE was first formalized by the International Union for the Conservation of Nature and Natural Resources (IUCN), in 1970 at a meeting in the United States (Nevada). At that meeting, EE was defined as: A process of recognizing values and classifying the concepts in order to develop skills and attitudes necessary to understand and appreciate the culture and his biophysical surroundings. EE also entails practice in decision-making and self-formulating of behavior about issues concerning environmental quality. The American States Conference on Education and Environment in America 1971 defined EE as follows: "Environmental Education involves teaching about value judgments and the ability to think clearly about complete problems – about the environment—which are as political, economical and philosophical as they are technical" (Krishnamacharyulu and Reddy, 2005). The term "environmental education" means the educational process dealing with man's relationship with his nature and man-made surroundings and includes the relation of population, pollution, resource allocation and depletion, conservation, transportation, technology for urban and rural planning to the total human development. At an International Conference of United Nations Educational, Scientific and Cultural Organisation (UNESCO) held at Tbilisi in 1977 defined EE as an integral part of the education process. It should be focused on practical problems and be of an interdisciplinary character. It should aim at building up a sense of value, contributing to public well-being and concern itself with the survival of the human species. Its focused should reside mainly in the initiative of the learners and their involvement in action and it should be guided by both immediate and future subjects of concern. By the analysis of these definitions, it is concluded that the definitions of EE encompassed a variety of concepts and approaches, with varying emphasis on each of them. The following points have been identified as the essential components of EE.

1. Knowledge and understanding of environment, its associated problems, and future consequences
2. Understanding the subtle relationship between man and man, man and nature, and its appreciation
3. Value clarification, development of attitudes, interest and awareness concerning quality of environment
4. Decision-making and problem-solving steps
5. Formulation of code of behavior

Considering all the abovesaid points, it is further inferred that EE is the education "about" the environment, "from" the environment and "for" the environment.

i.   Education "about" the environment means an understanding of the total environment.
ii. Education "from" the environment implies gathering concepts, knowledge, and skills referring to specific academic discipline.
iii. Education "for" the environment means the development of attitude, skills, and evaluation abilities for the proper use and the development of the environment.

The abovementioned definitions have a tilt toward cognitive aspects of learners. Perhaps, a considerable emphasis also needs to be put on aspects which relate to values, feeling, and attitude analysis in EE. Without such an emphasis, EE will be nothing more than a facile exercise in glibness.

14.2.2   NATURE AND SCOPE OF EE

EE is largely interdisciplinary and multilingual in nature as the environment problems and issues are not confined to any other disciplines such as physics, chemistry, biology, and so forth. It is an amalgamation of many of the conventional subjects. EE is considered both as an art (doing) and as science (understanding) organized from primary to university level. The objectives and the content of EE should be integrated with the curriculum varying from stage to stage, that is, primary to university level. EE is a process of creation of individual and collection commitment to improve the quality of the life through self-knowledge and an understanding of the physical, political, socio-economic, and behavioral concerns of man. Moreover, it is a continuous, individual, and community education process that is an integral part of the complicated web of communication for human understanding. EE is a perspective that is to be given to all the subjects in the curriculum, a second trend, which means that, without alterations of the subjects in the curriculum, teachers will give them an ecological or economic front toward environmental problems. If teachers are to do this, they will have to be given a lengthy and costly training, because unless they have a profound insight into the mechanisms of nature and the aggressive process of the economy and technology, they will not be able to carry out this difficult task successfully. EE should be broad, open to the internationalist spirit; it should not offer protection to narrow-minded chauvinism. It is not difficult to discern the propagandistic aims of the great industrial powers that have no hesitation in using hard technology in the areas of influence while recommending developing countries to see soft technology. EE seeks to develop the ability to assess environmental situations and the causal chain of relationships leading to environmental damage; the interaction among social, economic, and physical factors; mutually related and overlapping developments, networks and feedback; responsibility for the future generations' economy and care in the use of natural resources; respect for nature and life; recognition of the limits of nature, human action and self-restriction; and acquiring an ability to perceive nature. EE aims at ultimately for reaching and changing behavior in everyday life and workplace. The guiding principle and pedagogical ideal of EE is the environmentally responsible consumer, industrial producer, employee, citizen, policy-maker, traveler, athlete, tourist, and the far farmer—every individual who is aware of nature and lives in harmony with it.

The various aspects of environmental marketing and EE, which form its scope, are as follows:

1. **Pollution**: This includes air pollution, noise pollution, water pollution, land pollution, thermal pollution, and unclear pollution of social environment—their effects and prevention.
2. **Population education**: This includes importance, problems, and ways and means of promoting population education.
3. **Population explosion**: This includes causes and effects of population, explosion, and remedies for population explosion.
4. **Use of resources and conservation**: This includes use of resources, conservation of resources, land and soil conservation, water conservation, energy conservation, conservation of wildlife, forests, air, natural beauty, and other natural resources.
5. **Food and nutrition**: This includes production of food, supply and use of food, food adulteration and preservation, uses of food, value of food, balanced diet, malnutrition, and undernutrition.
6. **Health and hygiene**: This includes individual, family, community, health and hygiene, health hazards communicable and noncommunicable diseases, and their preservation and treatment.

14.3   NEED, CONTEXT, AND SIGNIFICANCE OF THE STUDY

Today, the global concern is to struggle against environment pollution and maintain the standard of human environment. Environment in developing countries like India have been threatened by problems such as poverty, over–pollution, and degradation and depletion of environment. In addition to the industrial revolution, unprecedented scientific and technological revolution in disastrous changes in the environment, leading to environmental degradation/crisis. The nature of environmental change (particularly, man-induced change) in recent years have brought about a series of environmental issues of global magnitude, including population explosion, energy resources and utilization, exploitation of raw materials and environmental pollution, leading to environmental degradation. This environmental degradation/crisis has become a serious issue as it threatens not only the tranquility of people's existence but also their health and lives as well. As such, the environmental protection and preservation has been an urgent need of the hour. No doubt, it is considered that education plays an important role in reducing the environmental degradation and protecting the environment and as such EE in India has emerged as a significant area of concern.

The National policy on Education 1986 (modified in 1992) states that "Protection of the Environment" is a value, like other values, must form an integral part of the curriculum at all stages of education. The national system on education as defined in the policy visualizes a national curricular framework, which contains direct bearing on the natural and social environment of the pupils. There is a considerable emphasis in the syllabus/textbooks on making teaching and learning more environment-oriented and socially relevant. While National Council of Educational Research and Training (NCERT) has integrated environmental concepts into various subjects taught at the school level under the central sector, State Councils of Educational Research and Training have attempted to integrate these concepts into the syllabus and textbooks at the state level. In addition, under the "Socially Useful Productive Work" there must be a provision made for the direct participation of the children in environment-related field programs such as planting and caring for trees, environmental sanitation, and so forth. It is very important to note that the EE started at the school level will have to become, over a period of time, a way of life. This is possible only when we aim at the development of certain values among children, youth, and adults, irrespective of their social, economic, and occupational status, which enable them to participate in solving real-life environmental problems. EE should create consciousness and this consciousness be translated into coherent behaviors in which collective action finds a fundamental solution for the problems related to the environment.

Today, environmental marketing and EE both are the important segments within the educational system. EE involves an interdisciplinary, integrated, and active approach to learning, mainly purports to create environmentally literate and active citizens and ensure that present and future enjoy a decent quality of life through the sustainable use of resources. Environmental literate citizens are able to consider the sustainability of development, to actively work to reverse environmental degradation, and to manage to use the country's natural resource base more wisely and democratically. The key to successful EE is marketing and advertising of environmental issues. They can use information, legislation, and community action to improve and protect human and environmental health. A proper advertising can make the young minds to know that within this finite system of earth, the air, the water, the soil, the landscape, and the biological organisms exist in very delicately balanced mutual relationships. As the balance is disturbed by any human intervention or technological innovations, the environment gets affected leading to environmental degradation. A committed teacher indeed can develop awareness about these and make students feel sorry for the undesirable act of human beings, help them incline their attitude toward aiming at benefiting the public trust properties. If teachers do not have knowledge, skills, and commitment to environmentalism, it is unlikely that an environmentally literate and active student will be the end product. For this, teachers need to have a positive approach and positive attitude toward EE.

There is a general feeling that teachers' attitude toward EE at different levels of education plays an important role in the development of the desired attitude, awareness, and behavior among students. Hence, there is a need to assess attitude and awareness of teachers about the environment, that is, environmental attitude and awareness and suggest measures to enhance them. An important reason for linking environmental attitude and awareness of classroom teachers with their attitude toward EE is that both can contribute significantly to transformation and development of individuals with utmost care for the environment. Environment pollution and degradation have been the universal phenomena and almost all the countries including India and Iran are attempting at minimizing the pollution, conserving the environment, and moving toward sustainable development. Ever since academic attention was drawn to the environmental issues and the adoption of EE as a strategy to combat it, considerable research has been conducted and adequate litera-ture is prepared/published. A thorough review of research studies related to

EE revealed that considerable research has been done on different aspects of EE with different variables, but only a few studies have been carried out to measure the existing levels of environmental awareness and attitude among teachers and students at different levels of education.

However, most of the studies carried out on EE in India and Iran have been only in the area of curriculum analysis in order to find out their adequacies or otherwise for integration of EE into school subjects. Minimum effort has been directed at finding out the level of awareness, attitude, and perception of environmental issues and EE among the school teachers and learners of the program and more over experience of successful countries, as revealed through comparative studies, could serve as sources of ideas for those involved in the Iranian and Indian education system.

Further, it is found that a considerable number of research studies have attempted to study environmental attitude and awareness and attitude toward EE in relation to certain variables such as sex, subject studied, qualification, teaching experience, and so forth. Some of those studies have reported the difference in the attitude and awareness between/among different categories on the selected variables and some have reported no differences. Hence, the findings of those studies are found to be inconsistent and therefore, it is an attempt here to consider the above-said variables to study the influence of them on teachers' environmental attitude and awareness.

Thus, the present chapter aiming at the comparison of environmental attitude, awareness and attitude toward EE in India is entitled "Attitude and awareness about environment and attitude towards EE in India."

14.4   APPLICATION STUDY OF THE INDIAN SOCIETY

The growing demand of society for greener products makes Indian corporate desires to meet this and to make a profit, seeking to a fascinating interaction with cultural change. Green advertising affects the culture of the community through mutual interaction between the needs of consumers and their desires and preferences, as well as culture conveyed by the advertising. When the advertising responds to the community, it will be able to continue to deliver the message and advertising in a compatible manner with community members, because the environmental requirements have become an urgent necessity with the growing environmental trends. Accordingly, Indian designer must develop mental and intellectual abilities to achieve the environmental requirements that the consumer needs.

14.4.1   ENVIRONMENTAL MESSAGE

Lord et al. (1995) reported that message agreement strongly influences the advertisement attitude. This finding was corroborated by Laczniak et al. (1999) who studied the influence of advertising message involvement, service and product involvement, and product knowledge on the way a consumer processes an advertisement. As a result, message involvement in environmental advertising had the strongest influence on advertisement processing through achieving consumers environmental needs as shown in Figure 14.1.

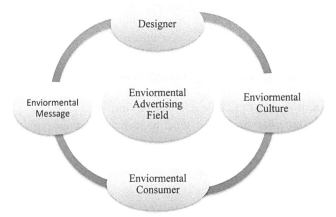

FIGURE 14.1   The relationship between factors affected in environmental advertising.

14.4.2   ENVIRONMENTAL CULTURE

Culture is defined as "The set of distinctive, spiritual, intellectual, material and emotional features of society or a social group, and that it encompasses, in addition to art and literature, value systems, lifestyles, ways of living together, traditions and beliefs" (UNESCO, 2002). The area of local culture has grown in salience in global development issues on account of the rising share of cultural goods, services, and intellectual property in global trade, as well as the threats to cultural diversities and identities associated with contemporary globalization. However, there is increasing awareness that the protection and promotion of cultural diversity to universal human rights, fundamental freedoms along with securing ecological and genetic diversity. A question that should be raised is whether former cultural values still have a meaning and validity. We must admit that only hypothetical thoughts that lie behind physical forms are desirable, such as the religious values, privacy, and so forth. The past is not very important in itself. It is the presence of the past and our attitude toward that matter (Bousaa, 1999). A number of following points are worth mentioning here:

- Not everything in our past is worth reintegration into present-day society. Only the cultural values that are significant and have a constructive function for society should be considered.
- Cultural heritage should be fostered in a creative way by extracting and analyzing lessons to be adapted to present.
- The demolition of traditional concepts does not erase them from the local memory; it tends to eliminate the recall of that memory, rendering less meaningful communication of that heritage to a new generation leading society. (Lowenthal, 1994)
- New development in the social and technological field has its merits and contributions to the quality of the present lifestyle. Soulless copies of the past exclude any social vitality.
- Visible tangible past may tend to culture, but current tends make economic profit from it as well. All cultural assets represent an economic value, which can be integrated in any planning development process.

14.4.3 ENVIRONMENTAL CONSUMER

The key element of environmental conscious consumption is a desire by consumers for information about the relationship between products and the environment with greater exposure to "green" information sources which influence consumer purchasing decisions (Peattie, 1995). The growth of environmentally conscious consumers has created a new trend in the market called "environmental consumerism" (Carlson et al., 1993). Gussow (1989) explained environmental consumerism as an orientation in which consumers' purchases, product usage, and disposal to preserve nature's ecological balance are included.

14.5 PRINCIPLES OF EE

The EE involves the task of creating an informed group of young citizens whose environmental issues will be a part of their daily lives. This will eventually bring about changes on a larger scale, creating a more socially just and ecologically sustainable society. This includes running EE programs in schools and colleges, conducting training and awareness-building workshops for teachers, developing and disseminating EE material to all segments in the society through the web and print media, and networking and coordinating among school communities across the country. In total, EE is a subject of a very practical nature and is also supported by sound pedagogical principles. Thus, it is very essential to consider the following principles in teaching EE.

EE should consider following things:

- Consider the environment in its totality and built, technological and social (economic, political, technological, cultural–historical, moral, and aesthetic).
- Be a continuous life-long process, beginning at the pre-school level and continuing through all formal and nonformal stages of life.
- Be interdisciplinary in its approach, drawing on the specific content of each discipline in making balanced perspective.
- Focus on current potential environmental situations while taking into account the historical perspective.
- Examine major environmental issues from local, national receive insights into environmental conditions in other geographical areas.
- Promote the importance and necessity of local, national, and international cooperation in the prevention and solution of environmental problems.
- Consider environmental aspects in plans for development and growth.
- Relate knowledge, problem-solving skills, values clarification, and environmental sensitivity but with special emphasis on participation of environmental sensitivity.
- Emphasize on environmental problems and the need to develop critical thinking and problem-solving skills.
- Utilize diverse learning environments and educational approaches to teaching/learning about and from the environment with due stress on practical activities.
- Helps in programming learning experiences from simplex to complex. It is this principle that makes EE as a medium of learning different subjects. It also helps a child proceed from indefinite ideas to definite ones. The first principle of thoughts which are vague will become clear later as it grows and EE helps in sharpening the development of observational skills for definiteness.
- Helps the ordering of learning experiences from empirical to rational as it is a very important educational maxim.

- Provides for the self-development of the child. Children are encouraged to conduct their own investigations and draw their own conclusions.
- Provide for self-instruction and self-discovery.
- Create delightful and pleasurable excitement in children because of the beauty and glory of the environment issued as teaching aid by the teachers. It makes the child's education problem based for understanding environment and has social relevance.

Generally, the aims of environmental marketing and EE fall into groups:

a) **Cognitive aims:** These include imparting knowledge about environment and an ability to think which will enable the individual and social group to work out political solution to the wide variety of problems related to the environment.
b) **Normative aims:** These relate to the inculcation of ecological awareness which will be conducive to the creation of value models, enabling the individual and the group to identify the factors that upset the environment equilibrium and protest against them.
c) **Technical and applicative aims**: This means planning collective practices which preserve, improve, and restore the quality of life as understood by the community in the light of formal and informal marketing and education in such a way that the demands made by economic development do not conflict with the biological rhythms of the ecosystem.

14.7　OBJECTIVES OF ENVIRONMENTAL MARKETING AND EE

The objectives of environmental marketing and EE are as follows:

1. **Awareness**: To help individuals and social groups to acquire an awareness of sensitivity for total environment and its allied problems.
2. **Knowledge**: To help individuals and social groups to acquire a basic understanding of the environment and problems associated with it and humanity's critically responsible presence and role in it.
3. **Attitude**: To help social groups and individuals to acquire social values, strong feelings of concern for the environment and motivation for actively participating in its protection and improvement.
4. **Skills**: To develop individuals and social groups acquire the skills to motivate for solving environmental problems.

5. **Evaluation ability**: To help individuals and social groups to evaluate environmental measures and marketing and education programs in terms of political, economic, social, aesthetic, educational, and ecological factors.

6. **Participation**: To help social groups and individuals to develop a sense of responsibility regarding environmental issues so as to ensure appropriate action to solve these problems.

Therefore, the objectives of the EE are to generate awareness, knowledge, attitude, skills, evaluation ability, and participation to help social groups and individuals. These objectives are further clarified in the flowchart (Larijani and Yeshodhara, 2003) given below in Flowchart 14.1.

Flowchart 14.1 Flowchart presenting the objectives of environmental marketing and education.

To sum up, the main objectives of EE are to generate and disseminate knowledge through research, teaching, and extension work, develop knowledge-based awareness, which will lead to responsible attitude toward the environment, without losing sight of the value system of society and individual, acquire skills for implementation of programs and policies conducive to solving immediate problems and overall development of the nation.

14.8   MAJOR ENVIRONMENTAL ISSUES

Major environmental issues are as follows:

1. **Forest and agricultural degradation of land**: Land degradation has accelerated during the 20th century due to increasing and combined pressures of livestock production and agricultural (over-cultivation, overgrazing, forest conversion), urbanization, deforestation, and extreme weather events.

2. **Resource depletion (such as water, mineral, forest, sand, and rocks)**: Natural resource depletion is another crucial current environmental problems. Fossil-fuel consumption results in emission of greenhouse gases, which is responsible for global warming and climate change. Globally, people are making efforts to shift to renewable sources of energy such as solar, wind, biogas, and geothermal energy.

3. **Environmental degradation**: Environmental degradation is known as the deterioration of the environment through depletion of resources such as water, air, and soil; the destruction of ecosystems; and the extinction of wildlife. It is defined as any change or disturbance to the environment perceived to be deleterious or undesirable.

4. **Public health:** The current environmental problems pose a lot of risk to health of humans and animals. Dirty water is the biggest health risk of the world and poses threat to healthy life. Run-off to rivers carries along toxins, chemicals, and disease-carrying organisms. These pollutants cause respiratory disease such as asthma and cardiovascular problems.

5. **Loss of biodiversity:** Human activity is leading to the extinction of species and habitats and the loss of biodiversity. Ecosystems, which took millions of years to perfect, are in danger situation when any species population is decimating. Balance of natural processes like pollination is crucial for the survival of the ecosystem and human activity threatens the same.

6. **Loss of resilience in ecosystems**: Ecosystem resilience refers to an ecosystem to recover from disturbance or withstand ongoing pressures. It is a measurement how well an ecosystem can tolerate disturbance without collapsing into a different state that is controlled by a different set of processes.

7. **Livelihood security for the poor**: A person's livelihood refers to their "means to secure the basic necessities—food, water, shelter, and clothing of life." Livelihood is defined as a set of activities, involving securing food, fodder, water, medicine, shelter, clothing, and the capacity to acquire above necessities for meeting the requirements of the self and his/her household on a sustainable basis with dignity. The activities are usually carried out repeatedly.

14.8.1   MAJOR ENVIRONMENTAL CONCERNS

14.8.1.1   LAND/SOIL DEGRADATION

Most of the land area in the country shows evidence of degradation, which affects the productive resource base of the economy. Erosion by water and wind is the most significant contributor to soil erosion with other factors such as salivation, water logging, and so forth. While soil erosion by river and rain in hill areas causes landslides and floods, deforestation, overgrazing, traditional agricultural practices, mining and development projects in forest areas have resulted in opening up of these areas to heavy soil erosion. The government strategy for preventing land degradation includes treatment of catchment areas, comprehensive watershed development, vegetative measures, survey and investigation of problem areas through remote sensing and geographic information system techniques, biomass production in reclaimed land, micro-level planning, and transfer of technology.

14.8.1.2   DEFORESTATION

Forests are a renewable resource and contribute substantially to the economic development by providing goods to the forest dwellers and forest-based industries, besides generating employment. Forests also play a vital role in enhancing the quality of environment by influencing the ecological balance and life-support requirement (checking soil erosion, maintaining soil fertility, conserving water, floods and regulating water cycle, balancing carbon dioxide and oxygen content in atmosphere, etc.). The country has very diverse forest ranging from the moist evergreen forests in the North-East, along the west coast of the Andaman and Nicobar Islands and alpine vegetation in the Himalayas. However, this forest wealth is depleting due to overgrazing, unsustainable practices, overexploitation, encroachments, forest fire, and indiscriminate sitting of development projects in the forest nearby areas. Withdrawal of forest products, including fuelwood, timber, and so forth, is much beyond the carrying capacity of our forests. To regulate unabated diversion of forest land for the use of non-forestry purposes, Forest (Conservation) Act, 1980 was enacted. It has resulted in reduction of diversion of forest area for non-forestry use considerably and the present rate of diversion is 16,000 ha annually. In recent past, there is no change in forest area because its diversion for non-forestry purposes has been more or less compensated by forestation and natural regeneration programs of the government.

14.8.1.3   BIODIVERSITY

The country's unique geographical and agroecological diversity endows it with a wide variety of agroclimatic zones that harbors a rich repository of biological resources. With only 2.4% of the total land area of the world, biodiversity from India contributes 8% to the known global biological diversity. It is one of the 12 mega biodiversity centers in the world. The biodiversity in forests, grasslands, wetlands and mountains, deserts, and marine ecosystems is under tremendous pressure. One of the major causes of the loss of biological diversity has been the cover in order to expand agriculture. From about 70% of the geographical area surveyed so far, 81,000 animal species and 46,000 plant species have been recorded by the Zoological Survey of India and the Botanical Survey of India, respectively.

14.8.1.4   ATMOSPHERIC POLLUTION

Air pollution and sound pollution are widespread in the country and regular monitoring is being carried out under the National Ambient Air Quality Monitoring System. A high level of suspended particulate matter (SPM) is the most prevalent form of air pollution. High concentration of sulfur dioxide (SO_2) and SPM occurs in the air. High domestic use of coal or biomass fuel is still a serious problem in high human exposure to SPM, SO_2, and carcinogenic agents. Vehicular traffic is the most important source of pollution in all the megacities. The number of vehicles in these cities has increased manifold. Central Pollution Control Board (CPCB) studies, on the ambient noise levels show that noise levels in most of the big cities exceed the prescribed standards. The major sources of noise are vehicles, diesel generator sets, loudspeakers, construction activities, and bursting of firecrackers. The toxic nature of air pollutants and their high concentrations in many industrialized areas are posing serious concerns in terms of human health. An attempt is being made to control the noise and air pollution by notifying the compliance through local authorities. The main factors contributing to urban air quality deterioration are growing industrialization, population, and increasing vehicular pollution.

14.8.1.5   WATER POLLUTION

According to an analysis of water quality by CPCB, the biochemical oxygen demand values, the quality is gradually declining down. This indicates that the water quality monitoring results indicate that the organic and bacterial pollution continues to be predominant source of pollution in our aquatic resources. The major sources of water pollution are discharge of domestic sewage and industrial effluents which contain harmful organic pollutants, chemicals, and heavy metals, and run-off from land-based activities such as agriculture and mining. The rising industrial and domestic wastes have resulted in heavy stress of the quality of water. The diseases commonly caused due to pollutant water are diarrhea, trachoma, intestinal worms, hepatitis, and so forth.

14.8.1.6   SOLID WASTES

Unregulated growth and improper collection, transportation, treatment, and disposal of solid wastes have resulted in increased pollution and health hazard from these wastes. Urban municipal waste (municipal solid waste) is a heterogeneous mixture of paper, plastic, cloth, metal, glass, organic matter, and so forth. Although our current per capita waste generation is very low as compared to advanced countries, the actual quantum of waste is large owing to the enormous size of our population. A plastics waste alone has increased tremendously from the last few years. The mode of waste disposal predominantly remains through landfilling, which is a conventional but unhygienic method. Alternative modes such as composting and other scientific approaches are sparsely used.

14.8.1.7   COASTAL AND MARINE POLLUTION

The coastal areas of India, with a coastline of over 7500 km harbor a variety of specialized marine ecosystems such as mangroves, coral reefs, salt lakes, and mudflats, which mainly are the habitat for endangered marine species. These coastal areas are exposed to environmental stress due to several reasons such as unplanned and improper development activities without appropriate coastal zone management plans, shipping and sea-based activities including sludge disposal, oil spills, and mining in coastal areas. An important impact on climate change is due to rise in sea level.

14.8.2   MAJOR ENVIRONMENTAL POLLUTION CONTROL ACTIVITIES

The major environmental pollution control activities should include:

• Policy initiatives to improve environment like the National Conservation Strategy and Policy Statement for Environment Protection and

Development, 1992; Policy Statement for Abatement of Pollution, 1992; and National Forest Policy, 1988.

- Notification and implementation of emission and effluent standards for air, water, and noise levels. Standards are formulated by the multidisciplinary group keeping in view the international standards, existing technologies, and impact on health and environment.
- Action plans for 141 polluted river stretches to improve quality of river water.
- For controlling vehicular pollution, progressive emission norms at the manufacturing stage have been notified, low sulfur diesel; cleaner fuels such as unleaded petrol and compressed natural gas are introduced.
- Identification of clean technologies for new large industries and clean technologies/processes for small-scale industries (SSIs).
- Setting up of common effluent treatment plants for clusters of SSI units.
- Implementation of an eco-mark scheme to encourage production/ consumption of environment-friendly products.
- Preparation of a zoning atlas, indicating status of the environment at district levels to guide environmentally sound location for setting the industries.
- Mandatory submission of annual environmental statement report which could be extended into environmental audit.
- Initiation of environmental epidemiological studies in critically polluted areas to study the impact of environment on health.
- Setting up of authorities like the environment pollution (prevention and control) in both rural and urban areas, authority for the national capital region and different zonal region for protecting and improving the quality of environment and controlling, preventing, and abating environmental pollution.
- Provision of fiscal incentives for installation of pollution monitoring equipment and also for shifting of industries from congested areas has been made to save country's ecology in general. Also, provisions should be made to monitor health effects associated with different types of air pollution.

14.9 CONCLUSION

From this chapter, following things should be done so as to ensure our ecosystem is secure:

1. Help to develop a sense of appreciation of natural resources.
2. Create a positive approach and relationship between healthy life and pure natural environment.
3. Give information about the pollutants so that the people will be aware of the diseases and consequences of water, air, and other pollutions.
4. Convince the society about the urgency of environmental protection.
5. Make people aware about their fundamental duties with regard to the environment as incorporated in Article 51, Part IV-A of Indian Constitution, to protect and improve the natural environment including forests, lakes, rivers, dams, and wildlife, and to have compassion for living creatures.

All the above-said things are possible if people are aware about the environment and have a positive attitude toward environment/EE.

KEYWORDS

- **environmental issues**
- **education**
- **marketing**
- **environmental concerns**

REFERENCES

Bousaa, J. Why the Past. Working Paper, University of Liverpool, 1999.

Carlson, L.; Grove, S. J.; Kangun, N. A Content Analysis of Environmental Advertising Claims: A Matrix Method Approach. *J. Advertising* **1993,** *22*(3), 27–39.

Carson's, R. *Silent Spring;* Houghton Mifflin Harcourt: USA, 1962.

Disinger, J. F.; Monroe, M. C. *Defining Environmental Education.* Environmental Protection Agency: Washington, DC, 1994, 1–36.

Gussow, A. Green Consumerism. *Business, 12* (November/December), 18–19. July/August 1989, 3–7.

Krishnamaracharyulu, V.; Reddy, G. S. *Environmental Education: Aims and Objectives of Environmental Education: Importance of Environmental Education.* Neelkamal Publications Pvt. Ltd: Hyderabad, 2005.

Laczniak, R.; Kempf, D. S.; Muehling, D. D. Advertising Message Involvement: the Role of Enduring and Situational Factor S. *J. Curr. Issues Advertising* **1999,** *21,* 51–61.

Lord, K. R.; Lee, M. S.; Sauer, P. L. The Combined Influence Hypothesis: Central and Peripheral Antecedents of Attitude Toward the Ad. *J. Advertising* **1995,** *24,* 73–85.

Lowenthal, D. *Our Past Before Us: Why Do We Save It?* Temple Smith: London, 1994.

Larijani, M.; Yeshodhara, K. An Empirical Study of Environmental Attitude among Higher Primary School Teachers of India and Iran. *J. Hum. Ecol.* **2008,** *24*(3) 195–200.

Peattie, K. *Environmental Marketing Management: Meeting the Green Challenge;* Pitman Publishing: London, 1995.

Stapp, W. The Concept of Environmental Education. *J. Environ.* **1969,** *1*(1), 33–36.

UNESCO. Universal Declaration on Cultural Diversity. In *Cultural Diversity Series No. 1*; Stenou, K., Ed.; UNESCO: Paris, France, 2002.

United Nations Conference. The Human Environment, June, Stockholm, Sweden, 1972, pp 5–16.

CHAPTER 15

GOING GREEN: TOWARD ORGANIC FARMING AND A PLASTIC-FREE ECO-FRIENDLY LIFESTYLE

SUMIT ROY*

MSW, Consultant, Jadavpur, Kolkata, West Bengal, India, Mob.: 9831024033

*Corresponding author. E-mail: sumitroyid@gmail.com

15.1 INTRODUCTION

It is a generally accepted belief and understanding that the whole world is moving toward a more advanced, developed, and civilized society. Human beings are evolving and so are their habits, behavior, thinking, attitude, and emotions. People are advancing toward a better, convenient, comfortable, modern, technological, and easy life. It is also called hi-tech life, as technology has made life very smooth. Communication, financial transactions, conveyance, construction, even presentation of imagination has reached new heights with the help of technology. Every day, some new inventions are being made which are said to help mankind. Resources which were luxuries a few decades ago have become necessities, without which life seems to be impossible. Human beings live in a world which was created by nature. The basic resources which have made life possible on this planet Earth were made available by nature. The mere survival and existence of human beings depend entirely on the environment which is a "natural self-sustaining system" and includes the soil, water, air, sunlight, plants and animals (other than human beings), weather, temperature, moisture, and so forth. The food that we eat grows on the soil or land and it requires sunlight, water, air, and a few members of the animal family for its nurturing. Human beings are probably the only creatures, which ignored the rest of the other elements of the "natural self-sustaining system" and built an "artificial easy system" for their own advancement, convenience, and comfort. Unfortunately, the "artificial easy system" cannot sustain unless it ensures the balance of the "natural self-sustaining system," so that the latter can function properly. In fact, "artificial easy system" has started affecting the human life adversely destroying habitats, disturbing the weather cycle, causing global warming, leading to various diseases, and taking human lives. However, initially, for a few decades it was not given any cognizance and other less relevant factors were considered to be responsible for the undesirable impacts. Fortunately, this has been recognized, acknowledged, and given priority very recently. Moreover, since the "artificial easy system" is very lucrative, in the majority of the cases, it has become very difficult for humans to retract back to the age-old practices which ensure the balance of the "natural self-sustaining system." Moreover, this "artificial easy system" has damaged the balance of the "natural self-sustaining system" so badly that unless a major thrust is given for awareness and sensitization to change the belief systems, attitude, behaviors, and practices, it is going to be a very difficult task to save the Earth and thus human beings.

This chapter will focus on how the "natural self-sustaining system" was replaced by the "artificial easy system" with respect to one of the most basic of life processes, but which has profound impact on the lives of all living beings on Earth. It will also put forth the existing methods and mechanisms which have been found to expand the "natural self-sustaining system" by making changes in the way humans consume and dispose the consumed product but are hardly implemented or put in use. It will present ways in which this "artificial easy system" can be broken or altered to incorporate mechanisms which can ensure sustainable consumption of natural resources and goods.

15.2 OBJECTIVE

This chapter is an effort to inform the readers the way in which the modern technologically advanced "artificial easy system" is jeopardizing the environment in general and all living beings in particular. The chapter illustrates how humans have changed their lives exploiting everything in and around them expecting to get more out of them, even from nature and in due course have started depleting the environment, their own lives and that of the future generations. The chapter focuses on these changes starting from something as basic and essential as human food to indiscriminate use of nonbiodegradable products, gadgets, and nonrenewable resources and how these are affecting the Earth, based on international studies. Certain sections bring to light the experiences and learning based on the consumer behavior and field exposures.

15.3   RESEARCH METHODOLOGY

This is basically a qualitative research with a combination of two qualitative approaches—grounded theory approach and field research approach. The data has been collected from the literature review and online investigation of various studies conducted by individuals and agencies across the globe. Broadly, this is an interpretation of facts and findings, but also includes experiences based on one to one interactions with farmers at the field in the rural areas as well as consumers in the urban market in West Bengal, India.

15.4   AGRICULTURE

One of the most primitive and basic processes without which the existence and mere survival of human life on Earth are impossible is agriculture. It was a "natural self-sustaining system," using the nature to produce what humans need to consume for their survival. Humans learned that certain crops yielded better during specific weather and soil conditions and started growing the crops accordingly. The crop yield depended on inputs of nature sunlight, water, and nutrients from the soil.

India is one of the most ancient agricultural civilizations of the world, being an agricultural society since more than 40 centuries (Randhawa, 1980). Traditionally, Indian farmers had developed the tools, techniques, and other input resources to improve the crop quality and yield. Each region had indigenous varieties to suit the region-specific climatic, soil, and water conditions. These varieties included saline resistant, drought resistant, flood resistant, and so forth. Since the seeds were indigenous, the capacity of the crops to fight the pests based on the region-specific climatic changes was also very high. Some indigenous high yielding varieties (HYV) also evolved which were identified, selected, and improved by peasants and tribal (Shiva, 1989). Most of the farmers used wooden plough, which was not dependent on nonrenewable fuel and households had cows and bullocks. The cow dung manure and cow urine were used extensively as fertilizer, pesticide, and herbicide. In fact, Dr. John Augustus Voelcker, who was commissioned by the Secretary of State to India in 1889 to advise the

Imperial Government on the application of agricultural chemistry to Indian agriculture, had stated in his report to the Royal Agricultural Society of England: "I do not share the opinions which have been expressed as to Indian agriculture being, as a whole, primitive, and backward, but I believe that in many parts there is little or nothing that can be improved." (Voelcker, 1893, p 6). Again he says "I make bold to say that it is a much easier task to propose improvements in English agriculture than to make really valuable suggestions for that of India (Voelcker, 1893, p 10)." Further he says that "At his best the Indian *raiyat* or cultivators quite as good as, and, in some respects, the superior of, the average British farmer, whilst at his worst it can only be said that this state is brought about largely by an absence of facilities for improvement which is probably unequalled in any other country, and that the *raiyat* will struggle on patiently and uncomplainingly in the face of difficulties in a way that no one else would" (Voelcker, 1893, p 11). It is thus evident that in spite of the fact that whatever little scientific research was going on at that point of time in India, was not focused on agriculture, but the inherent techniques, methods, and processes followed in Indian agriculture were quite improved and advanced. Though most of the small and marginal farmers of rural India were unaware of Integrated Pest Management, they had means and ways through which they protected their crops and managed to get the required yield. Most of the households used wooden ploughs on the land, which only helped the soil to breathe but did not lead to the death of vermin, which made the soil more fertile. The crop cycle and mixed cropping were followed to ensure natural nutrient retention and regeneration within the soil, the rice or wheat cultivation was followed by oilseed, pulses, and millets, which were highly nutritious and resource prudent crops (Randhawa, 1980). Mixed and diverse cropping also helped to ward off pests or minimize the damage caused due to pests. Farmers always saved a portion of their yield as seed for the next year. Seed stocking, protection, and sharing were common practices adopted among farmers (Randhawa, 1980). Hence, there was no need to buy seeds each year.

After Independence, the newly formed government did focus on agriculture, village, and farms with a major focus on self-reliance and ecological sustainability. In the first 5 year plan (1951–1956), land reforms and improvement in agriculture production were given priority, containing a full chapter for promotion of organic farming, encouraging the use of organic manure for increasing soil fertility and productivity (Planning Commission Government of India, n.d., para 21). The focus was on farming strategies at the village and in some cases at the individual field level.

15.4.1 HOW GREEN WAS THE GREEN REVOLUTION: A CRITICAL ANALYSIS

By the end of World War II, the US companies which were producing nitrogen (ammonia) for ammunition in the war had no use of the product. They were producing 730,000 tons of ammonia each year and had the capacity to produce 1.6 million tons (Herger et al., 2015, para 10). Suddenly, there were no takers for the nitrogen. By that time, it was known to the world that plants need three main nutrients called macronutrients—nitrogen, phosphorous, and potassium. Organophosphate which was used as nerve gas during the war was now available but was of no use. It was decided that the nitrogen and other such products such as organophosphate will be used for increasing the yield of the crops and protect the crop from pests (Biddle, 1984).

In India, the import of food grains in 1951–1953 was 3.41 million tons annually which reduced to 0.94 million tons annually in 1954–1956, but again rose to 3.93 million tons in 1957 due to shortage of food grains, high prices, hoarding, and speculation by traders (Chapter III-Case Studies-1: The Food Problem, n.d.). The West was trying to introduce green revolution in India with the miracle seeds of Norman E. Borlaug since 1963, but the Government of India was stressing on ecologically balanced measures of agriculture ensuring self-reliance of the farmers. In 1950, India's food grain import was only 5% of the total food grain available in the country which increased in 1966 to 10 million tons with domestic food grain production of 72 million tons (Abrol, n.d., p 2). Moreover, with the drought in 1966 and the major food crisis, India had to import 10 million tons of wheat and had no other option but to explore the possibility of green revolution, expecting that it would yield better results than the existing agricultural system.

The miracle HYVs of wheat was first introduced in Punjab followed by HYVs of rice, which subsequently spread to other parts of India. The agriculture extension workers of the government reached to each farmer in the villages and provided them with the new seeds. These seeds initially were available at a subsidized rate and the basic inputs required were chemical fertilizers and pesticides, which were also highly subsidized. The initial yield was quite high and the farmers were happy with the yield. Something that was not taken into consideration was the fact that these HYVs did not yield high by themselves; they needed the inputs of more water and a good amount of chemical fertilizers to give the higher yield (Agricultural Production, n.d.). The primary focus was on big farmers with large landholdings and organic manures took a backstage as specific N, P, and K fertilizers became the basic input for these new HYV seeds. The farmers did achieve higher yields but not only for HYVs but because of better irrigation facilities, more supply of chemical fertilizers and pesticides, and better research and extension (Abrol, n.d., p 2).

However, with time it is becoming more and more evident that this initiative which had created a notion that use of these HYV seeds, fertilizers, pesticides, and herbicides will increase crop yield and ensure food security leading to prosperity of farmers is actually a mirage (Khajuria, 2016). Not only has the yield decreased over the years but also the farmers had to take loan at very high-interest rates from money lenders or the retailers of chemical inputs to buy seeds, fertilizers, pesticides, and herbicides and in majority of the cases the yield of the crop is not sufficient enough to repay the debt; in fact, the farmers became more and more poor and indebtedness of the farmers kept on increasing. This indebtedness even led to the mass suicide of farmers in many parts of India. The younger generation, realizing the future of agriculture, lost all interest in the occupation and started looking for other opportunities, which led to mass migration of rural population either temporarily or permanently to the urban area which further led to a population explosion in the urban areas.

15.4.2 THE YIELD OF GREEN REVOLUTION

15.4.2.1 INCREASED USE OF CHEMICAL FERTILIZERS AND PESTICIDES

As a basic prerequisite of these HYVs, in order to get higher yield, a huge quantity of chemical fertilizers must be applied to the soil before sowing and at different stages of growth of the crop. As a result of the loss of micronutrients, the yield of these HYVs started decreasing. Expecting to get better results by increasing the dose of the chemical fertilizers, the resource crunch poor farmers kept on buying and increasing the use of these, but the yield continued to decrease over the years (Khajuria, 2016).

Farmers were using organic pesticides before the HYV seeds were introduced. But with HYV the scenario changed. "… rice varieties like PR106, which was considered to be resistant to white back planthopper and stem rot at the time of its introduction in 1976, has been found to be susceptible to both the diseases in addition to other pests like rice leaf-folder, hispa, stem borer, and several other pests" (Shiva, 1991). "The introduction of HYV has brought about a marked change in the status of insect pests such as gall midge, brown planthopper, leaf-folder, whore maggot, and so forth. Most of the HYV released so far are susceptible to major pests with a crop loss of 30–100 %" (Dogra, 1984). In cases, where new varieties were bred to ensure resistance to various diseases, it was found that "breakdown in resistance can occur rapidly and in some instances, replacement varieties may be required every 3 years or less" (CGIAR, 1979). Moreover, the price of these diseases or pest-resistant varieties which were bred afterward also increased over the years, which failed and simultaneously the price of pesticides to control these also skyrocketed.

15.4.2.2   LOSS OF RESOURCES

When artificial nitrogen is given to plants through fertilizers, their carbon–nitrogen balance gets disturbed as a result of which plants uptake more amount of water (Shiva, 1991). Since almost all of the lands were planted with the HYVs, there was excessive use of water leading to water shortage.

As a result of monocropping, the land got depleted of two major resources—soil and micronutrients. Compared to rice and wheat, legumes which were grown after these crops, such as oilseeds, pulses, and so forth, had a better capacity to hold the soil. The area under wheat has almost been doubled and that of rice being increased five-fold, whereas that under legumes decreased to half (Shiva, 1991). This has resulted in land erosion. Since the land is being depleted of the same micronutrients again and again due to the planting of the same crop round the year, the land became devoid of micronutrients such as zinc, iron, copper, manganese, magnesium, molybdenum, boron, and so forth (Shiva, 1991). Micronutrient deficiencies resulted in new diseases in plants.

Moreover, with peasants expecting better yield, the forests were cut and the area of cultivation was increased, which led to the loss under forest land. For example, 84% of the land of Punjab is under cultivation, whereas it is only 42% for India as a whole (Shiva, 1991). One of the striking facts is that of the total land in Punjab, only 4% is forest, and that too Eucalyptus which has been planted under afforestation program (Kang, 1982).

These HYVs yielded lesser straw as compared to traditional varieties which lead to loss of fodder and organic matter. The straw was also used for thatching the roof of houses of poor peasants, which also got affected.

Before the introduction of green revolution, farmers in India used to grow 100,000 varieties of rice (Naik et al., 2012) many of which were high yielding, and all of them were efficient users of water and had higher resistance to pests and diseases (Shiva, 1997, p 212). Farmers have lost the seeds of many of these indigenous varieties. With loss of indigenous varieties, focus on only staple food (rice and wheat), ignoring the fodder and plant biomass through plantation of other crops and drastic reduction in the cultivation of legumes such as oilseeds, pulses, and millets, there has been loss of biodiversity. In addition, the increase in cultivation of only wheat and rice, whose varieties too belonged to a narrow genetic base, further reduced the biodiversity.

15.4.2.3   IMPACT ON HUMAN LIFE

With no stringent norms or regulations on the use of the pesticides, fertilizers, and weedicides/herbicides, and no existing mechanism to aware and supervise the farmers on the use of these chemicals, farmers have used these in excessive amounts in the fields expecting to increase the yield and reduce the damage by protecting the crop from various diseases and pests. However, it has not yielded much result. It has led to seeping of these deadly chemicals into the air, water (surface as well as ground table including water table), and soil leading to environmental calamity affecting human lives adversely. The following interesting facts are self-explanatory:

1.   There is a train of Indian Railways, Train No. 339—Abohar Jodhpur Passenger, which runs via Punjab, once called the "wheat bowl of India," commonly known as Cancer Express. The cancer patients from Bhathinda of Punjab travelled to Bikaner, Rajasthan, where Acharya Tulsi Regional Cancer Treatment and Research Institute is situated, to get cancer treatment at a very subsidized cost (Donthi, 2010). Malwa region of Punjab is the most affected area (Aggarwal et al., 2015) and has now become a factory of cancer patients— ranging from throat cancer, nasal cavity cancer, paranasal sinus cancer, blood cancer, testicular cancer, breast cancer, intraocular cancer, and so forth. As per government statistics, cancer is taking 300 lives every year in Malwa region since 2008 (Kumar and Kaur, 2014). According to the Report of the Department of Health Affairs, Government of Punjab, there are 136 cancer patients per lakh of population in the Malwa region and 18 people are dying of cancer in Punjab everyday (Tribune News Service, 2013).

2.   In India, there is a regulatory body—Central Insecticide Board and Registration Committee that decides which pesticide has to be

used for which crop. Food Safety and Standards Authority of India (FSSAI) regulates the Maximum Residue Limit/Levels (MRLs) of the each of these pesticides in the final product which is available to the consumer. The MRL is the measure to ensure that the product has minimal ill effect of the pesticide after consumption. The MRL is available in the Prevention of Food Adulteration Act (Rule 65). However, it must be noted that the MRL must match the Acceptable Daily Intake (ADI) (Centre for Science and Environment, n.d.). Moreover, there are pesticides which are not to be used in fruit plantation and some which are not to be used in crops which are pollinated by honey bees. Interestingly, there are no means to measure the level of awareness regarding all of these among the farmers and regarding the MRL and ADI among the consumers or for that matter regarding the laboratories where the MRL can be tested. This can be better understood with the following facts:

- Of the 234 enlisted pesticides, FSSAI has not fixed the MRL of 59 pesticides (Centre for Science and Environment, n.d.).
- The Kerala University of Agriculture, in one of its studies, found a significant quantity of a dangerous banned pesticide—profenofos in vegetables such as cabbage, onion, and tomato (Martin, 2013).
- Delhi High Court informed that the fruits and vegetables available in the market in and around Delhi are not suitable for human consumption as it contains pesticides at dangerous levels (Delhiites Eating Food, 2014).
- Greenpeace India conducted a study to point out the pesticides used in tea cultivation in India. It was found that a pesticide namely monocrotophos, a suspected mutagen and neurotoxicant (Shah, 2014), the contamination of which lead to the death of 23 school children in Bihar and which was recommended to be phased out by Food and Agriculture Organization (FAO, 2013), is used in excess quantities in tea plantation.
- In Kerala, commonly known as "God's own country," endosulfan, a pesticide, is responsible for the death of 500 people as per the records of Government of India, though according to a study by a nongovernmental agency the figure goes up to 4000. As per United Nations Organisation, endosulfan is a dangerous pesticide and has been banned in 81 countries and not permitted but not banned in 12 countries (Government of Kerala, 2011).

However, it is being used for cultivation of cotton and cashew nuts in Kerala, Karnataka, Andhra Pradesh, Maharashtra, and Punjab. In 2000–2001, a study was conducted to understand the adverse effects of endosulfan, which found that it can cause—cancer, physical deformity, congenital disorder, cerebral diseases, disease of the central nervous system, mental disorder, and infertility (Government of Kerala, 2011).

- Though dichlorodiphenyltrichloroethane (DDT) is banned in many countries, it is used abundantly in India to protect the population from malaria and other vector (mosquito) borne diseases. United Nations Environment Program (UNEP) has directed the prohibition of 12 chemicals [persistent organic pollutants (POPs) DDT is one of these] exposure to which in humans can cause—death, cancers, allergies, hypersensitivity, developmental changes, damage to the central and peripheral nervous systems, disruption of the endocrine, reproductive, and immune systems (Simard and Spadone, 2012) and as per a study published in 2006, an increased level of POPs in human blood serum can be linked to diabetes (Lee et al., 2006).

- About 17 scientists from 11 countries across the globe met at International Agency for Research on Cancer and have identified Glyphosate, a weedicide, which is available as the brand "round up" to be "probable carcinogenic for humans" and two pesticides, namely tetrachlorvinphos and parathion as "possibly carcinogenic to humans" (IARC Monographs, 2015). Moms across America, a US-based organization, in one of its studies have found glyphosate in mother's milk (Zen, 2014). Each of these chemicals is used in large quantities in India.

- Carbide is being used to artificially ripen fruits in India. Both FSSAI and Prevention of Food Adulteration Act Rules, 1955, have found it to be highly dangerous for human consumption and have banned it. Carbide contains arsenic and phosphorous hydride which is capable of causing cancer, skin disease, vision disorder including loss of vision, diseases of central nervous system such as loss of memory, dizziness, mental imbalance, cerebral edema, and even seizure (National Centre for Cold Chain Development, n.d.).

- An unborn child receives these chemicals through the placenta, after birth, initially through mother's milk and later in life through the food s/he consumes.

- It has been proved through studies that once these chemicals enter the food cycle, a lot of chemical changes take place within them and in many cases their capacity to adversely affect human life increases (Beg, 2017, p 150–151).
- There are 67 pesticides which are banned (51) or restricted (16) in the world but are used in India (Press Trust of India, 2016).

15.4.3 RELEVANCE OF GREEN REVOLUTION TODAY

It has become quite evident today that the world has lost more than it has gained by the introduction of the artificial chemical fertilizers, pesticides, and herbicides. These chemicals have polluted the soil, water, and air: the three major resources the humans depend on for living. Studies have proved that the only benefactors of the introduction of these chemicals are the corporations manufacturing these agrochemicals. The HYVs have lost their significance as neither are they high yielding nor do they have the resistance to pests and diseases. The following studies give a better picture of the relevance of green revolution in the present scenario or the fallacy of it:

- Cornwell University conducted a study based on which it was estimated that 99.9% of all the chemical pesticides reach and stay in the environment (soil, water, and air), whereas only 0.1% kill the targeted pests (Pimental and Levitan, 1986).
- Indigenous agriculture based on biodiversity needs only 5 units of input to produce 100 units of food, whereas industrial agriculture requires 200 units of input to produce the same 100 units of food (Shiva, 1997, p.212).
- In 2000, International Rice Research Institute, Manila, had concluded based on a study that use of fertilizers in rice in Asian subcontinent is a "waste of time and effort." The study was conducted in India and some other Asian countries, where it was found that farmers got bumper yield without using the chemical fertilizers and pesticides (Sharma, 2006).
- Considering the labor and energy investment, comparative analysis of 22 rice growing varieties have confirmed that the indigenous process of agriculture is far more efficient (Baylis-Smith, 1984).
- Based on a study, Kang has concluded that this chemical agriculture "implies a downward spiraling of agricultural land used from legume to wheat to rice to wasteland" (Kang, 1982).

- In Iran, it has been concluded based on a study that with same and similar inputs, the local improved varieties gave better yield than the HYVs (Vann, 2012).
- Comparing small farms with large farms with similar soil composition and crop, the net production per unit of land is more in small farms irrespective of traditional or HYVs seeds (Vann, 2012).
- As per international estimates, the total food produced in the world is enough to feed 12 billion people. However, the population of the world is 7 billion. Hence, starvation deaths are taking place due to two reasons—irregular distribution of food and inability of people to buy the food. As a result, it can be concluded that if chemical fertilizers, pesticides, and herbicides are not used and it affects the yield (if at all it does), it would still be sufficient enough to feed the hungry in the world.
- As per the report of US Department of Agriculture, 1980, the concerns raised in relation to chemical-intensive agriculture are summarized as

  - Sharply increasing costs and uncertain availability of energy and chemical fertilizer, and our heavy reliance on these inputs
  - Steady decline in soil productivity and tilth from excessive soil erosion and loss of soil organic matter
  - Degradation of the environment from erosion and sedimentation, and from the pollution of natural waters by agricultural chemicals
  - Hazards to human and animal health and to food safety from heavy applications of pesticides
  - Demise of the family farm and localized marketing systems (USDA, 1980)

- In the preface of the National Program For Organic Production (NPOP), Government of India, it is clearly mentioned—"Growing consciousness of health hazards due to the possible contamination of farm produce from the use of chemical fertilizers have immensely contributed to the revival of this (organic) form of farming during the last 5 years." It further states that "Agro-climatic condition in India and our agricultural biodiversity are conducive to organic agriculture and, hence, offers tremendous scope for cultivation of a wide range of organic products" (Government of India, 2005).
- A study being carried out in test plots of Rodale Institute since last 27 years has found that the yield of conventionally and organically grown soybean is same. A meta-analysis of 115 comparative studies between conventional and organic yield by researchers at University

of California, Berkley, has found that organic farming yielded 19.2% less than conventionally grown crop which can further reduce to 9% with multi-cropping and crop rotation. Moreover, there was no difference in yield in case of legumes such as peas, beans, and lentils (Mehta, 2015).

15.5   ARTIFICIAL EASY SYSTEM: DAY TO DAY LIVING

15.5.1   LESS BRIGHT TO SLEEP TIGHT

Instead of looking at the way majority of the educated middle-class population starts the day in the morning, it would be interesting to look at how they go to bed at night. There was a time not so long ago when sun was the only major source of light, either before the artificial light was invented or before electricity reached to all across the globe as a basic essential service. In the absence of sunlight, evenings used to be dark or dimly light with candles or lamps. But with the dawn of the era of electricity and electric artificial lights, humans have tried their best to replace the sunlight (as far as illumination is concerned) with all sorts of electrical lamps and make sure that they put on most of them as soon as the sun sets, regardless of their actual need. It has been found that people keep lights on even during the day. However, taking security into consideration, it is very crucial to keep light on in the streets and passages, stairways, and other outdoor locations. However, people want the indoors to be so well—lit as if its daytime. It is a common practice that people switch off lights before going to sleep (with few exceptions who wants lights on even at night) and hence it can be logically concluded without any technical research that for the body to rest it needs minimal or no light. Nature also wants the beings to go to sleep after the sunset, may be not immediately after, but within a few hours after sunset, but of course not a few hours before sunrise, the normal practice followed by many these days. These artificial bright lights disturb the body's natural biological clock—the circadian rhythm, as a result of which sleep get disturbed (Dodson and Zee, 2011). The bright lights not only make the body not realize that it must rest within a few hours but also the use of electricity (which in majority of the cases is being generated using nonrenewable sources) is also consumed massively after the sunset. Hence, by using more lights during evening, late evening, and night, human beings are consuming or rather wasting a lot of nonrenewable energy source in the form of electricity. Research studies have shown that people suffering from insomnia has increased over the years. The link between health, maintaining sun timetable, and artificial lights can be better understood with the help of following studies:

- In 1981, Dr. Charles Czeisler of Harvard Medical School showed that daylight keeps a person's internal clock aligned with the environment (Harvard Health Publications, 2012).
- Several studies have linked working the night shift and exposure to light at night to several types of cancer (breast, prostate), diabetes, heart disease, and obesity. The basic reason being, exposure to the light during night time suppresses the secretion of melatonin, a hormone that influences the circadian cycle rhythms. A study has shown that lower secretion and hence lower levels of melatonin can lead to obesity and even cancer (Pepis, 2016).
- As per the studies of Harvard University, exposure to blue light has more capacity to affect the circadian cycle and melatonin secretion than the white or yellow light emitted by the filament bulbs and mercury gas fluorescent tube lights used earlier (Dougherty, n.d.). However, the curlicue compact fluorescent lights and the light-emitting diode (LED) emit blue lights. Stephen Lockley, a Harvard sleep researcher, conducted a study using a light of eight lux, which is less bright than most of the table lamps and twice of that of a standard night lamp and found that even it affected the circadian cycle. Based on his study, he concluded that the exposure to light during the night is one of the reasons why a lot many people suffer from lack of sleep (Harvard Health Publications, 2012).
- The exposure to this light at night has been linked to increased risk of depression, diabetes, and cardiovascular diseases (Chepesiuk, 2009).
- When Harvard researchers compared the effect of exposure to blue light with green light of comparable brightness for 6.5 h, they found that the blue light suppressed the melatonin secretion for double the time period and shifted the circadian cycle by double the time (3 h vs. 1.5 h). Blue light emitted by the compact fluorescent lamps and LEDs are thus more harmful (Harvard Health Publications, 2012).

15.5.2 LOW ON HIGH TECH

The usual trend is that people keep their internet, Wi-Fi, laptop, palmtop, tablet, and other gadgets on till they go to bed. Surprisingly, the mobile and its Wi-Fi/internet are kept on even when people go to sleep. Irrespective of whether people are late in going to bed, they need alarms to get up in the morning and suffer from morning sickness. The usual excuse put forth by people for keeping the mobile (s) on during the night is that they need the alarm. Even if they need alarm, there are alarms clocks still available in the market. Moreover, in all mobiles, the alarm works even if the mobile is switched off. Since it is known that tumor takes almost 30 years to develop, it is very difficult to link cancer with Wi-Fi or radio frequency (RF). The following studies have clearly shown a link between the Wi-Fi/broadband, gadget used, and the health hazards associated with it:

- A study which appeared in *Journal of Microscopy and Ultrastructure*, funded by Saudi Society of Microscopes, suggests that compared to adults, young adults, children, and babies have higher risk from certain RF signals, the reason being, thinner skull and more absorbent brain tissues (Morgan, 2014).
- Agency for Research on Cancer (a wing of World Health Organization) lists RF and electromagnetic frequency as a Class 2B carcinogenic which means that under specific conditions and levels of exposure these may cause cancer (WHO, 2014).
- Two different studies, one conducted by Rodney Croft of the Brain Science Institute, Swinburne University of Technology in Melbourne, Australia, and the other by James Horne and colleagues at the Lough-borough University Sleep Research Centre in England, have found that the cell phone affected the functioning of the alpha and delta waves of the brain even when the phone is on standby and the effect continues for hours after the phone has been switched off. These studies concluded that though these findings do not suggest possible health hazards, inducing disturbance to brain waves cannot be considered healthy. However, both concluded that it affects sleep (Fields, 2008).
- There is a 20-cm safe distance rule for laptops and tablets but it contradicts with the normal operating position. There have been two cases in the US, where women keeping their smartphones under bras have developed breast cancer exactly at the position where they used to keep their phone (Safe Tech for Schools Maryland, 2016).
- As per a study, the use of laptop resulted in decreased motility of sperm and increased DNA defragmentation (Avendano, 2011)
- These gadgets also emit blue light which has been found to deter the melatonin secretion and circadian cycle (Parson, 2015).
- The Apple iPhone 6 recommends that the phone should be kept 5 mm away from the body. (iPhone 6, n.d.)

• Besides the traditional motion sickness, a new age disease has been identified to be digital motion sickness or cybersickness or virtually induced motion sickness. Cyriel Diels, a cognitive psychologist and human factors researcher at Coventry University, Centre for Mobility and Transport, England, calls it "a natural response to an unnatural environment" (Saner, 2015). Dr. Steven Raunch, medical director of Massachusetts Eye and Ear Balance and Vesticular Centre and professor of otolaryngology at Harvard University states that this disease stems from a basic mismatch between sensory inputs—the motion that the patient sees when s/he is looking at the screen of the gadget lying very close to the eyes but does not feel it (Bratskeir, 2015).

15.5.3 WHY STICK TO PLASTIC?

The "t-shirt polythene bags" emerged as one of most convenient option for grocery bags. By 1985, the use of polythene bags was widespread in all developed countries, but environmental activists raised concerns regarding its environmental impact. By 1990s, these polythene bags had entered Indian market and consumers were happy with this alternative which could carry 1000 times its own weigh (Doucette, 2011). For the convenience with regard to the soft, transparent, flexible, lightweight, cheap, disposal space (foldable), nonseepage of wet contents, and so forth, people stopped using the normal grocery bags and started using these polythene bags. Over the years, the normal polythene bag has seen two major changes—the thickness and development of recyclable plastics. Quite similar to the use of polythene bags, a new product emerged which was found to be very lightweight, ductile, cheap, and impermeable, it was thermocol, which was soon molded to be used for making single-use plates, cups, glasses, bowls replacing the traditional ones made from *saal* (*Shorea robusta*) leave mats and earthen pots. However, the world over, the disposal of plastics and thermocols have become one of the major problems as these take many years to biodegrade and hence lies in the soil or under the sea/ocean releasing various chemicals during the degrading phase and harming the environment (Gewert et al., 2015). An estimated 46,000 pieces of plastic occupy each square mile of ocean (Ferguson, 2006). According to U.S. National Park Service; Mote Marine Lab, Sarasota, FL., the average time for the biodegradation of few commonly used products are: plastic beverage bottle—450 years, disposable diaper—450 years, foamed plastic bouy—80 years, foamed plastic cups—50 years, and plastic bag—10–20 years (Government of New

Hampshire, n.d.). Considering just one of these products as an example—disposable diaper, in the United States more than 300 pounds of wood, 50 pounds of petroleum feedstock, and 20 pounds of chlorine are used to produce disposable diapers for one regular baby for 1 year (Kelly, 2012). There are other products which take longer to biodegrade but once they are compared with the plastic and foam products in respect of quantity disposed per second, it would be very evident that the disposal of the plastic and foam products poses a greater threat to the existence of life on Earth. There have been numerous studies around the world which have proved that the plastic use harms human health, inappropriate plastic disposal affects the lives of humans and other living beings on Earth and water and since it does not degrade easily, it affects all the other elements that sustain life on earth. The following facts and studies will illustrate the ways plastic is damaging the environment and life on earth:

- To improve the performance, the virgin plastic polymer is usually mixed substantially with a few additives-bisphenol-A (BPA), brominated flame retardants, antimicrobial agents and phthalates, to name a few which are of main concern. "BPA and phthalates are found in many mass-produced products including medical devices, food packaging, perfumes, cosmetics, toys, flooring materials, computers, and compact discs and can represent a significant content of the plastic" (Thompson et al., 2009). These are dangerous because of the combination of their persistence in environment and toxicity. Phthalates can constitute a substantial proportion, by weight, of PVC (Oehlmann et al., 2009), while endocrine-disrupting BPA is the monomer used for production of polycarbonate plastics as well as an additive used for production of PVC. Phthalates can leach out of products because they are not chemically bound to the plastic matrix, and environmentalists are more concerned about these because of the magnitude of their production volumes and range of products on which these are used (Wagner and Oehlmann, 2009, p 278; Talsness et al., 2009). Phthalates and BPA are detectable in aquatic environments, in dust and, because of their volatility, in air (Rudel et al., 2001, 2003). These chemicals have very dangerous effect on humans themselves as well as on other living creatures (Meeker, 2009; Oehlmann et al., 2009).
- Around 4% of the world oil production is used for feedstock and approximately 3–4% more for providing energy for the manufacture of plastic (Hopewell et al., 2009). Hence by manufacturing plastics,

we are polluting the environment in terms of the pollution associated with more extraction of oil and its refinement.

- Studies have shown that the plastic debris laced with chemicals, often ingested by marine animals lead to injury or death. About 100,000 marine creatures die each year due to plastic entanglement and 1 million seabirds die by entanglement or ingesting plastic (Ocean Crusaders, n.d.).
- As the plastic buried deep in landfill slowly degrades, very harmful chemicals seep into the ground and reach the groundwater (Knoblauch, 2009).
- Some 299 million tons of plastic was being produced in 2013 (Gourmelon, 2015), which are getting disposed without proper management in majority of the cases leading to clogging of drains, convergences of rivers, the sea/ocean floors, apart from filling the upper layer of the soil making it unfavorable for any living being, including plants. It is being estimated that considering the multiplying increase in the use of plastics, the amount of plastic produced during the first decade of the century is almost equal to the total produced in the last century (Knoblauch, 2009).

15.6   THE WAY FORWARD–TOWARD SELF-SUSTAINING NATURAL SYSTEM

15.6.1   ORGANIC FARMING AND ORGANIC FOOD

It is high time that governments of countries across the globe setup plans to help farmers to reduce and subsequently stop the use of chemical fertilizers and pesticides for agriculture. For further regulation, ban on the use of these artificial chemicals can be imposed by the government in order to ensure that farmers do not have access to these.

By the time, the chemical-intensive agriculture was in use in full swing all across India, the United States had realized the ill effects of this form of farming. In 1980, US Department of Agriculture defined organic farming as

"Organic farming is a production system which avoids or largely excludes the use of synthetically compounded fertilizers, pesticides, growth regulators, and livestock feed additives. To the maximum extent feasible, organic farming systems rely upon crop rotations, crop residues, animal manures, legumes, green manures, off-farm organic wastes, mechanical cultivation, mineral-bearing rocks, and aspects of biological pest control to maintain soil productivity and tilth, to supply plant nutrients, and to control insects, weeds, and other pests." (USDA, 1980)

The NPOP, Government of India defines organic agriculture as

"It is a system of farm design and management to create an ecosystem, which can achieve sustainable productivity without the use of artificial external inputs such as chemical fertilizers and pesticides" (Department of Commerce, Government of India, 2014).

However, all the programs that are running to promote and implement organic farming have incorporated a certification of the land and/or the product by enlisted authorized certifying agencies as one of the prerequisites for calling any product as "organic." It is evident that the certification process has been introduced in order to ensure that the product is authentically organic. However, it is really interesting that farmers are free to cultivate crops using the artificial chemicals and poison the soil, water, air, and the food of the consumers and they do not need any penalty or fine, whereas farmers who make efforts to grow safe food have to pay a premium for labeling and selling it as "organic."

The process of organic certification is very resource intensive. Those farmers who are converting their land from chemical to organic have to wait for at least a certain period of time, in the majority of cases almost 3 years, during which they are not allowed to call their products organic. The certification process involves payment to the certifying agency at a rate specified by the agency per square unit of land. Besides this, the farmer has to bear the cost of the travel and accommodation of the consultants of the agency for and during the certification process. The certification has to be renewed each year, the renewal is also charged by the agency per unit area. Further, every plot where the organic product is being grown, a buffer or boundary has to be maintained in order to ensure that the prohibited substances used in the adjoining plots do not reach the products in the organic plot through the air (spray), water (waterlogging or seepage), and so forth. All these costs, add up to something that the small and marginal farmers cannot afford. Though there is the provision of a form of certification which is not much resource intensive, called the group certification or Participatory Guarantee System and where the certification is provided by farmer groups and farmers cooperatives. This is based on internal control among different farmers groups or cooperatives. Generally, it is very difficult to convince the consumers that the products are genuinely organic even with a certificate from an enlisted certifying agency; hence if the certificate is provided by farmers groups, the majority of the consumers may not consider the certificate to be authentic.

This is not only making this process of group certification null and void, the small and marginal farmers who can afford only this form of certification are losing interest in growing organically as it requires more effort, energy, and money (in the form of labor lost). Since almost three quarters of the world's farmers cultivate small plots and of the total operational area in India, 44.58% is operated by small and marginal farmers (Press Information Bureau, 2015) which account for almost 67% of the total farmer (Indo-Asian News Service, 2015), they play a very crucial role in agriculture. More so, because they contribute 41% of the total food grains produced in the country (FAO, 2002). These small and marginal farmers spend more time on mulching, trellising, weeding, removing rocks stones, soil conservation, and building natural irrigation systems. They mostly engage their family members and rarely local labors, for the work on their farms. They usually plant as soon as they harvest and grow multiple crops, from paddy to legumes, to vegetables. Hence, this process of conversion of agriculture to organic must include the small and marginal farmers and governments should take special initiatives to encourage and involve them. In fact, greater thrust needs to be put now to include them in organic farming than that which was implemented when the green revolution was introduced, as this time it is not only about food security of human beings, but about the survival of all living beings on earth.

Special initiatives have to be undertaken to identify, collect, and conserve the indigenous seeds of various crops and provide these seeds free of cost to the small and marginal farmers. This, of course, will be a one-time investment, in most of the cases, as unlike the HYV seeds, most of the indigenous seeds can be then preserved by the farmers themselves for subsequent years. However, they have to be given appropriate training on seed conservation (an age-old practice which most of them have completely forgotten). Initiatives similar to the one taken by Dr. Debal Deb of Basudha/Vrihi to conserve and create a seed bank of indigenous varieties of rice seeds, which presently has 1200 rice landraces (Vrihi, n.d.), have to be adopted by government and nongovernment agencies to revive the agriculture.

The way the HYV seeds and chemical fertilizers, pesticides, and herbicides, including the machinery such as tractors and tillers have been provided to farmers at subsidized rates, it is time to invest in the production of organic fertilizers, pesticides and provide these at subsided rates to the farmers. Small and marginal farmers also need to be supported to buy the traditional breeds of cattle including bullocks, so that the farming can be revived to the old traditional method of ploughs which do not require any machinery and fuel (nonrenewable resource) and which will ensure that the farm-friendly worms can survive in the soil.

In order to ensure that optimal yield is achieved, it is very crucial that the farmers are trained on integrated pest management, mixed cropping, multiple cropping, integrated farming, crop rotation, and so forth, and such trainings have to be provided repeatedly and consistently at regular intervals along with reinforcements.

15.6.2   HEALTHY LIFESTYLE

Human beings, the most advanced of all the creatures on earth must safeguard themselves first and in order to do that they must protect their environment and use their resources optimally. It has been proved scientifically that in order to live healthy or to become healthy, full deep sleep which may vary from 7–9 h plays a vital role. Furthermore, it has been proved that the best time to sleep is between 10:00 pm and 4:00 a.m. Of course, individuals have their own circadian cycles and need different duration of sleep, if they can bring it to a time close to this ideal period, it will help in regenerating the health. It is better to have your meals 2 h before sleep. In order to get a good sleep, it is very essential that we use dim light and stay away from our gadgets at least an hour or two before going to bed. These will not only protect humans from the ill effects of the exposure to these gadgets and lights, it will also save electricity (which mostly uses nonrenewable energy source). Initiatives need to be taken by the government and internet service providers, to ensure that rates for internet use during the night is increased which will deter consumers to use internet during the night. Messages warning the common mass regarding the ill effects of overexposure to bright lights, gadgets, and Wi-Fi/internet should be spread through newspapers, radio (frequency modulation channels), and television.

15.6.3   RETHINKING THE USE AND RECYCLING THE PLASTIC WASTE

Landfilling, incineration, recycling, and biodegradation are four key processes which can be employed for disposal of plastics (North and Halden, 2014). Landfilling leads to toxicity reaching the soil and the groundwater, incineration leads to releasing toxins in the air. As far as recycling is concerned, developed countries such as the United States in 2012 was able to recycle only 9% of the total 32 million tons of plastic waste generated (Sparling, 2016). Biodegradation requires too long time and hence humans have to opt for the only other available option—replace the use of plastics with aluminum, steel, wood, glass, bagasse (the pulp leftover when juice is extracted from sugarcane and beets and can be used for disposable plates, cups, or take out containers), and bioplastics (made from vegetable fats and oils or cornstarch instead of fossil fuel). Special drives need to be undertaken by governments to promote bagasse and bioplastics so that these are available in abundance and are used to replace plastic.

There are two major deterrents that can be put in use to ensure that the consumer stops, at least, asking if not using, polythene bags from the retailers, which over the years have been found to be successful:

1. Retailer stops providing polythene bags to the consumers, except for the first-timers who were not aware of the unconventional rule of the retailer. Even if the first timer is provided with a polythene bag, that has to be an old one to ensure reuse of the bag. Just like a "no smoking zone" signboard, the retailer should put "no plastic zone"/"no plastic bags" sign at different positions in and around the shop. It has been found that subsequently, consumers visiting these retailers bring bags from their home. It becomes the responsibility of the retailer to encourage the customer by thanking him/her for bringing the bag. Further, it has been found that the customers who are sensitized on the "reuse" issue, bring a bunch of used polythene bags from their home and give these to the retailer so that the retailer can provide these to the first time customers, who are new to the practice.

2. Charge a good amount of money for the plastic bag. In general, it is a practice that the retailers charge a nominal amount for the bags, which has not been found to be a deterrent. It has to be explained to the consumer that the price the consumer is paying, is not only for the plastic bag, but also for the damage that bag would cause to the environment by not degrading, clogging the drains, and injuring other living creatures.

Since 2005, when UNEP initiated the Social Economic Environmental Design (SEED) award for promoting entrepreneurship for sustainable development, youths from all the six focus countries have come up with innovative enterprises which are making significant contribution in eradicating poverty and environmental sustainability. One of the sectors under the SEED is waste management. Innovative plastic or polyethylene terephthalate recycling projects have been receiving awards under waste management sector. The waste plastic is being recycled to convert it into utility products ranging from road construction material (used along with bitumen) to handbags, table mats, laptop cases, clothing, accessories, jewelry, housewares, and so forth. Plastic poles, which are cost-effective, less maintenance intensive and durable alternative to timber or iron poles have also been manufactured by recycling the plastic waste.

However, in order to implement any such project for recycling the plastic waste, it is very crucial that the population must sort the garbage into biodegradable and nonbiodegradable waste. In order to achieve the same, it is a prerequisite that the population is sensitized on the environmental cost of the indifference toward garbage disposal and the significant impact such project can have on the environment and the lives of people living at present and for future generations. Government and nongovernment organizations need to make sure that the common mass is made aware of the gravity of the situation through messages in newspapers, television, and radio channels.

15.7   CONCLUSION

The present living generation has probably witnessed the peak of techno-logical development in almost every aspect of life. This is prompting them to adapt their lives to this "artificial easy system" which is giving them a notion that they can become more and more superior by engaging their lives and their minds into this new system. On the other hand, in spite of this, the present generation is also witnessing how helpless this technological advancement is in tackling the new health concerns, whether it is the development of treatment for the new emerging diseases or curing the chronic lifestyle diseases. It is also witnessing a good number of natural and man-made calamities. Unfortunately, it is this present generation which is turning a blind eye to anything else they are witnessing, which is pointing out the ill effects this new system is having on Earth, whether it is global warming or extinction of different species.

The primary focus should be on the health of human beings, to take measures so that healthy living styles are adopted instead of the mere provision of treatment of diseases. To start with, this should include organic food that is free from any genetic modification and efforts to ensure that local people eat locally grown seasonal food except for spices and beverages which do not grow everywhere. The primary hindrance, of course, is the high price of organic food (because of the resource intensive organic certification) and the limited farmland under organic agriculture. Steps have to be initiated to encourage farmers to grow organic food by subsidizing on seeds, saplings, and organic fertilizers and pesticides and simplifying the process and reducing the cost involved for certification. Deterrents have to be planned for those continuing conventional agriculture using chemicals. Governments and Corporates can come up with disincentives and incentives for employees for adopting green practices at household and community level. "Swachh Bharat Abhiyan" (Clean India Initiative) has already sown the seed for garbage disposal. Electricity use at the household level can be charged heavily as the night progresses in order to ensure that people use less electricity at night, which will compel them to sleep early and thereby ensure that the natural cycle of the sun is adopted and people use dim lights at night. Most of the internet and mobile service providers offer "happy hours" at night which prompt people to use these gadgets heavily during those hours. Instead, stricture should be imposed on such corporate companies who are putting profits before the health of the users.

The onus lies on those who are aware, sensitive and have the authority and capacity to take steps in order to curb the situation before it goes out of bounds. The government, educational institutions, teachers, professors, and civil society organizations/NGOs have to come forward to take the lead in planning and implementing steps, either by adopting the existing successful models or developing new ones based on the learning of other's experiences. This will obviously require lots of effort in bringing all stakeholders in one platform, brainstorming, planning, and executing the projects. However, one of the prerequisites is to shift the focus from individual achievement to individual contribution, not only for those who will be executing the projects but also for the population who are busy earning a living and are not habituated to think beyond the daily earning.

KEYWORDS

- green revolution
- high yielding varieties
- exposure to blue light
- circadian cycle
- radio frequency
- plastic disposal
- organic farming
- recycling of plastic waste

REFERENCES

Abrol, I. P. *Agriculture in India*. (n.d.). planningcommission.nic.in/reports/sereport/ser/vision2025/agricul.doc

Aggarwal, R.; Manuja; Aditya, K.; Singh, G. P. I. Pattern of Cancer in a Tertiary Care Hospital in Malwa Region of Punjab, in Comparison to Other Regions in India. *J. Clin. Diagn. Res.* **2015**, *9*, 3. DOI: 10.7860/JCDR/2015/11171.5685.

Agricultural Production. (n.d.). http://indiabudget.nic.in/es1969-70/2%20Agricultural%20Production.pdf

Apple Inc. *IPhone 6 RF Exposure Information*. (n.d.). http://www.apple.com/legal/rfexposure/iphone7, 2/en/

Avendano, C.; Mata, A.; Sanchez Sarmiento, C. A.; Doncel, G. F. *Fertil. Steril.* **2011**, *91, 1*. DOI: http://dx.doi.org/10.1016/j.fertnstert.2011.10.012

Bayliss-Smith, T.; Wanmali, S.; Ed. *Understanding Green Revolutions*. Cambridge University Press: Cambridge, UK, 1984.

Beg, M. A. A. *Pesticides Toxicity Specificity and Politics;* Jan 2017. DOI: 10.13140/RG.2.2.35594.67520. https://www.researchgate.net/publication/312042352_Book_Pesticides_Toxicity_Specificity_Politics_Chapter_5_MECHANISM_OF_BIOTRANSFOR-MATION_OF_PESTICIDES_TOXICITY_TRAIL

Biddle, W. Nerve Gases and Pesticides: Links are Close. *New York Times*. March 30, 1984. http://www.nytimes.com/1984/03/30/world/nerve-gases-and-pesticides-links-are-close.html

Bratskeir, K. 8 Physical Risks of Too Much Screen Time. *The Huffington Post*. Nov 18, 2015. http://www.huffingtonpost.in/entry/technology-health-physical-effects_us_564a1df4e4b045bf3df03368

Centre for Science and Environment. *State of Pesticide Regulations in India*. (n.d.). http://www.cseindia.org/userfiles/State%20of%20Pesticide%20Regulations%20in%20India_CB.pdf

Chapter III: Case Studies-1 The Food Problem (a) India. n.d. http://pacific.unescap.org/publications/survey/surveys/survey1964-3.pdf

Chepesiuk, R. Missing the Dark: Health Effects of Light Pollution. *Environ. Health Perspect.* **2009**, *117*, A20–A27. https://www.ncbi.nlm.nih.gov/pmc/articles/PMC2627884/

Consultative Group of International Agricultural Research. *Integrative Report*. Washington D.C. 1979, p. 13. https://library.cgiar.org/bitstream/handle/10947/5411/cgint79.pdf?sequence=1 (accessed April 11, 2017).

Delhiites Eating Food Unfit for Humans: Delhi High Court. *India Today*. March 6, 2014. http://indiatoday.intoday.in/story/delhiites-unhygenic-food-delhi-high-court-fruits-vegetables-pesticide-residue/1/347018.html

Department of Commerce, Ministry of Commerce and Industry, Government of India. *National Program for Organic Production*. Sept 2005, http://www.apeda.gov.in/apedawebsite/organic/ORGANIC_CONTENTS/English_Organic_Sept05.pdf

Department of Commerce, Ministry of Commerce and Industry. *Revised National Program for Organic Production*. 2014, p. 8. http://cgcert.com/source/Download/RevisedNPP2014.pdf

Department of Environmental Services, Government of New Hampshire. *Time it Takes for Garbage to Decompose in the Environment*. (n.d.). https://www.des.nh.gov/organization/divisions/water/wmb/coastal/trash/documents/marine_debris.pdf

Department of Health and Family Welfare, Government of Kerala. *Endosulfan The Kerala Story*. April 20, 2011. http://www.cseindia.org/userfiles/endosulfan_kerala_story.pdf

Dodson, R. D.; Zee, P. C. Therapeutics for Circadian Rhythm Sleep Disorders. *Sleep Med. Clin.* **2011,** *5,* 4. DOI: http://dx.doi.org/10.1016/j.jsmc.2010.08.001 (accessed April 12, 2017).

Dogra, B. *Empty Stomachs and Packed Godowns*. New Delhi, India, 1984.

Donthi, P. Cancer Express. *Hindustan Times.* Jan 17, 2010. http://www.hindustantimes.com/india/cancer-express/story-G0i4G2mnoQnLDRry0kiV6L.html

Doucette, K. The Plastic Bag Wars. *Rolling Stone.* July 25, 2011. http://www.rollingstone.com/politics/news/the-plastic-bag-wars-20110725

Dougherty, E. *Blue Cues.* (n.d.). https://hms.harvard.edu/news/harvard-medicine/blues-cues.

FAO. *Smallholder Farmers in India: Food Security and Agricultural Policy*. March 2002. ftp://ftp.fao.org/docrep/fao/005/ac484e/ac484e00.pdf

FAO. *Highly Hazardous Pesticides should be Phased Out in Developing Countries*. July 30, 2013. http://www.fao.org/news/story/en/item/180968/icode/

Ferguson, N. Fact: 46,000 Pieces of Plastic Float on Each Square Mile of Sea. *The Telegraph.* Aug 6, 2006. http://www.telegraph.co.uk/comment/personal-view/3626914/Fact-46000-pieces-of-plastic-float-on-each-square-mile-of-sea.html

Fields, R. D. Mind Control by Cell Phone. *Scientific America.* May 7, 2008. https://www.scientificamerican.com/article/mind-control-by-cell/#

Gewert, B.; Plassmann, M. M.; Macleod, M. Pathways for Degradation of Plastic Polymers Floating in the Marine Environment. *Environ. Sci. Process. Impacts* **2015,** *17.* DOI: 10.1039/c5em00207a.

Gourmelon, G. *Global Plastic Production Rises, Recycling Lags*. World Watch. Jan 18, 2015. http://www.worldwatch.org/global-plastic-production-rises-recycling-lags-0

Harvard Health Publications. *Blue Light has a Dark Side*. Harvard Health Letter. May 2012. http://www.health.harvard.edu/staying-healthy/blue-light-has-a-dark-side

Herger, G.; Nielsen, R.; Margheim, J. *Fertilizer History P3: in WWII Nitrogen Production Issues in Age of Modern Fertilizers*. April 10, 2015. http://cropwatch.unl.edu/fertilizer-history-p3

Hopewell, J.; Dvorak, R.; Kosior, E. Plastics Recycling: Challenges and Opportunities. *R. Soc. Publ.* **2009,** *364,* 1526. DOI: 10.1098/rstb.2008.0311.

Indo-Asian News Service. Nearly 70 Percent of Indian Farms are Very Small, Census Shows. *Business Standard.* Dec 9, 2015. http://www.business-standard.com/article/news-ians/nearly-70-percent-of-indian-farms-are-very-small-census-shows-115120901080_1.html

International Agency for Research on Cancer. *IARC Monographs Volume 112: Evaluation of Five Organophosphate Insecticides and Herbicides.* March 20, 2015. http://www.iarc.fr/en/media-centre/iarcnews/pdf/MonographVolume112.pdf

Kang, D. S. Environmental Problems of the Green Revolution with a Focus on Punjab, India. In *International Dimensions of the Environmental Crisis;* Barett, R., Ed.; Westview: Boulder, Colorado, 1982; p. 204.

Kelly, A. *The Dirt on Disposable Diapers*. April 4, 2012. https://www.therapidian.org/dirt-disposable-diapers

Khajuria, A. Impact of Nitrate Consumption: Case Study of Punjab, India. *J. Water Resour. Prot.* **2016,** *8,* 211–216. DOI: 10.4236/jwarp.2016.82017.

Knoblauch, J. A. *The Environmental Toll of Plastics*. July 2, 2009. http://www.environmentalhealthnews.org/ehs/news/dangers-of-plastic

Kumar, G.; Kaur, A. Factors Responsible for Cancer in Bhatinda: Socio-Economic Impacts. *Int. J. Adv. Res. Manage. Soc. Sci.* **2014,** *3,* 8. http://www.garph.co.uk/IJARMSS/Aug 2014/8.pdf (accessed March 10, 2017).

Lee, D. H.; Lee, I. K.; Song, K.; Steffes, M.; Toscano, W.; Baker, B. A.; Jacobs, D. R., Jr. A Strong Dose-Response Relation between Serum Concentrations of Persistent Organic Pollutants and Diabetes: Results from the National Health and Examination Survey 1999–2002. *Diabetes Care* **2006,** *29,* 7. DOI: 10.2337/dc06-0543.

Martin, K. A. Banned Pesticide Residues Found in Vegetables. *The Hindu*. June 18, 2013. http://www.thehindu.com/news/cities/Kochi/banned-pesticide-residues-found-in-vegetable-samples/article4824152.ece

Meeker, J. D.; Sathyanarayana, S.; Swan, S. H. Phthalates and Other Additives in Plastics: Human Exposure and Associated Health Outcomes. *R. Soc. Publ.* **2009,** *364,* 1526. DOI: 10.1098/rstb.2008.0268

Mehta, S. *Sustainable Agricultural Development for Food Security and Nutrition, Including the Role of Livestock.* Global Forum on Food Security and Nutrition for High Level Panel of Experts. Jan 13, 2015. http://www.fao.org/fsnforum/cfs-hlpe/node/689

Morgan, L. L.; Kesari, S.; Davis, D. L. *J. Microsc. Ultrastruct.* **2014,** *2,* 4. DOI: http://dx.doi.org/10.1016/j.jmau.2014.06.005.

Naik, G. H.; Balachandran, C.; Subash Chandran, M. D.; Ramachandra, T. V. *In Situ Conservation of Traditional Rice Varieties of Uttara Kannada.* Proceedings of the LAKE 2012: National Conference on Conservation and Management of Wetland Ecosystems, Nov 6–9, 2012; School of Environmental Sciences, Mahatma Gandhi University: Kottayam, Kerala. 2012. http://wgbis.ces.iisc.ernet.in/energy/water/paper/lake2012_traditional_rice/index.htm (accessed Feb 23, 2017).

National Centre for Cold Chain Development. *NCCD-Skill Development Program-Fruit Ripening–Huge Scope for Employment Generation.* (n.d.). http://www.nccd.gov.in/PDF/Report%20South%20India.pdf

North, E. J.; Halden, R. U. Plastics and Environmental Health: The Road Ahead. *De Gruyter* **2014,** *28,* 1. DOI: https://doi.org/10.1515/reveh-2012-0030

Ocean Crusaders. *Plastic Statistics in Plastic Ain't so Fantastic.* (n.d.). http://oceancrusaders.org/plastic-crusades/plastic-statistics/

Oehlmann, J.; et al. A Critical Analysis of the Biological Impacts of Plasticizers on Wildlife. *R. Soc. Publ.* **2009,** *364,* 1526. DOI: 10.1098/rstb.2008.0242.

Parson, J. Blue Light from Smart Phones and Laptop Causes Insomnia but this App Lets you Keep Reading without Preventing Sleep. *Mirror.* Nov 5, 2015. http://www.mirror.co.uk/news/technology-science/science/blue-light-smartphones-laptop-causes-6772588

Pepis, S. *Fat is Our Friend.* 2QT Limited. 2016. https://books.google.co.in/books?id=igzXCwAAQBAJ&pg=PT48&lpg=PT48&dq=false

Pimental, D.; Levitan, L. Pesticides: Amounts Applied and Amount Reaching Pests. *Bio Science* **1986,** *36,* 2. http://www.beyondpesticides.org/assets/media/documents/mosquito/documents/Pimentel%201985%20crop%20spray%20effectivness.pdf

Planning Commission, Government of India. *Chapter 18: Some Problems of Agricultural Development.* (n.d.). http://planningcommission.nic.in/plans/planrel/fiveyr/1st/1 planch18.html

Press Information Bureau. *Highlights of Agriculture Census 2010-11.* Dec 9, 2015. http://pib.nic.in/newsite/PrintRelease.aspx?relid=132799

Press Trust of India. Use of 51 Pesticides Banned Elsewhere Allowed in India, Centre Tells High Court. *The Hindu Business Line.* Dec 7, 2016. http://www.thehindubusinessline.com/economy/agri-business/use-of-51-pesticides-banned-elsewhere-allowed-in-india-centre-tells-high-court/article9416251.ece

Randhawa, M. S. *A History of Agriculture in India Volume One, Beginning to 12th Century.* Indian Council of Agricultural Research, New Delhi, India. Dec 1980. https://archive.org/details/HistoryAgricultureIndia1

Rudel, R. A.; Brody, J. G.; Spengler, J. C.; Vallarino, J.; Geno, P. W.; Sun, G.; Yau, A. Identification of Selected Hormonally Active Agents and Animal Mammary Carcinogens in Commercial and Residential Air and Dust Samples. *J. Air Waste Manag. Assoc.* **2001,** *51*, 499–513.

Rudel, R. A.; Camann, D. E.; Spengler, J. D.; Korn, L. R.; Brody, J. G. Phthalates, Alkylphenols, Pesticides, Polybrominated Diphenyl Ethers, and Other Endocrine-Disrupting Compounds in Indoor Air and Dust. *Environ. Sci. Technol.* **2003,** *37*, 4543–455.

Safe Tech for Schools Maryland. Top Ten Facts About Laptops [Web log Comment]. March 20, 2016. http://safetechforschoolsmaryland.blogspot.in/2015/07/top-ten-facts-about-laptops.html

Saner, E. Why Staring at Screens is Making us Feel Sick. *The Guardian.* Nov 18, 2015. https://www.theguardian.com/technology/shortcuts/2015/nov/18/why-staring-at-screens-is-making-us-feel-sick

Shah, S. Ek Cup Chai-Sans Pesticides Please [Web log comment]. Aug 11, 2014. http://www.greenpeace.org/india/en/Blog/Campaign_blogs/ek-cup-chai-sans-pesticides-please/blog/50247/

Sharma, D. *Similarities and Differences between Indian and American Agriculture and need for the KIA.* Paper Presented at Indo-US Knowledge Initiative on Agriculture-Whither Indian Farmer, National Workshop, Hyderabad, India, Dec 8–9, 2006. http://www.kicsforum.net/workshop/Indo-US-knowledge-initiative/Similarities-&-Differences.htm

Shiva, V. *The Violence of Green Revolution: Ecological Degradation and Political Conflict in Punjab.* Zed Books Ltd and Third World Network. 1989. http://www.trabal.org/courses/pdf/greenrev.pdf

Shiva, V. The Green Revolution in the Punjab. *Ecologist* **1991,** *21,* 2. file:///E:/Books/CHAPTER/Ref/The%20Green%20Revolution%20in%20the%20Punjab.html

Shiva, V. "Letter Forum". *Ecologist* **1997,** *27,* 5. http://exacteditions.theecologist.org/print/307/308/6182/3/46 (accessed Dec 10, 2016).

Simard, F.; Spadone, A. (Eds.). *An Ecosystem Approach to Management of Seamounts in the Southern Indian Ocean: Volume2: Anthropogenic Threats to Seamount Ecosystems and Biodiversity.* IUCN. 2012. http://www.mu.undp.org/content/dam/mauritius_and_seychelles/docs/Seamount%20report%20Vol%202.pdf (accessed March 4, 2017).

Sparling, D. W. *Eco Toxicity Essentials: Environmental Contaminants and Their Biological Effects on Animals and Plants.* Academic Press. 2016. https://books.google.co.in/books?id=kiDfCQAAQBAJ&pg=PA278&lpg#v=onepage&q&f=false

Talsness, C. E.; Andrade, A. J.; Kuriyama, S. N.; Taylor, J. A.; vom Saal, F. S. Components of Plastic: Experimental Studies in Animals and Relevance for Human Health. *R. Soc. Publ.* **2009,** *364*, 1526. DOI: 10.1098/rstb.2008.0281

Thompson, R. C.; Moore, C. J.; vom Saal, F. S.; Shanna, H. S. Plastics, the Environment and Human Health: Current Consensus and Future Trends. *R. Soc. Publ.* **2009,** *364*, 1526. DOI: 10.1098/rstb.2009.0053.

Tribune News Service. 18 Die of Cancer in Punjab Everyday 33, 318 Deaths Reported in last Five Years, Says First State-Wide Survey. Jan 29, 2013. http://www.tribuneindia.com/2013/20130129/main3.htm

United Stated Department of Agriculture. *Report and Recommendation on Organic Farming.* July 1980. https://naldc.nal.usda.gov/download/CAT80742660/PDF

Vann, A. (Ed.). *Ecological Effects of Current Development Processes, Human Ecology and World Development: Proceedings of a Symposium.* Organized Jointly by the Commonwealth Human Ecology Council and the Huddersfield Polytechnic, Huddersfield, Yorkshire, England in April 1973, 2012. https://books.google.co.in/books?id=4oHgBwAAQBAJ&pg=PA80&lpg=PA80&dq=false

Voelcker, J. A. *Abstract of the Report.* In Report on the Improvement of Indian Agriculture. Eyre and Spottiswoode. 1893. https://archive.org/details/cu31924001039324

Vrihi. *Introduction.* (n.d.). http://cintdis.org/vrihi/

Wagner, M.; Oehlmann, J. Endocrine Disruptors in Bottled Mineral Water: Total Estrogenic Burden and Migration from Plastic Bottles. *Environ. Sci. Pollut. Res.* **2009,** *16*(3), 278–286. DOI: 10.1007/s11356-009-0107-7.

WHO. *Electromagnetic Fields and Public Health: Mobile Phones.* Oct 2014. http://www.who.int/mediacentre/factsheets/fs193/en/

Zen, H.; Rowlands, H. *Glyphosphate Test Results.* April 7, 2014. http://www.momsacrossamerica.com/glyphosate_testing_results

CHAPTER 16

EFFECTIVE UTILIZATION OF RENEWABLE BIOMATERIALS FOR THE PRODUCTION OF BIOETHANOL AS CLEAN BIOFUEL: A CONCEPT TOWARD THE DEVELOPMENT OF SUSTAINABLE GREEN BIOREFINERY

GEETIKA GUPTA[1,*], PINAKI DEY[2], and SANDEEP KAUR SAGGI[1,3]

[1]Department of Biotechnology, Thapar Institute of Engineering and Technology, Bhadson Road, Patiala, Punjab 147004, India, Mob.: 9815700175

[2]Department of Biotechnology, Karunya Institute of Technology and Sciences, Karunya Nagar, Coimbatore, Tamil Nadu 641114, India, Mob.: 9893355139, E-mail: saspinaki@gmail.com

[3]Mob.: 7888356712, E-mail: sandeepsaggi4@gmail.com

*Corresponding author. E-mail: geetika_12_gupta@yahoo.com

16.1 INTRODUCTION

Owing to the enhanced worldwide energy demands, rising prices of petroleum-based feedstocks and mounting global warming-based environmental pollution issues, research efforts are now more focused on the development of sustainable green energy, based on some renewable feedstocks. Recently, second generation feedstock, more especially nonfood lignocellulosic biomasses are gaining huge importance as a potential feedstock for sustainable green biofuel production. Being polysaccharides, cellulose and hemicellulose which are part of lignocellulosic material can be easily hydrolyzed to sugars and then fermented to bioethanol. Although agriculture-based lignocellulosic bioresources in terms of waste are largely available in this world but development of economic, sustainable, and environment friendly bioethanol production process through selecting such feedstock has been still the major challenge persisting in this area. Consolidated bioprocessing (CBP), which integrates enzyme production, saccharification and fermentation, simultaneous saccharification and fermentation (SSF), separate hydrolysis and fermentation (SHF) are the promising strategies for effective ethanol production from such lignocellulosic materials. The application of thermotolerant yeast strains or thermophilic bacteria with thermostable enzymes to the process would overcome the drawback by performing hydrolysis and fermentation at elevated temperature. To overcome all the existing complicacies and problems associated with second-generation biofuel production, process integration of pretreatment processes with saccharification processes or saccharification process with fermentation process or enzyme production. Saccharification processes with fermentation process plays an instrumental role to make the overall process realistic and commercially sustainable. Development of some advanced membrane reactor-based green processes facilitate SSF, SHF, simultaneous saccharification, filtration, and fermentation (SSFF) schemes in more intensified and simplified way.

Demand of petroleum-based fuels has been highly raised in last two decades due to increasing industrialization and transportation throughout the world. Among 80% of the fossil fuel-based primary energy consumption, 58% is alone involved in the transport sector (Escobar et al., 2009). Such high dependency on fossil fuel-based energy sources contributed in large amount greenhouse gas (GHG) emissions and hence caused to many negative effects including climate change, receding of glaciers, rise in sea level, loss of biodiversity, and so forth. (Gullison et al., 2007).Owing to increasing energy demand and high GHG emissions, the significance of alternative source of energy has been raised globally (John et al., 2011). Hike in price of fossil-based energy resources raised various concerns such as the subject of depletion of energy resources in near future and generates risk of uncertainty. Campbell and Laherrere (1998) predicted that annual global oil production would deteriorate from the present 25 billion barrels to around 5 billion barrels in 2050 and can last for around 35 years. As green fuel, biofuels are gaining acceptance, as they are believed to substantially reduce GHG emission. Bioethanol is the most widely used biofuel among all other liquid biofuels, used for the transport sector (Demirbaş, 2005). Bioethanol is a clean biofuel that does not play any role in global warming as the carbon dioxide is formed by the combustion of bioethanol and it can be easily consumed by green plant and hence it can be considered as a zero carbon energy source. Presently, most of the bioethanol is being formed from food crops such as corn starch (in USA) and sugar cane (Brazil, South Africa), which leads the disturbance in food security and creates food versus fuel controversy. Hence, lignocellulosic biomass such as agricultural residues or wastes, energy crops, paper industry waste, kitchen waste, and municipal waste which are rich in carbohydrates influenced researcher's to promote lignocellulosic bioethanol production (Sarkar et al., 2012). Utilization of this lignocellulosic biomass as raw material for production of biofuel needs some processing prior to fermentation process by microorganism because microorganisms cannot convert complex lignocellulosics to biofuels. Owing to such complicacies in the production of second-generation bioethanol, researchers are now directing their attention from past agricultural substrates and waste vegetable oils to microscopic organism-based biofuel. Based on the current scientific knowledge and technology projections, third-generation biofuels specifically derived from microbes and microalgae are considered to be a viable alternative energy resource and free from the major drawbacks associated with first and second-generation biofuels. Whatever the materials can be used for such third-generation biofuel, economic viability of the overall process will be dependent on its intensification and simplification strategies.

16.2 UTILIZATION OF DIFFERENT BIOLOGICAL FEEDSTOCK FOR PRODUCTION OF BIOETHANOL

The broad classification for liquid biofuels includes "first-generation" and "second-generation," and third-generation of biofuels. Based on the feedstock used and microorganism utilized they were separated.

16.2.1 FIRST-GENERATION BIOETHANOL

For the production of first-generation bioethanol, different sugar sources such as grains, seeds have been utilized as raw materials (Love et al., 1998; Banat et al., 1992). Ethanol produced by fermenting sugar which was extracted from crop plants and starch composed of maize kernels or other starchy crops (Larson,2008). Grain alcohol is produced with the help of food crops such as barley, corn, sweet sorgum and wheat, and so forth (Table 16.1). Generally, bioethanol is produced from the organic matter with high sugar content which is then fermented by yeast. By fermentation, yeast converts six carbon sugars into ethanol. First of all, raw material, sugar is separated after that with the help of fermentation glucose is converted into ethanol. Last steps involved the distillation and dehydration for fulfillment of the desired concentration (anhydrous or hydrated ethanol) that can be blended with fossil fuels or used directly as a fuel (Nigam and Singh, 2011). When grains are used as raw materials, hydrolysis is done for conversion of starch into glucose (IEA, 2004). The conventional procedure utilized the germs of the seeds or grains for production of ethanol that serve a small percentage of the total mass of the plant, generating a significant amount of residue (Escobar et al., 2009).

First-generation bioethanol is in continuation and being produced in significant economic quantity in a number of countries. The survival of the first-generation biofuels production is, however, questionable because of the strife with food supply (Patil et al., 2008).The usage of only a small portion of total plant biomass reduced the land use productivity. The first-generation bioethanol has tremendous production expenditure due to competition with food. The rapid advancement of global bioethanol production from grain, sugar, and oilseed crops has raised the price of certain crops and food stuffs. These limitations have led to the discovery of nonedible biomass for the production of bioethanol (Nigam and Singh, 2011).

TABLE 16.1 Feedstocks That Can Be Used for First-Generation Bioethanol Production.

Feedstock	Crop
Sugar	Beet root, fruits, sugar cane, palm juice, wheat and so forth
Starch	Grain such as barely, corn, rice, sweet sorgum, wheat, and so forth, and root plants such as cassava, potato, and so forth

Sugarcane is the main feedstock for production of ethanol (Lee and Lavoie, 2013). In Brazil, 79% of ethanol is produced from fresh sugar cane juice and remaining from cane molasses (Wilkie et al., 2000). In this process, sugarcane is first crushed in water to expel sucrose than purified to produce raw sugar or ethanol (Lee and Lavoie, 2013). Detailed descriptions of feedstocks which can be used for first-generation bioethanol production are tabulated in Table 16.1.

Starch consists of long-chain polymers of glucose. Starch being macro-molecular cannot be fermented to ethanol by conventional fermentation technology; therefore, its macromolecular structure first broke down into simpler and smaller glucose. In this way, starch feedstocks are processed and infused with water to prepare a mash which contains 15–20% starch. With two steps enzyme preparation mash is then cooked at or above its boiling point. The first enzyme "amylase," hydrolyses starch molecules into glucose. The second enzyme "pullulanase" and "glucoamylase" further hydrolyses oligosaccharides and the process is known as saccharification. After that mash is cooled at 30°C and yeast is inoculated for fermentation (Lee, 2007).

16.2.2 SECOND-GENERATION BIOETHANOL

The second-generation liquid biofuels involved two different approaches, that is, biological or thermochemical processing, from agricultural ligno-cellulosic biomass, which are either nonedible whole plant biomass (e.g., grasses or trees specifically grown for production of energy) or nonedible residues of food crop production (Nigam and Singh, 2011).

The major achievement of the production of second-generation biofuels from nonedible feedstocks is that it limits the direct food versus fuel conflict associated with first-generation biofuels. Feedstock involved in the process can be raised precisely for energy purposes, enabling higher production per unit land area, and a higher amount of aboveground plant material can be transformed and used to produce biofuels. As a result this will increase land use efficiency compared to first-generation biofuels.

16.2.2.1 BIOETHANOL FROM LIGNOCELLULOSIC BIOMASS

Lignocellulosic biomass such which contains cellulose involves biological conversion technologies based on microbial and enzymatic process for producing sugars which are later on converted into ethanol (Naik et al., 2010). Agricultural wastes, forest wastes, postharvest processing of industrial food crops generate huge amounts of carbohydrate containing lignocellulosic waste (Singh et al., 1995). Lignocellulosic biomass comprises of three main structural units: cellulose, hemicelluloses, and lignin. Cellulose is crystalline glucose polymer, hemicelluloses is amorphous polymers of xylose, arabinose, and lignin a large polyaromatic compounds. As compared to starch-based feedstocks, the conversion of complex lignocellulosic biomass to ethanol is more difficult, comprises of three steps, that is, pretreatment of biomass, acid or enzymatic hydrolysis, and fermentation/distillation (Naik et al., 2010).

Lignocellulosic material contains three different types of polymers, namely cellulose $(C_6H_{10}O_5)_n$, hemicellulose $(C_5H_8O_4)_m$, lignin $[C_9H_{10}O_3(OCH_3)_{0.9-1.7}]$

$_x$, along with pectins, extractives, glycosylated proteins, and several inorganic materials (Sjöström,1993) which are combined which each other to form the structural framework of the plant cell wall.

Hemicelluloses is an amorphous and irregular structure composed of heteropolymers such as pentoses (D-xylose and L-arabinose), hexoses (D-glucose, D-galactose, and D-mannose), and sugar acids (D-glucuronic, D-galacturonic, and methylgalacturonic acids) (McMillan 1993,Saha 2003). The backbone chain is made up of xylan β (1→4) linkages that consist of D-xylose (approximately 90%) and L-arabinose (nearly 10%) (Girio et al., 2010). Depending upon the nature and feedstocks branch frequencies varies. Softwood hemicelluloses are made up of glucomannans while hardwood is composed mostly of xylan (McMillan1993).

Cellulose is the most extensive organic polymer and it constitutes 30% of the plant composition. Purest sources of cellulose such as cotton, flax, and chemical pulp constitutes (80–95% and 60–80%, respectively), whereas soft and hardwoods contains about 45% cellulose (Demirbaş, 2005; Lugar and Woolsey1999; Pettersen1984). Cellulose is a linear structural component of plant cell wall composed of a long chain of glucose monomers linked β (1→4) glycosidic bonds. Owing to broad hydrogen linkages molecules lead to a crystalline and strong matrix structure (Ebringerova et al., 2005). Functional groups, hydroxyl, and glycosidic bonds, present in the cellulose are responsible for chemical reactivity (Fan et al., 1987).Cross-linkages of various hydroxyl groups comprise of microfibrils which is responsible for high strength and compactness in a molecule (Limayem and Ricke, 2012).

After cellulose and hemicelluloses, lignin can be characterized as a polyphenolic material growing mainly from enzymatic dehydrogenetive polymerization of three phenylpropanoid (p-hydroxycinnamyl alcohols) units, namely trans-coniferyl alcohol, trans-sinapyl alcohol, and trans-p-coumaryl alcohol (Saggi et al., 2016a). Lignin is a rigid and aromatic biopolymer having a molecular weight of 10,000 Da bonded along covalent bonds to xylans conferring high level of compactness and rigidity to the plant cell wall (Mielenz, 2001). In plant cell wall, lignin is responsible for the resistant against microbes and chemicals (Himmel et al., 2007). Forest woody biomass mainly consists of cellulose and lignin polymers. Softwood barks have the maximum content of lignin (30–60%) followed by the hardwood barks (30–55%), whereas grasses and agricultural residues contains minimum content of lignin (10–30% and 3–15%, respectively) (Demirbaş, 2005; Pettersen,1984). During the process of degradation, lignin may form furan (furfural and hydroxymethyl-furfural) compounds that could inhibit fermentation (Ran et al., 2014).

16.2.2.2 PRETREATMENT PROCESS

Pretreatment is needed to change the biomass microscopic and macroscopic size and structure furthermore as its submicroscopic structure and chemical composition so breakdown of carbohydrate fraction to monomeric sugars may be achieved earlier and with larger yields (Sun and Cheng, 2002; Mosier et al., 2005a). Pretreatment strategies attribute to the solubilization and separation of one or a lot of those factors of biomass. It makes the remaining solid biomass further accessible to any chemical or biological treatment (Demirbaş, 2005). Pretreatment has been studied as the one of the overpriced step in biomass to possible sugar conversion with price as high as30 cents/gal alcohol created (Mosier et al., 2005a). To assess the price and performance of pretreatment strategies, technoeconomic analysis have been achieved recently (Eggeman and Elander, 2005). Enzymatic digestion is recalcitrant to native lignocellulosic biomass. Therefore, a lot of thermochemical pretreatment strategies are developed to improve digestibility (Wyman et al., 2005b). Recent studies have clearly proved that there is an on the spot correlation between the removal of polymer and hemicellulose on polyose edibleness (Kim and Holtzapple, 2005). Thermochemical process options seem more convincing than biological choices for the conversion of lignin of cellulosic biomass, which may have an adverse effect on enzyme hydrolysis. It can function as process energy and potential coproducts that have great benefits during a life cycle context (Sheehan et al., 2003). Pretreatment should meet the subsequent requirements: (1) improve the ability or formation of sugars by enzymatic hydrolysis; (2) avoid the loss or degradation of carbohydrate; (3) avoid the by-products formation inhibitory to hydrolysis and fermentation processes; and (4) be cost-efficient. Pretreatment strategies are classified in to physical, physicochemical, chemical, and biological pretreatment of lignocellulosic materials prior to enzymatic hydrolysis or digestion (Fan et al., 1982; Berlin et al., 2006).

16.2.2.2.1 Physical Pretreatment

In order to increase the specific surface and reduce the degree of polymerization, the mechanical pretreatment helps in reduction of size of particles and crystallinity of lignocellulosic biomass (Alvira et al., 2010). Depending on the final particle size of the material (10–30 mm after chipping and 0.2–2 mm after milling or grinding, it can be achieved by a combination of chipping, grinding, or milling) (Sun and Cheng, 2002). To improve the enzymatic hydrolysis of lingocellulosic biomass different milling processes (ball milling, two-roll milling, hammer milling, colloid milling, and vibro energy milling) can be used (Taherzadeh and Karimi, 2008). Depending on the final particle size and the biomass characteristics the power requirement of this pretreatment is relatively high (Alvira et al., 2010). This process is not economically feasible due to high energy requirements of milling and the continuous rise of energy prices (Hendriks and Zeeman, 2009).

16.2.2.2.2 Chemical Pretreatment

On the basis of alkaline pretreatment, some bases are effective on lignocellulosic biomass, depending on lignin content present in the biomass. Lime reduces steric hindrance of enzymes and enhancing cellulose digestibility by removing acetyl groups from hemicelluloses (Mosier et al., 2005a). NaOH increases the internal surface of cellulose and decreases the degree of polymerization and crystallinity by causing swelling in the biomass, which provokes the disruption of the structure of lignin (Taherzadeh and Karimi, 2008). At room temperature, alkali pretreatment time ranges from second to days. Alkali pretreatment is more effective on agricultural residues than on wood materials and it caused less sugar degradation than acid pretreatment (Kumar and Wyman, 2009a). For alkaline pretreatment ammonium hydroxide, calcium, potassium, and sodium are suitable alkalis. NaOH has been reported to reduced lignin content from 24–55 to 20% which increase hardwood digestibility from 14 to 55% (Kumar and Wyman, 2009a). Alkali pretreatment are more effective for lignin solubilization, exhibiting minor cellulose and hemicellulose solubilization which increases cellulose digestibility more than acid or hydrothermal processes (Carvalheiro et al., 2008).

Acid pretreatment with the help of acids such as H_2SO_4 and HCl solubilize the hemicellulosic fraction of the lignocellulosic biomass to make the cellulose more accessible to enzymes. To make the process economically feasible, after hydrolysis concentrated acid must be recovered (Von Sivers and Zacchi, 1995). Pretreatment of olive tree biomass with 1.4% H_2SO_4 at 210°C yields 76.5% of hydrolysis (Cara et al., 2008). Acids such as nitric acid and phosphoric acid have also been tested (Mosier et al., 2005b). Saccharification produced as high as 74% was shown when wheat straw was given pretreatment with 0.75% v/v of H_2SO_4 at 121°C for 1 h (Saha et al., 2005).

Ozonolysis has been applied on different agricultural residues such as wheat straw and rye straw, increasing in both cases the enzymatic hydrolysis yield after ozonolysis pretreatment (García-Cubero et al., 2009). Ozone is a powerful oxidant as it shows high delignification efficiency (Sun and Cheng, 2002). The main disadvantage in this process is the requirement of large amount of ozone, which can make the process economically unviable (Sun and Cheng, 2002).

16.2.2.2.3 Physicochemical Pretreatment

Steam explosion technology has been showed for production of ethanol from extensive range of raw materials such as olive residues (Cara et al., 2006), poplar (Oliva et al., 2003), and wheat straw (Ballesteros et al., 2006) and herbaceous residues as corn stover (Varga et al., 2004). In combination with the partial hemicellulose solubilization and hydrolysis, the lignin is redistributed and to some extent removed from the material (Pan et al., 2005). Steam explosion process offers various interesting features when compared to other pretreatment technologies. These include the potential for significantly low capital investment, low environmental impact, high potential for energy efficiency, less hazardous process chemicals and conditions, and complete sugar recovery (Avellar and Glasser, 1998). Although utilization of acid in steam explosion has some disadvantages, many pretreatment approaches (SO_2-explosion) have involved external acid addition for solubilization of the hemicellulose, lower the optimal pretreatment temperature, and give a partial hydrolysis of cellulose (Brownell et al., 1986; Tengborg et al., 1998). With the aim of high sugar recoveries, some researchers have suggested a two-step pretreatment (Tengborg et al., 1998). The main disadvantage of steam explosion pretreatment includes the partially hemicellulose degradation and the production of some toxic compounds that could affect the hydrolysis and fermentation steps (Oliva et al., 2003).

In herbaceous feedstocks, such as sugarcane bagasse (Laser et al., 2002), corn stover (Mosier et al., 2005b), and wheat straw (Pérez et al., 2008), liquid hot water has been studied to eliminate up to 80% of the hemicellulose and to increase the enzymatic digestibility of pretreated material. Flow through systems has been believed to eliminate more hemicellulose and lignin than batch systems from some materials. During the flow through process addition of external acid has been studied (Wyman et al., 2005a).

On agricultural residues and herbaceous crops, the ammonia fiber expansion (AFEX) pretreatment is more effective with controlled effectiveness exposed on high lignin and woody biomass (Wyman et al., 2005a). After AFEX pretreatment, digestibility of biomass is increased (Galbe and Zacchi, 2007). AFEX eliminates acetyl groups from several lignocellulosic materials (Kumar and Wyman, 2009b). AFEX pretreatment resulted in only a small amount of the solubilzation of solid and removal of less hemicellulose and lignin content from biomass (Wyman et al., 2005a).

In the pretreatment of wet oxidation, air or oxygen acts as a catalyst. It is used for production of ethanol followed by SSF (Martín et al., 2008). In the pretreatment of wheat straw by using Na_2CO_3, it resulted in 96% recovery of cellulose (65% converted to glucose) and yield of 70% of hemicelluloses (Klinke et al., 2002) was achieved. After wet oxidation, pretreatment of corn stover and spruce, high yields have been achieved (Palonen et al., 2004).

Although less research has been carried out on ultrasound pretreatment from lignocellulose, some investigators demonstrated that saccharification of cellulose is increased by ultrasonic pretreatment (Yachmenev et al., 2009). Ultrasound pretreatment helps in extraction of hemicelluloses, cellulose, and lignin but little research has been carried out to study the susceptibility of lignocellulosic materials to hydrolysis.

16.2.3 THIRD-GENERATION BIOETHANOL

The third-generation biofuel mainly focus on the direct production of vegetable oil or production of oil from agricultural substrates and wastes and microalgae-based biofuel production. Third-generation biofuels particularly derived from microbes and microalgae are considered to be suitable alternative energy resource without the major disadvantages associated with first and second-generation bioethanol (Nigam and Singh, 2011). But still second-generation biofuel is more preferred than microalgae-based third-generation biofuel production as biofuel produced from microalgae is comparatively less stable, production will not take place for a long time and extraction of such fuel is comparatively difficult and costly.

16.2.3.1 BIOETHANOL FROM ALGAE

Bioethanol could be very significant to encourage energy independence and lower GHG emissions. A substantial contest on the gradual substitution of petroleum with the use of renewable alternatives such as biofuels influences the political and economic agenda worldwide (Demain, 2009). Different bioethanol production methods from microalgae and cyanobacteria need to be developed so that the costs associated with the land, labor, and time of conventional fermented crops can be circumvented. Ueda et al. (1996)

have patented a two-stage process for fermentation of microalgae. In the first stage, microalgae sustain fermentation in anaerobic environment to produce ethanol. In the fermentation process, the CO_2 yield can be recycled in algae cultivation as a nutrient. The second stage involves the application of remaining algal biomass for production of methane, by anaerobic digestion process, which can be converted to generate electricity. However, Bush and Hall (2006) pointed out the patented process of Ueda et al. (1996) which was not commercially adaptable due to the confinement of single cell free-floating algae. They patented a mutated fermentation process wherein yeasts, *Saccharomyces cerevisiae* and *Saccharomyces uvarum*, were combined to algae fermentation broth for production of ethanol. Harun et al. (2010) reported the suitability of microalgae (*Chlorococum* sp.) as a substrate, by using yeast for the production of bioethanol with the help of fermentation. They produced a yield of approximately 38% weight, which supports the suitability of microalgae as a promising substrate for production of bioethanol.

16.3 UTILIZATION OF DIFFERENT GROUP OF EFFICIENT MICROORGANISM FOR PRODUCTION OF BIOETHANOL, BASED ON DIFFERENT FEEDSTOCKS

Lignocellulosic biomass is the largest biorenewable source of carbohydrates on earth and cellulose is its main component. Cellulose is a homopolymer of β-1,4 linked glucose molecule which is organized in linear form as recalcitrant crystalline-like structures. In the plant cell wall, cellulose is tightly compacted with the components such as hemicellulose, lignin, and pectin, making the whole structure extremely intractable to microbial attack.

16.3.1 *LIGNASE*

Lignases employ low molecular and diffusible reactive compounds to affect initial changes to the lignin substrate (Call and Mücke, 1997) that is why lignases affect lignin degradation, as enzymes are too large to penetrate the cell wall of biomass. Fungi with highest activity for lignases are *Stropharia coronilla, Botrytis cinerea, Phanerochaete chrysosporium* but the most efficient and highest degraders are white-rot fungi belonging to the *Basidiomycetes* (Akin et al., 1995; Gold and Alic, 1993) with *P. chrysosporium* being the best-studied lignin degrading fungus and least known white-rot fungi such as *Daedalea flavida, Phlebia fascicularia, Paralepetopsis floridensis, and P. radiata* have been known to selectively degrade lignin in wheat straw. Other less lignin degraders microorganism belongs to the genera are *Cellulomonas, Pseudomonas, Actinomycetes, Thermomonospora, Microbispora, Clostridium thermocellum, and Ruminococcus* (Howard et al., 2003).

16.3.2   CELLULASE

Cellulases are usually a mixture of several enzymes for the degradation of biomass. Three major groups of cellulases are involved in the hydrolysis process: endoglucanases (EG, endo-1,4-d-glucanohydrolase or EC 3.2.1.4.), attacking the regions of low crystallinity in the structure of cellulose fiber which makes free chain ends in the structure; the second cellulase enzyme is exoglucanase or cellobiohydrolase which degrades the molecule further by removing cellobiose units from the free chain ends; the most popular third enzyme is glucosidase (EC 3.2.1.21.), which hydrolyses cellobiose to produce glucose molecules (Coughlan and Ljungdahl, 1988). In addition to these cellulase enzymes there are several number of ancillary enzymes that attack hemicellulose, such as acetylesterase, xylanase, xylosidase, galactomannanase, glucuronidase, and glucomannanase (Duff and Murray, 1996). During hydrolysis, cellulose component is degraded by the enzyme cellulases and it converts cellulose to reducing sugars, which can be further fermented by yeasts or bacteria to bioethanol (Sun and Cheng, 2002; Prasad et al., 2007). *Trichoderma reesei* are widely employed for the commercial production of enzymes such as hemicellulases and cellulases (Esterbauer et al., 1991; Jørgensen et al., 2003; Nieves et al., 1998). *T. reesei* might be a good producer of hemicellulolytic and cellulolytic enzymes but is quite unable to degrade high prolific lignin. Bower (2005) introduced several bacterial endoglucanases into *T. reesei*. One of them *Acidothermus cellulolyticus*, was fused to *T. reesei* cellobiohydrolase CBH1. Cellulases often contain carbohydrate-binding modules (CBMs) to facilitate the interaction between the enzyme and the substrate surface. CBMs target their cognate catalytic domains to specific substrates and enhance the catalytic efficiency by increasing the effective concentration at the surface. Some of bacteria with the highest specificity for cellulose are *Bacillus subtilis, Streptomyces murinus, C. thermocellum, Bacillus* sp., *Bacillus macerans*. Most of fungi with the highest specific activity for cellulose are *Sclerotium rolfsii, Orpinomyces* sp., *Aspergillus, Achlya bisexualis, Rhizopus chinensis*, and *Penicillium brefeldianum* (Howard et al., 2003).

Pentose-fermenting *Escherichia coli* (Ingram et al., 1987) and *Klebsiella oxytoca* (Burchhardt and Ingram, 1992) have been generated by introducing ethanologenic genes from *Zymomonas mobilis*. At the same time, the first xylose fermenting *S. cerevisiae* strain was generated through the introduction of genes for xylose metabolizing enzymes from *Pichia stipitis* (Kotter and Ciriacy, 1993). Later introducing the genes encoding xylose isomerase from the bacterium *Thermus thermophilus* (Walfridsson et al., 1996) and the anaerobic fungus *Piromyces* sp.(Kuyper et al., 2003), respectively, xylose-fermenting strains of *S. cerevisiae* were constructed. Using *S. cerevisiae* for high ethanol yields from xylose also require metabolic engineering strategies to enhance the xylose flux (Hahn-Hägerdal et al., 2006). Microbes with the highest specific activity for hemicellulose are *Aspergillus nidulans, Aspergillus niger, Trichoderma, Phanerochaete, Mortierella vinacea melibiose, Humicola insolvens, Sclerotium, and Schizophyllum commune* (Howard et al., 2003). Detail description of microbes which are involved in the process is tabulated in Table 16.2.

16.4 DEVELOPMENT OF PROCESSES TO ACHIEVE GREEN AND CLEAN BIOFUEL

Most important and final stage of bioethanol production is the fermentation and it can be carried out through batch, fed-batch, or continuous process. The most suitable process can be selected based on the kinetic properties of microorganisms and type of lignocellulosic hydrolysate in addition to process economics aspects (Chandel et al., 2007). Batch culture is generally considered as a closed culture system which is consist of an initial, limited amount of nutrient and inoculated with suitable microorganisms at the beginning of the process to allow the fermentation (Abtahi,2008). During such fermentation process, nothing can be added after inoculation except possibly acid or alkali for pH control or air for aeration during the processes. Another production approach is fed-batch operation which is widely used in industrial applications as it is somewhat advantageous from both batch and continuous processes (Saarela et al., 2003). The major advantage of fed-batch operation with respect to batch operation is the ability to increase maximum viable cell concentration for certain cases and hence allow product accumulation to a higher concentration (Frison et al., 2002). This process is highly applicable for bioethanol production. In the continuous process, fermentation medium with all required nutrients is passed continuously into an agitated vessel where the microorganisms are active. The continuous

TABLE 16.2 Microorganism and their Species with Genus and Substrate.

Major group	Genus	Species	Substrate
Fungi	Aspergillus	Aspergillus niger	Cellulose
		Aspergillus nidulans	Wheat bran/corn cob
		A. Oryzae	Wheat straw/wheat bran
		F. Saloni	Carboxymethyl cellulose/wheat Bran/
	Fusarium	F. Oxysporum	xylose/sorbose
		H. Insolens	
	Humicola	H. Grisea	
		P. Brasilianum	
	Penicillium	T. Ressei	
	Trichoderma		
Bacteria	Acidothermus	A. cellulolyticus	Carboxymethyl cellolose/glycerol
	Bacillus(facultative anaerobes)	B. subtilus	Bannna waste
	Clostridium(anaerobes)	C. acetobutylicum	Carboxymethyl cellulose
	Pseudomonas(aerobic)	C. thermocellum	
	Rhodothermus	P. cellulosa	
		R. marinus	
Actinomycetes	Cellulomonas(facultative anaerobes)	C. fimi	Caxboxymethyl cellulose
	Streptomyces	C. uda	Wheat straw
	Thermomonospora	S. lividans	
		S. drozdowiczii	
		T. usca	
		T. curvata	

fermentation is carried out in a specific dilution rate and in a steady-state condition. After achieving the steady state condition, the product can be collected from the bioreactor throughout the process. One of the first advanced approaches taken to improve the yeast bioethanol fermentation process was operating the fermenters in a continuous mode rather than the conventional batch mode and hence it results in increasing the productivity about three-fold (Lawford,1988). Continuous cultivation with high cell densities can be achieved using membrane cell recycle bioreactor systems or through cell immobilization system which were highly effective to increase the productivity of bioethanol.

16.4.1   BATCH FERMENTATION

Traditionally, batch fermentation is used for the production of ethanol using microorganism in each industry. In batch fermentation, the microorganism works in high substrate concentration initially to get high product concentration (Olsson and Han-Hagerdal, 1996).The batch process is a multivessel process, allows flexible and easy method approach. Generally, batch fermentation process is performed by low productivity substrate with an intensive labor (Shama, 1988). For this fermentation process, the elaborate preparatory procedures are needed. Owing to discontinuous start-up also with shut down operations, high labor costs are incurred. This leads to disadvantage and the low productivity offered by the batch process have led many commercial operators to consider the other fermentation methods.

16.4.2   FED-BATCH FERMENTATION

To increase the ethanol production in fed-batch fermentation, substrate is used at low level concentration. Fed-batch cultures fermentation provide better productivities than batch cultures for the production of microbial metabolites. The low feed rate of substrate solution containing its high concentration is some time useful to prevent substrate inhibition effect during ethanol production. However, the inhibitory effect of these compounds to yeast can be reduced. The specific ethanol productivity has been reported to decrease with enhancing the cell mass concentration (Palmqvist et al., 1996; Lee and Chang, 1987).The cell density should be kept at a level to give maximum ethanol productivity.

16.4.3   CONTINUOUS FERMENTATION

Continuous fermentation process performed in different kind of bioreactors (1) stirred tank reactors and (2) plug flow reactors. Continuous fermentation is much better than batch fermentation process to give the higher rate of production of ethanol at low dilution rates. Unproductive time associated with cleaning, recharging, adjustment of media, and sterilization are removed in this continues fermentation process. A high cell substrate concentration of microbes give the high productivity in this exponential phase by locking in this phase and we can say overall processing takes 4–6 h for the ethanol production which is much lesser time than conventional batch fermentation process (24–60 h). This continuous fermentation process saved labor part and in the lesser cost which achieves a high goal and production level with much smaller plant.

16.4.4   IMMOBILIZED CELLS

The use of immobilized cells makes easy part in the difficulty of fermentation process. Immobilization by adhesion to a surface by this mechanism either electrostatic or covalent, entrapment in polymeric matrices and retention by membranes has been successful for ethanol production from hexoses (Godia et al., 1987). The application of immobilized cells has made significant changes and advances in the fuel ethanol production technology. Immobilized cells offer rapid fermentation rates with high productivity and large fermenter volumes of mash put through per day without risk of cell washout and remove. Cells can be kept inside of bioreactors in suspension as free cells or immobilized in various supports. There are four main immobilization techniques (1) attachment to a surface, (2) entrapment within a porous matrix, (3) cell aggregation as flocculation, and (4) containment behind barriers.

16.4.5   SEPARATE HYDROLYSIS AND FERMENTATION

SHF is basically a two-stage process, first by using cellulolytic enzymes, cellulose which is contained in a solid phase of pretreated lignocellulosic material, is hydrolyzed to glucose or starch-based material can be converted to glucose by amylolytic enzymes (Sreenath et al., 2001). In some cases, directly microorganism which can produce cellulolytic or amylolytic enzymes can be directly grown in that media to produce simple sugars such as glucose. In the next stage, released glucose is then converted into ethanol by fermentation process by selecting suitable microbial. Both processes (i.e., enzymatic hydrolysis and fermentation) can be performed under their optimal conditions (temperature, pH, nutrient composition, solid loading), which is considered as the main advantage of this configuration as the optimum operating temperature of both the process differs considerably. Cellulolytic enzymes works at optimum temperature around 50°C while the microbial strains which are employed for bioethanol production (yeast *S. cerevisiae* and the bacterium *Z. mobilis*, although many other strains) produce ethanol most efficiently at 28–37°C. Moreover, the overall performance of the process not gets hampered as two processes are conducted separately in different vessels and cellulolytic enzymes are not influenced by the presence of ethanol (Wingren et al., 2003). On the other hand, the biggest drawback of the process is inhibition of cellulolytic enzyme activity by increasing concentrations of released glucose in the vessel which is considered as end product inhibition problem. It slows down the rate of cellulose hydrolysis. There are many reports suggesting that the investment is normally increased for SHF due to the use of more than one vessel at different times. Another biggest drawback associated with SHF process is when it is performed under high solid loading condition, activity of cellulolytic enzymes get reduced due to inhibition problem. The use of various surfactants such as Tweens, polyethene glycols, or ionic liquids can play a vital role in the process by reducing enzyme attachment to lignin which ultimately improves the rate of cellulose hydrolysis. The complete description of the process is presented in Figure 16.1.

16.4.6   SIMULTANEOUS SACCHARIFICATION AND FERMENTATION PROCESS (SSF)

There are always three steps present in bioethanol production from lignocellulosic waste materials which include (1) biomass pretreatment (2) enzymatic hydrolysis (3) fermentation. Based on these three individual steps, three different process can be configured which are SHF, SSF, and CBP. CBP approach which combines step 2 and 3 together while utilizing only one microbial community that can produce cellulases and ferments sugars to ethanol simultaneously. For SSF and CBP processes, hydrolysis and fermentation are carried out in the same reactor. In one-point SSF, pretreatment step can be combined with SSF. It actually merges all three steps together. However, the major drawbacks lies with the SSF and CBP process is the optimum temperature required for the saccharification and fermentation

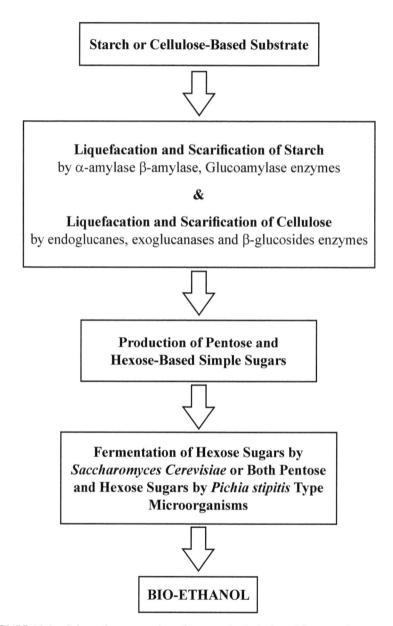

FIGURE 16.1 Schematic presentation of separate hydrolysis and fermentation process.

stages are different. Saccharification with cellulolytic enzymes is normally carried out around 50°C while the temperature normally maintained for ethanol fermentation is between 28 and 37°C for most fermenting microbes. Lowering the optimum temperature of cellulases through protein engineering is a difficult task in practice (Szczodrak and Targonski, 1988). Accordingly, high-temperature fermentation has been the major priority for SSF process and thermotolerant yeast strains can play a big role in that process (Lin and Tanaka, 2006)

Because of many advantages, bioethanol production at elevated temperature through SSF process has received bigger attention. The reasons are reduction in cooling costs associated with cooling after sterilization, continuous and easy evaporation of ethanol from broth under reduced pressure, a reduced risk of contamination, and suitability for application in tropical countries. SSF can be mostly preferred because it can involve less equipment and time which could lead to reduction in the investment cost. The operation can be also simplified by the integration of saccharification and fermentation. The major reason behind the selection of SSF is it can avoid cellulase inhibition by glucose and hence can be useful in increasing the saccharification rate and ethanol yield. However, the disadvantages associated with the SSF process is the lower efficiency of hydrolysis carried out at lower temperature to be compatible with yeast fermentation compared to SHF systems (Hasunuma and Kondo, 2012).

Therefore, thermotolerant microbial strains are used for the production of substantial amount of ethanol at temperatures more favorable for saccharification which is actually necessary for the improvement of SSF efficiency (D'Amore et al., 1989). Thermotolerant yeast strains such as *Kluyveromyces*, *Saccharomyces*, and *Fabospora* genera that can produce more than 5% (w/v) ethanol at elevated temperature (>40°C) which are already identified. *Candida glabrata* is expected to be an another useful microbes in the development of an ideal SSF process, since the yeast has higher stress tolerance to both acid and high temperature in addition to effective ethanol production capability. Additionally, microbes such as *Kluyveromyces marxianus* can offer extra benefits including a high growth rate and it has the ability to utilize a wide variety of sugar substrates (e.g., arabinose, galactose, mannose, and xylose) at elevated temperatures (Hasunuma and Kondo, 2012). As a result, it is highly preferred microorganism in industry. For industrial use, *S. cerevisiae* is considered as a good ethanol producer and shows high ethanol tolerance but it is not thermotolerant. Thus, mutation screening and genetic engineering approaches were performed to make it thermotolerant (Gírio et al., 2010). On the other hand, thermotolerant *S.*

cerevisiae strains have been isolated from tropical regions, which shows a high-temperature (41°C) growth phenotype with higher production abilities (Hasunuma and Kondo, 2012).

16.4.7 MEMBRANE BIOREACTOR-BASED GREEN TECHNOLOGY

Membrane reactors are specifically well suited for bioethanol production as they can play a number of key roles, such as intensifying the contact between the reactants and the catalyst, reduces the chances of product inhibition by selective removal of products from the reaction mixture and controlling the addition of reactants to the reaction mixture, formulating low energy demanding continuous mode fermentation, maintaining high product yield and productivity. With addition to that it also offer a number of other environment friendly advantages over conventional systems including these operation at moderate temperature and pressure conditions, low-energy consumption, increased safety, simple operation, elimination of the need for wastewater treatment, easy scale-up; higher mechanical, thermal, and chemical stability; and resistance to corrosion. As a result, membrane bioreactor is comparatively considered as an effective tool to develop sustainable green technology to produce bioethanol. The most useful strategy for effective bioethanol production is cell recycle batch fermentation, as cell recycling system significantly reduces production time. Maintaining high concentration of alcohol producing strain specifically yeast concentration in the bioreactor is advantageous to achieve high concentration of ethanol and reduces the chances of contamination. Membrane-based process is considered economical and low energy involving as conventional methods such as evaporation to make sugar concentration, solvent extraction, and activated charcoal-based processes can be avoided. The microorganism can be retained by different means to achieve high cell concentrations inside the bioreactor, for example, by recycling or immobilizing methods. Cell retention not only increases the biomass amount in the reactor but also improves the sugar utilization and in situ detoxification in bioethanol production process.

16.4.8 MEMBRANE DISTILLATION-BASED GREEN TECHNOLOGY

Membrane distillation is a process which separates simple ethanol and water solution using membrane. Ethanol productivity can be improved when membrane distillation operation for ethanol recovery can be coupled with fermentation process as the inhibitor was removed. Considering the partial pressure of ethanol which is higher than the water, ethanol vapor can transfer preferentially through the membrane pores. Membrane distillation is also considered as sustainable alternative green technology approach as because conventional distillation technique to separate ethanol from water requires high temperature, electricity, and labor. Ethanol production increased by 15.5% due to facilitation of the continuous process and more complete fermentation of sugars in integrated membrane distillation process was already demonstrated. It was reported that the separation efficiency was achieved by lowering the osmotic pressure in the fermentation broth, decreasing glycerol synthesis level, and increasing yeast cells number and viability. The membrane flux increases as the feed temperature increases due to membrane swelling. It was found feed temperature make large influences on the separation process.

14.4.9   MEMBRANE-BASED GREEN TECHNOLOGY FOR ENZYMATIC SACCHARIFICATION

Using enzymes in free form in continuous enzymatic saccharification process is usually not economical and not recommended. The main reason behind such decision is free expensive enzymes leaves with the effluent and can be wasted. Such thing can be avoided by recycling and reusing the enzyme. Most usual way to reuse the enzyme is to make it in an immobilized form. The biggest drawback of using enzymes in immobilized form is that it restricts its ability to penetrate the solid substrate and less saccharification efficiency can be achieved. Ultrafiltration-based membrane bioreactor can be used for ideal separation of unused enzymes after saccharification process and it can be effectively recycled back to bioreactor for next-stage saccharification process. Such kind of membrane bioreactor is ideal for enzymetic saccharification process.

16.4.10   MEMBRANE-BASED SUSTAINABLE SSFF SCHEME

Most common processes for bioethanol production are SHF and SSF-based schemes. SHF allows the saccharification and fermentation process to be conducted at separate conditions. The biggest drawback of such process is it suffers from product inhibition problem as produced sugars will not be instantly separated from the reactor. Similarly, SSF-based concepts also suffer from stability of the reaction as both saccharification and fermentation must be carried out in a single environmental condition which is difficult to optimize. To avoid the disadvantages associated with both the schemes, membrane-based SSFF scheme is developed to bypass such product inhibition problem while instantly separating the simple sugars and carrying out both saccharification and fermentation processes simultaneously in different optimized conditions.

16.5   TECHNOECONOMIC EVALUATION OF THE PROCESS TO ACHIEVE CLEAN BIOFUEL

For commercial feasibility, the production cost for conversion of biomass to liquid bioethanol must be lower than the current gasoline prices (Wayman and Parekh, 1990). Owing to the increasing research efforts toward the development of most efficient biomass conversion technologies, the production process seems to be feasible and attractive. According to trend of research investigations, still huge opportunity is present to bring down the production cost of biomass-based bioethanol production. The most influencing parameters for low-cost bioethanol production are the nature of the feedstock and cellulolytic enzymes. Biomass feedstock cost normally contributes around 40% of the ethanol production cost, whereas downstream processing technology contributes more than 60% of the ethanol production cost. To make the overall process economically viable, industrially acceptable, integrated approaches of combining feedstock processing stages, enzymatic action, and downstream processing stages are accepted and encouraged. Feasibility of long-term supply of cheapest feedstock and uses of potent cellulases could make the process economically sustainable (Sun and Cheng, 2002; Dien et al., 2006) and for that the feedstock should be available throughout the year at bulk amount and at lowest cost. Therefore, the availability and selection of the raw material is the biggest issue as it should not create food versus fuel controversy.

For bioethanol production, agroresidues such as wheat straw, sorghum, and barley straw are not preferable as they are normally used as animal fodder, whereas sugarcane bagasse, rice straw, rice bran, groundnut shell, corn stover, soya bean stalks, and so forth can be used directly because these sources are not preferably used as fodder for livestock. Organic waste and municipal solid waste that contain significant amount of cellulose could contribute significantly for the development of cost-effective bioethanol production process. Other parameters which can significantly reduce the bioethanol production cost are plant size, continuous process operation, and developing the process by integrating with existing plant. The overall production cost per unit will be reduced by increasing the plant size. It was investigated that a 10-fold increase in size reducing the unit cost to less than one-half (Wayman and Parekh, 1990). Similarly, continuous nature of the operation makes significant contribution in lowering the production cost. To further improve the economy of ethanol production, judicious integration of the ethanol production can be done with already existing plants such as pulp and paper plants. It was projected that the cost of bioethanol can be reduced from US$1.22/l to about US $0.31/l on the basis of continuous development of pretreatment of biomass, enzyme application, and fermentation process. Further economic analysis of bioethanol ($0.78/gal) was done by Wooley et al. (1999) and he suggested a projected cost of as low as $0.20/l by 2015 if enzymatic processing and biomass improvement processes are targeted to be combined. Economic evaluation has been done by Wingren et al. (2003) for the SHF and SSF-based processes using cellulase enzymes. In both configuration, SSF-based approach was proved to be less expensive by about 10%; and estimated ethanol production cost was 0.56–0.67 $/l. Ethanol production cost of 20 cents/l can be possible in another 15 years from lignocellulose biomass employing designer cellulases and simultaneous saccharification and cofermentation process, was suggested by National Renewable Energy Laboratory (Colorado, USA). How cell recycle and vacuum fermentation processes for continuous ethanol production can contribute in production capacity of 78,000 gal ethanol/day with estimated ethanol production cost 82.3 and 80.6 cent/gal from molasses-based medium was clearly demonstrated by Cysewski and Wilke (1976). Another main contributor of bioethanol production process is the distillation cost which can be substantially higher at low ethanol concentrations. Therefore, the idea of concentrating sugar solutions prior to fermentation (Cyweski et al., 1976; Oh et al., 2000; Iraj et al., 2002) has been developed by the researchers. Ethanol distillation cost can be further improved by using membrane distillation process which is lowest in operational cost, flexible, simple to use, and is easy to maintain.

16.6 CONCLUSION

Bioethanol can be produced from renewable sources of feedstock such as wheat, sugar beet, corn, straw, and wood. It can be produced from various domestic, agricultural wastes, and similarly from municipal and industrial waste streams. Production of bioethanol from biomass is one way to lower both environmental pollution and the consumption of crude oil. Selection of relevant raw material for bioethanol production is the major concern regarding its worldwide commercial sustainability. Considering the different process for the biotransformation of substrates to bioethanol, different research challenges are being faced by the researchers to finally commercialize the bioethanol and hence it needs critical evaluation. Despite of specific advantages associated with the different process, still large-scale production facilities are limited. In spite of some technoeconomic challenges, second-generation biofuel is more preferred and advantageous than first and third-generation biofuel production. Knowing the huge availability of lignocellulosic biomass resources, researchers are working toward to achieve realistic solutions on existing complicacies and problems associated with second generation biofuel production. Process integration of pretreatment processes with saccharification processes or saccharification process with fermentation process (SSF) or enzyme production, saccharification processes with fermentation process (CBP) helps to overcome those existing complicacies up to certain extent. Development of some advanced equipments such as membrane reactor to facilitate SSF, SHF, SSFF, and distillation processes with process intensification and process simplification strategy can play a major role to resolve those issues in the same direction.

KEYWORDS

- second-generation biofuel
- lignocellulosic biomasses
- saccharification
- fermentation
- hydrolysis

REFERENCES

Abtahi, Z. Ethanol and Glucose Tolerance of *M. indicus* in Aerobic and Anaerobic Conditions. Master Thesis, University College of Boras School of Engineering, Boras, Sweden, 2008, *47*(9), 1287–1294.

Akin, D. E.; Rigsby, L. L.; Sethuraman, A. Alterations in the Structure, Chemistry, and Biodegradation of Grass Lignocellulose Treated with White Rot Fungi *Ceriporiopsis subvermispora* and *Cyathus stercoreus*. *Appl. Environ. Microbiol.* **1995,** *61,* 1591–1598.

Avellar, B. K.; Glasser, W. G. Steam-Assisted Biomass Fractionation. I. Process Considerations and Economic Evaluation. *Biomass Bioenergy* **1998,** *14*(3), 205–218.

Ballesteros, I.; Negro, M. J.; Oliva, J. M.; Cabañas, A.; Manzanares, P.; Ballesteros, M. Ethanol Production from Steam-Explosion Pre-Treated Wheat Straw. In *Twenty-Seventh Symposium on Biotechnology for Fuels and Chemicals;* McMillan, J. D., Adney, W. S., Mielenz, J. R., Klasson, K. T., Eds.; Humana Press: Totowa, NJ, USA; 2006; pp 496–508.

Banat, I. M.; Nigam, P.; Marchant, R. Isolation of Thermotolerant, Fermentative Yeasts Growing at 52°C and Producing Ethanol at 45 and 50°C. *World J. Microbiol. Biotechnol.* **1992,** *8*(3), 259–263.

Berlin, A.; Balakshin, M.; Gilkes, N.; Kadla, J.; Maximenko, V.; Kubo, S.; Saddler, J. Inhibition of Cellulase, Xylanase and β-glucosidase Activities by Softwood Lignin Preparations. *J. Biotechnol.* **2006,** *125*(2), 198–209.

Bower, B. S. (Genencor International IU): Fusion Proteins of an Exocellobiohydrolase and an Endoglucanase for use in the Saccharification of Cellulose and Hemicellulose. Patent 2,005,093,073, 2005.

Brownell, H. H.; Yu, E. K. C.; Saddler, J. N. Steam-Explosion Pre-Treatment of Wood: Effect of Chip Size, Acid, Moisture Content and Pressure Drop. *Biotechnol. Bioeng.* **1986,** *28*(6), 792–801.

Burchhardt, G.; Ingram, L. O. Conversion of Xylan to Ethanol by Ethanologenic Strains of *Escherichia Coli* and *Klebsiella oxytoca. Appl. Environ. Microbiol.* **1992,** *58,* 1128–1133.

Bush, R. A.; Hall, K. M. U.S. Patent 7,135,308, Washington, DC: U.S. Patent and Trademark Office, 2006.

Call, H. P.; Mücke, I. History, Overview and Applications of Mediated Lignolytic Systems, Especially Laccase-Mediator-Systems (Lignozyme®-Process). *J. Biotechnol.* **1997,** *53,* 163–202.

Cara, C.; Ruiz, E.; Ballesteros, I.; Negro, M. J.; Castro, E. Enhanced Enzymatic Hydrolysis of Olive Tree Wood by Steam Explosion and Alkaline Peroxide Delignification. *Process Biochem.* **2006,** *41*(2), 423–429.

Cara, C.; Ruiz, E.; Oliva, J. M.; Sáez, F.; Castro, E. Conversion of Olive Tree Biomass into Fermentable Sugars by Dilute Acid Pre-Treatment and Enzymatic Saccharification. *Bioresour. Technol.* **2008,** *99*(6), 1869–1876.

Carvalheiro, F.; Duarte, L. C.; Gírio, F. M. Hemicellulose Bio Refineries: A Review on Biomass Pre-Treatments. *J. Sci. Ind. Res.* **2008,** 849–864.

Chandel, A. K.; Es, C.; Rudravaram, R.; Narasu, M. L.; Rao, L. V.; Ravindra, P. Economics and Environmental Impact of Bioethanol Production Technologies: An Appraisal. *Biotechnol. Mol. Biol. Rev.* **2007,** *2,* 14–32.

Cysewski, G. R.; Wilke, C. R. Utilization of Cellulosic Materials through Enzymatic Hydrolysis. I. Fermentation of Hydrolysate to Ethanol and Single Cell Protein. *Biotechnol. Bioeng.* **1976,** *18,* 1297–1313.

Demain, A. L. Bio Solutions to the Energy Problem. *J. Ind. Microbiol. Biotechnol.* **2009,** *36*(3), 319–332.

Demirbaş, A. Bioethanol from Cellulosic Materials: A Renewable Motor Fuel from Biomass. *Energy Sources* **2005,** *27*(4), 327–337.

Dien, B. S.; Jung, H. J. G.; Vogel, K. P.; Casler, M. D.; Lamb, J. A. F. S.; Iten, L.; Mitchell, R. B.; Sarath, G. Chemical Compositionand Responseto Diluteacid Pre-Treatment and Enzymatic Saccharification of Alfalfa, Reed Canarygrass and Switchgrass. *Biomass Bioenergy* **2006,** *30*(10), 880–891 (in Press).

Duff, S. J. B.; Murray, W. D. Bioconversion of Forest Products Industry Waste Cellulosics to Fuel Ethanol: A Review. *Bioresour. Technol.* **1996,** *55,* 1–33.

Ebringerova, A.; Hromadkova, Z.; Heinze, T.; Hemicellulose, T. H. Polysaccharides 1: Structure, Characterization and use. Springer-Verlag: Berlin, 2005; Vol. 186, pp 1–67.

Eggeman, T.; Elander, R. T. Process and Economic Analysis of Pre-Treatment Technologies. *Bioresour. Technol.* **2005,** *96*(18), 2019–2025.

Escobar, J. C.; Lora, E. S.; Venturini, O. J.; Yáñez, E. E.; Castillo, E. F.; Almazan, O. Biofuels: Environment, Technology and Food Security. *Renewable Sustainable Energy Rev.* **2009,** *13*(6), 1275–1287.

Esterbauer, H.; Steiner, W.; Labudova, I. Production of Trichoderma Cellulase in Laboratory and Pilot Scale. *Bioresour. Technol.* **1991,** *36,* 51–65.

Fan, L. T.; Lee, Y. H.; Gharpuray, M. M. The Nature of Lignocellulosics and their Pre-Treatments for Enzymatic Hydrolysis. In *Microbial Reactions;* Springer: Berlin Heidelberg, 1982; pp 157–187.

Fan, L. T.; Gharpuray, M. M; Lee, Y. H. Design and Economic Evaluation of Cellulose Hydrolysis Processes. In *Cellulose Hydrolysis*; Aiba, S., Fan, L. T., Fiechter, A., Klein, J., Schügerl, K., Eds.; Springer: Berlin Heidelberg, 1987; pp 149–187.

Frison, A.; Memmert, K. Fed-Batch Process Development for Monoclonal Antibody Production with Cellferm-Pro. *Gen. Eng. News.* **2002,** *22,* 66–67.

Galbe, M.; Zacchi, G. Pre-treatment of Lignocellulosic Materials for Efficient Bioethanol Production. *Adv. Biochem. Eng. Biotechnol.* **2007,** *108,* 41–65.

García-Cubero, M. T.; González-Benito, G.; Indacoechea, I.; Coca, M.; Bolado, S. Effect of Ozonolysis Pre-Treatment on Enzymatic Digestibility of Wheat and Rye Straw. *Bioresour. Technol.* **2009,** *100*(4), 1608–1613.

Godia, F.; Casas, C.; Sola, C. A Survey of Continuous Ethanol Fermentation Systems using Immobilized Cells. *Process Biochem.* **1987,** *22*(22), 43–48.

Gold, M. H.;Alic, M. Molecular Biology of the Lignin-Degrading Basidiomycetes Phanerochaete Chrysosporium. *Microbiol. Rev.* **1993,** *57*(3), 605–622.

Gullison, R. E.; Frumhoff, P. C.; Canadell, J. G.; Field, C. B.; Nepstad, D. C.; Hayhoe, K.;et al. Tropical Forests and Climate Policy. *Science* **2007,** *316,* 985–986.

Hahn-Hägerdal, B.; Galbe, M.; Gorwa-Grauslund, M. F.; Lidén, G.; Zacchi, G. Bio-Ethanol—TheFuel of Tomorrow from the Residues of Today. *Trends Biotechnol.* **2006,** *24*(12), 549–556.

Harun, R.;Danquah, M. K.;Forde, G. M. Microalgal Biomass as a Fermentation Feedstock for Bioethanol Production. *J. Chem. Technol. Biotechnol.* **2010,** *85*(2), 199–203.

Hasunuma, T.;Kondo, A. Consolidated Bioprocessing and Simultaneous Saccharification and Fermentation of Lignocellulose to Ethanol with Thermotolerant Yeast Strains. *Process Biochem.* **2012,** *47*(9), 1287–1294.

Hendriks, A. T. W. M.; Zeeman, G. Pre-treatments to Enhance the Digestibility of Lignocellulosic Biomass. *Bioresour. Technol.* **2009,** *100*(1), 10–18.

Himmel, M. E.; Ding, S. Y.; Johnson, D. K.; Adney, W. S.; Nimlos, M. R.; Brady, J. W.; Foust, T. D. Biomass Recalcitrance: Engineering Plants and Enzymes for Biofuels Production. *Science* **2007,** *315*(5813), 804–807.

Howard, R. L.; Abotsi, E.; Van Rensburg, E. J.; Howard, S. Lignocellulose Biotechnology: Issues of Bioconversion and Enzyme Production. *Afr. J. Biotechnol.* **2003,** *2*(12), 602–619.

Ingram, L. O.; et al. Genetic Engineering of Ethanol Production in Escherichia Coli. *Appl. Environ. Microbiol.* **1987,** *53*, 2420–2425.

International Energy Agency. *Biofuels for Transport: An International Perspective;* International Energy Agency (IEA): Paris, France, 2004; http://www.iea.org/textbase/nppdf/free/2004/biofuels2004.pdf

Iraj, N.; Giti, E.; Lila, A. Isolation of Flocculating *Saccharomyces cerevisiae* and Investigation of its Performance in the Fermentation of Beet Molasses to Ethanol. *Biomass Bioenergy* **2002,** *23*, 481–486.

John, R. P.; Anisha, G. S.; Nampoothiri, K. M.; Pandey, A. Micro and Macroalgal Biomass: A Renewable Source for Bioethanol. *Bioresour. Technol.* **2011,** *102*(1), 186–193.

Jørgensen, H.; Erriksson, T.; Börjesson, J.; et al. Purification and Characterisation of Five Cellulases and One Xylanases from *Penicillium brasilianum* IBT20888. *Enzyme Microb. Technol.* **2003,** *32*, 851–861.

Kim, S.; Holtzapple, M. T. Lime Pre-Treatment and Enzymatic Hydrolysis of Corn Stover. *Bioresour. Technol.* **2005,** *96*(18), 1994–2006.

Klinke, H. B.; Ahring, B. K.; Schmidt, A. S.; Thomsen, A. B. Characterization of Degradation Products from Alkaline Wet Oxidation of Wheat Straw. *Bioresour. Technol.* **2002,** *82*(1), 15–26.

Kotter, P.; Ciriacy, M. Xylose Fermentation by *Saccharomyces cerevisiae.* *Appl. Microbiol. Biotechnol.* **1993,** *38*, 776–783.

Kumar, R.; Wyman, C. E. Effects of Cellulase and Xylanase Enzymes on the Deconstruction of Solids from Pre-Treatment of Poplar by Leading Technologies. *Biotechnol. Prog.* **2009a,** *25*(2), 302–314.

Kumar, R.; Wyman, C. E. Does Change in Accessibility with Conversion Depend on both the Substrate and Pre-Treatment Technology? *Bioresour. Technol.* **2009b,** *100*(18), 4193–4202.

Kuyper, M.; et al. High-Level Functional Expression of a Fungal Xylose Isomerase: The Key to Efficient Ethanolic Fermentation of Xylose by *Saccharomyces cerevisiae*? *FEMS Yeast Res.* **2003,** *4*, 69–78.

Larson, E. D. Biofuel Production Technologies: Status, Prospects and Implications for Trade and Development. United Nations Conference on Trade and Development, New York and Geneva, 2008.

Laser, M.; Schulman, D.; Allen, S. G.; Lichwa, J.; Antal, M. J.; Lynd, L. R. A Comparison of Liquid Hot Water and Steam Pre-Treatments of Sugar Cane Bagasse for Bioconversion to Ethanol. *Bioresour. Technol.* **2002,** *81*(1), 33–44.

Lawford, H. G. A New Approach to Improving the Performance of Zymomonas in Continuous Ethanol Fermentations. *Appl. Biochem. Biotechnol.* **1988,** *17*, 203–219.

Lee, C. W.; Chang, H. N. Kinetics of Ethanol Fermentations in Membrane Cell Recycle Fermentors. *Biotechnol. Bioeng.* **1987,** *29*, 1105–1112.

Lee, R. A.; Lavoie, J. M. From First-to Third-Generation Biofuels: Challenges of Producing a Commodity from a Biomass of Increasing Complexity. *Anim. Front.* **2013,** *3*(2), 6–11.

Lee, S. Chapter 4 Coal Slurry Fuel. In *Handbook of Alternative Fuel Technologies;* Lee, S., Speight, J. G., Loyalka, S. K., Eds.; CRC Press: New York, USA; 2007;pp 129–156.

Lee, S. H.; Doherty, T. V.; Linhardt, R. J.; Dordick, J. S. Ionic Liquid-Mediated Selective Extraction of Lignin from Wood Leading to Enhanced Enzymatic Cellulose Hydrolysis. *Biotechnol. Bioeng.* **2009,** *102*(5), 1368–1376.

Limayem, A.; Ricke, S. C. Lignocellulosic Biomass for Bioethanol Production: Current Perspectives, Potential Issues and Future Prospects. *Prog. Energy Combust. Sci.* **2012,** *38*(4), 449–467.

Love, G.; Gough, S.; Brady, D.; Barron, N.; Nigam, P.; Singh McHale, A. P. Continuous Ethanol Fermentation at 45°C using *Kluyveromyces marxianus* IMB3 Immobilized in Calcium Alginate and Kissiris. *Bioprocess Eng.* **1998,** *18*(3), 187–189.

Lugar, R. G.; Woolsey, R. J. New Petroleum. *Foreign Aff.* **1999,** *78,* 88.

Martín, C.; Thomsen, M. H.; Hauggaard-Nielsen, H.; BelindaThomsen, A. Wet Oxidation Pre-Treatment, Enzymatic Hydrolysis and Simultaneous Saccharification and Fermentation of Clover–Ryegrass Mixtures. *Bioresour. Technol.* **2008,** *99*(18), 8777–8782.

McMillan, J. D. Pre-Treatment of Lignocellulosic Biomass, Chapter 15. In *Enzymatic Conversion of Biomass for Fuel Production;* Himmel, M. E., Baker, J. O., Overend, R. P., Eds.; National Renewable Energy Laboratory: Golden, CO, 1993; pp 292–324.

Mielenz, J. R. Ethanol Production from Biomass: Technology and Commercialization Status. *Curr. Opin. Microbiol.* **2001,** *4*(3), 324–329.

Mosier, N.; Wyman, C.; Dale, B.; Elander, R.; Lee, Y. Y.; Holtzapple, M.; Ladisch, M. Features of Promising Technologies for Pre-Treatment of Lignocellulosic Biomass. *Bioresour. Technol.* **2005a,** *96*(6), 673–686.

Mosier, N.; Hendrickson, R.; Ho, N.; Sedlak, M.; Ladisch, M. R. Optimization of pH Controlled Liquid Hot Water Pre-Treatment of Corn Stover. *Bioresour. Technol.* **2005b,** *96*(18), 1986–1993.

Naik, S. N.; Goud, V. V.; Rout, P. K.; Dalai, A. K. Production of First and Second Generation Biofuels: A Comprehensive Review. *Renewable Sustainable Energy Rev.* **2010,** *14*(2), 578–597.

Nieves, R. A.; Ehrman, C. I.; Adney, W. S.; et al. Technical Communication: Survey and Commercial Cellulase Preparations Suitable for Biomass Conversion to Ethanol. *World J. Microbiol. Biotechnol.* **1998,** *14,* 301–304.

Nigam, P. S.; Singh, A. Production of Liquid Biofuels from Renewable Resources. *Prog. Energy Combust. Sci.* **2011,** *37*(1), 52–68.

Oh, K. K.; Kim, S. W.; Jeong, Y. S.; Hong, S. I. Bioconversion of Cellulose Into Ethanol by Nonisothermal Simultaneous Saccharification and Fermentation. *Appl. Biochem. Biotechnol.* **2000,** *89,* 15–13.

Oliva, J. M.; Sáez, F.; Ballesteros, I.; González, A.; Negro, M. J.; Manzanares, P.; Ballesteros, M. Effect of Lignocellulosic Degradation Compounds from Steam Explosion Pre-Treatment on Ethanol Fermentation by Thermotolerant Yeast *Kluyveromyces marxianus*. In *Biotechnology for Fuels and Chemicals;* Davison, B. H., Evans, B. R., Finkelstein, M., McMillan, J. D., Eds.; Humana Press: Totowa, NJ, USA; 2003; pp 141–153.

Palmqvist, E.; Hahn-Hägerdal, B.; Galbe, M.; Zacchi, G. The Effect of Water-Soluble Inhibitors from Steam-Pretreated Willow on Enzymatic Hydrolysis and Ethanol Fermentation. *Enzyme Microb. Technol.* **1996,** *19,* 470–476.

Palonen, H.; Thomsen, A. B.; Tenkanen, M.; Schmidt, A. S.; Viikari, L. Evaluation of Wet Oxidation Pre-Treatment for Enzymatic Hydrolysis of Softwood. *Appl. Biochem. Biotechnol.* **2004,** *117*(1), 1–17.

Pan, X.; Xie, D.; Gilkes, N.; Gregg, D. J.; Saddler, J. N. Strategies to Enhance the Enzymatic Hydrolysis of Pretreated Softwood with High Residual Lignin Content. In *Twenty-Sixth Symposium on Biotechnology for Fuels and Chemicals;* Davison, B. H., Evans, B. R., Finkelstein, M., McMillan, J. D., Eds.; Humana Press: New York City, USA; 2005; pp 1069–1079.

Patil, V.; Tran, K. Q.; Giselrød, H. R. Towards Sustainable Production of Biofuels from Microalgae. *Int. J. Mol. Sci.* **2008,** *9*(7), 1188–1195.

Pérez, J. A.; Ballesteros, I.; Ballesteros, M.; Sáez, F.; Negro, M. J.; Manzanares, P. Optimizing Liquid Hot Water Pre-Treatment Conditions to Enhance Sugar Recovery from Wheat Straw for Fuel-Ethanol Production. *Fuel* **2008,** *87*(17), 3640–3647.

Pettersen, R. C. The Chemical Composition of Wood, Chapter 2. In *The Chemistry of Solid Wood;* Rowell, R., Ed.; American Chemical Society: Washington, DC, 1984; Vol. 207, pp 57–126.

Prasad, S.; Singh, A.; Joshi, H. C. Ethanol as an Alternative Fuel from Agricultural, Industrial and Urban Residues. *Resour. Conserv. Recycl.* **2007,** *50*(1), 1–39.

Ran, H.; Zhang, J.; Gao, Q.; Lin, Z.; Bao, J. Analysis of Biodegradation Performance of Furfural and 5-Hydroxymethylfurfural by *Amorphotheca resinae* ZN1.*Biotechnol. Biofuels* **2014,** *7*(1), 1.

Saarela, U.; Leiviska, K.; Juuso, E. *Modelling of a Fed-Batch Fermentation Process;* Control Engineering Laboratory, Department of Process and Environmental Engineering, University of Oulu, Report A No. 21, Finland; June 2003.

Saggi, S. K.; Gupta, G.; Dey, P. Biological Pre-treatment of Lignocellulosic Biomaterials. In *Advances in Biofeedstocks and Biofuels:Bio Feedstocks and their Processing;* 2016a; Vol.1, p 97.

Saha, B. C. Hemicellulose Bioconversion. *J. Ind. Microbiol. Biotechnol.* **2003,** *30*(5), 279–291.

Saha, B. C.; Iten, L. B.; Cotta, M. A.; Wu, Y. V. Dilute Acid Pre-Treatment, Enzymatic Saccharification and Fermentation of Wheat Straw to Ethanol. *Process Biochem.* **2005,** *40*(12), 3693–3700.

Sarkar, N.; Ghosh, S. K.; Bannerjee, S.; Aikat, K. Bioethanol Production from Agricultural Wastes: An Overview. *Renewable Energy* **2012,** *37*(1), 19–27.

Shama, G. Developments in Bioreactors for Fuel Ethanol Production. *Process Biochem.* **1988,** *10,* 138–145.

Sheehan, J.; Aden, A.; Paustian, K.; Killian, K.; Brenner, J.; Walsh, M.; Nelson, R. Energy and Environmental Aspects of using Corn Stover for Fuel Ethanol. *J. Ind. Ecol.* **2003,** *7*(3-4), 117–146.

Singh, D.; Dahiya, J. S.; Nigam, P. Simultaneous Raw Starch Hydrolysis and Ethanol fermentation by Glucoamylase from *Rhizoctonia solani* and *Saccharomyces cerevisiae. J. Basic Microbiol.* **1995,** *35*(2), 117–121.

Sjöström, E. *Wood Chemistry: Fundamentals and Applications;* Gulf Professional Publishing: USA, 1993.

Sreenath, H. K.; Moleds, A. B.; Koegel, R. G.; Straub, R. J. Lactic Acid Production by Simultaneous Saccharification and Fermentation of Alfalfa Fiber. *J. Biosci. Bioeng.* **2001,** *92,* 518–523.

Sun, Y.; Cheng, J. Hydrolysis of Lignocellulosic Materials for Ethanol Production: A Review. *Bioresour. Technol.* **2002,** *83*(1), 1–11.

Szczodrak, J.; Targonski, Z. Selection of Thermotolerant Yeast Strains for Simultaneous Saccharification and Fermentation of Cellulose. *Biotechnol. Bioeng.* **1988,** *31,* 300–303.

Taherzadeh, M. J.; Karimi, K. Pre-Treatment of Lignocellulosic Wastes to Improve Ethanol and Biogas Production: A Review. *Int. J. Mol. Sci.* **2008,** *9*(9), 1621–1651.

Tengborg, C.; Stenberg, K.; Galbe, M.; Zacchi, G.; Larsson, S.; Palmqvist, E.; Hahn-Hägerdal, B. Comparison of SO2 and H2SO4 Impregnation of Softwood Prior to Steam Pre-Treatment on Ethanol Production. In *Biotechnology for Fuels and Chemicals*; Humana Press: USA; 1998, pp 3–15.

Ueda, R.; Hirayama, S.; Sugata, K.; Nakayama, H. U.S. Patent No. 5,578,472. Washington, DC: U.S. Patent and Trademark Office, 1996.

Varga, E.; Réczey, K.; Zacchi, G. Optimization of Steam Pre-Treatment of Corn Stover to Enhance Enzymatic Digestibility. *Appl. Biochem. Biotechnol.* **2004,** *114*(1–3), 509–523.

Von Sivers, M.; Zacchi, G. A Techno-Economical Comparison of Three Processes for the Production of Ethanol from Pine. *Bioresour. Technol.* **1995,** *51*(1), 43–52.

Walfridsson, M.; et al. Ethanolic Fermentation of Xylose with *Saccharomyces cerevisiae* Harbouring the *Thermus thermophilus* Xyla Gene which Expresses an Active Xylose (Glucose) Isomerase. *Appl. Environ. Microb.* **1996,** *62*, 4648–4651.

Wayman, M.; Parekh, S. R. *Biotechnology of Biomass Conversion: Fuels and Chemicals from Renewable Resources;* Open University Press: Milton Keynes; 1990.

Wilkie, A. C.; Riedesel, K. J.; Owens, J. M. Stillage Characterization and Anaerobic Treatment of Ethanol Stillage from Conventional and Cellulosic Feedstocks. *Biomass Bioenergy* **2000,** *19*(2), 63–102.

Wingren, A.; Galbe, M.; Zacchi, G. Techno-Economic Evaluation of Producing Ethanol from Softwood: Comparison of SSF and SHF and Identification of Bottlenecks. *Biotechnol. Prog.* **2003,** *19,* 1109–1117.

Wooley, R; Ruth, M.; Sheehan, J.; Ibsen, K.; Majdeski, H.; Galvez, A. *Lignocellulosic Biomass to Ethanol Process Design and Economics Utilizing Co-Current Dilute Acid Prehydrolysis and Enzymatic Hydrolysis: Current and Futuristic Scenarios;* NREL/TP-580–26157, 1999; National Renewable Energy Laboratory: Golden, CO; 1999.

Wyman, C. E.; Dale, B. E.; Elander, R. T.; Holtzapple, M.; Ladisch, M. R.; Lee, Y. Y. Comparative Sugar Recovery Data from Laboratory Scale Application of Leading Pre-Treatment Technologies to Corn Stover. *Bioresour. Technol.* **2005a,** *96*(18), 2026–2032.

Wyman, C. E.; Dale, B. E.; Elander, R. T.; Holtzapple, M.; Ladisch, M. R.; Lee, Y. Y. Coordinated Development of Leading Biomass Pre-Treatment Technologies. *Bioresour. Technol.* **2005b,** *96*(18), 1959–1966.

Yachmenev, V.; Condon, B.; Klasson, T.; Lambert, A. Acceleration of the Enzymatic Hydrolysis of Corn Stover and Sugar Cane Bagasse Celluloses by Low Intensity Uniform Ultrasound. *J. Biobased Mater. Bioenergy* **2009,** *3*(1), 25–31.

Further Reading

Alvira, P.; Tomás-Pejó, E.; Ballesteros, M.; Negro, M. J. Pre-Treatment Technologies for an Efficient Bioethanol Production Process Based on Enzymatic Hydrolysis: A Review. *Bioresour. Technol.* **2010,** *101*(13), 4851–4861.

Campbell, C. J.; Laherrère, J. H. The End of Cheap Oil. *Sci. Am.* **1998,** *278*(3), 60–65.

Coughlan, M. P.; Ljungdahl, L. G. Comparative Biochemistry of Fungal and Bacterial Cellulolytic Enzyme Systems. In *Biochemistry and Genetics of Cellulose Degradation;* Aubert, J. P., Btguin, P., Millet, J.,Eds.; Academic Press: Londonand San Diego; 1988; *43*, 11–30.

D'Amore, T.; Celloto, G.; Russel, I.; Stewart, G. G. Selection and Optimization of Yeast Suitable for Ethanol Production at 40°C. *Enzyme Microb. Technol.* **1989,** *11,* 411–416.

Gírio, F. M.; Fonseca, C.; Carvalheiro, F.; Duarte, L. C.; Marques, S.; Bogel-Łukasik, R. Hemicellulose for Fuel Ethanol: A Review. *Bioresour. Technol.* **2010,** *101*, 4775–4800.

Lin, Y.; Tanaka, S. Ethanol Fermentation from Biomass Resources: Current State and Prospects. *Appl. Microbiol. Biotechnol.* **2006,** *69,* 627–642.

Olsson, L.; Hahn-Hägerdal, B. Fermentation of Lignocellulosic Hydrolysates for Ethanol Production. *Enzyme Microb. Technol.* **1996,** *18,* 312–331.

INDEX

α

α-amylase, 110, 111
α-carotene, 65, 67, 69, 70, 75
α-pinene, 133

β

β-carotene, 64–70, 72, 74–77
β-cryptoxanthin, 64, 69, 70
β-pinene, 133, 135

A

Abiotic stress, 13
Absorption, 45, 92, 172, 209
Acetaldehyde, 97
Acetate groups, 96
Acids, 5, 6, 14, 21, 22, 47, 53, 92, 97, 103,
    107, 113, 115, 116, 118–120, 122–124,
    133, 161, 162, 168, 191
Acinetobacter, 22
Acoustic cavitation, 76
Activated carbon, 231
Active
    coatings, 28
    compounds, 22, 103, 137, 142
    edible coatings, 18
    packaging, 17, 21, 90, 92, 132, 142
Acylglycerols, 5
Adhesion, 12
Adsorbent, 209, 212, 213, 231
Adsorption, 122, 213
Aeromonas hydrophilla, 22
Aesthetic, 3, 6, 91
Agro-industry, 108
Alcaligenes, 22
Alcohols, 123, 133
Aldehydes, 94, 133
Algae, 19, 69, 95, 200, 209, 211
Alginate, 3, 5, 12, 18, 23, 27, 44, 49, 54, 55,
    102, 103, 139, 140, 199–201, 203, 205,
    206, 209–212, 214, 215

fenugreek, 44, 55
Alginic acid, 211
Aliphatic, 231
Alkaline condition, 5
Alkalis, 5
Aloe mucilage, 8
Alpha tocopherol protection, 54
Alternative materials, 55
Alveolar inflammation, 214
Ambient temperature, 12
Amines, 133
Amino acid, 4, 19
Amphiphilic nature, 4
Amylopectin, 95, 96
Amylose, 95
Analytical test, 12
Anatomic analyses, 188
Anethole, 133
Animal
    collagen, 97
    proteins, 7, 9
    slaughtering, 7
Anthocyanidins, 107, 113–115
Anthocyanins, 114, 121, 124
Anthropocene, 213
Antibacterial activities, 12
Antifungal activity, 11, 139
Antimicrobial, 2, 5, 6, 9, 10, 15–18, 21–23,
    25, 26, 28, 89–92, 97, 99, 101, 103,
    131–137, 139, 142, 170, 205, 209, 212
    activity, 25, 103
    compound, 22
Anti-nutritional compounds, 109, 112
Antioxidant, 2, 5, 6, 8–10, 12, 17, 20–23,
    28, 63, 64, 66–68, 89, 91, 97, 101, 107,
    116, 124, 131, 137, 138, 142, 143, 159,
    160, 162, 168, 170, 208
    activity, 8, 12, 63, 170
    agents, 2
Anti-tumor immunity, 48
Apoptosis, 48

Appetite regulation, 47
Aqueous solution, 97, 204, 207, 211
Arabic gum, 6, 12
Arabinoxylan, 45
Arcobacter butzleri, 22
Aroma, 7, 8, 17, 20, 160
    aromatic compounds, 110, 114, 115, 123,
        133, 200, 231
    deterioration, 20
Arsenic, 221–228, 230–237, 239–241
    groundwater, 221–223, 233, 235
Arsenicosis, 234, 235
Ascorbic acid, 8, 18, 90, 119, 152, 153
Aspergillus, 22, 120, 123, 170, 209, 213
Assessment of persian lime quality, 11
Astaxanthin, 64, 66
Auricularia, 111
Autonomic imbalance, 214

B

Bacillus cereus, 22
Bacteria, 19, 21, 22, 44, 45, 47–50, 53–56,
    65, 69, 91, 97, 99, 101, 132, 133, 139,
    167, 191, 200, 209, 212
Bacterial
    endophthalmitis, 210
    infection, 48
Bacteriocins, 56
Barrier, 2, 4, 6, 7, 9, 13, 14, 17–20, 28, 47,
    50, 91, 95, 100–103, 132, 137, 140
    properties, 4, 14, 18, 19, 28, 100, 102,
        137, 140
Beneficial microflora, 48
Benzoic acid, 19
Beta-cyclodextrin, 18
Bifidobacterial populations, 47
Bifidobacterium (BF), 46, 48, 49, 53–55
    bifidum, 46, 55
    lactis BB12 (L), 46
    longum BB 536 (L), 46
Bilateral symmetry, 65
Bioactive
    components, 6, 52, 135
    compounds, 9, 19, 43, 50, 52, 64, 76,
        101, 124, 160, 162, 163, 166–168,
        170–172, 202
    ingredients, 44, 49
    proteins, 95

substances, 47, 90, 92, 93
Bioavailability, 68, 172, 214
Biochothrix thermospacta, 23
Biocompatibility, 9, 44, 96, 202, 212–214
Biocompatible, 51, 93, 95, 99, 103, 200,
    214
Biodegradability, 7, 9, 44, 96, 99, 101, 213,
    214
Biodegradable, 3–6, 9, 11, 17, 24, 28, 91,
    93, 95–97, 99–104, 131, 200, 212, 214
Bioethanol, 110, 111, 205
Biofertilizer, 110
Biofunctionality, 51
Biogas, 110
Biological systems, 214
Biomacromolecules, 94, 101
Biomaterials, 97
Biomolecules, 94, 95, 101, 170
Biopolymers, 4, 6–8, 15, 22, 52, 91, 96,
    100–103, 105, 199–201, 203, 205, 208,
    209, 212–215
Biopreservation technology, 16
Biosorption, 236
Biosynthesis, 72, 73
Biosynthetic genes, 69
Biotechnology, 77, 96, 99
Biotransformation, 107–109, 111, 170
Bitter vetch protein, 9
Bladder, 68, 234
Blakeslea trispora, 65
Bovine
    hides, 19
    serum albumin (BSA), 213
    spongiform encephalopathy, 7
Breba figs, 8
Briquettes, 110
Brochothrix thermosphacta, 22, 136
Butylated hydroxytoluene (BHT), 160
Butyrate, 48, 49

C

C. arabica, 108, 117
C. canephora, 108, 117, 121
C. jambhiri, 188
Cactaceae, 149, 150, 152, 186
Caffeic, 116, 119, 120, 122–124
    acid phenethyl ester (CAPE), 123, 124
Calcium, 64, 90, 99, 102, 211

Campylobacter spp., 22

Cancer, 48, 63, 64, 68, 201, 205, 206, 211, 212, 214, 234, 235
  cells, 48, 212, 214

Candelilla, 12, 13

Candida strains, 22

Capsanthin, 64, 70

Capsicum anuum, 70

Capsorrubina, 70

Carbohydrates, 5, 6, 51, 56, 112, 153, 161

Carbon, 2, 63, 65, 72, 76, 78, 110, 114, 201, 231
  dioxide, 2, 12, 76
  nanotubes, 231

Carboxyl, 65, 115, 122, 200, 207

Carboxylate groups, 99

Carboxymethyl
  cellulose, 5, 15
  starch, 15

Carbures, 133

Carcinogenic, 48, 91

Carcinogens, 47, 48

Cardiovascular
  diseases, 63, 64, 68, 79
  health, 66

Carica papaya, 70

Carnauba, 13
  wax, 12

Carotenoid, 8, 18, 63–79, 161, 168
  profile, 69
  recovery, 74
  triplet, 72

Carrageenan, 5, 18, 21, 27, 44, 51, 138, 199–201, 203, 206, 209, 212, 213, 215

Carvacrol, 23, 27, 133–135, 140

Casein, 7, 52

Caseinates, 102

Cassava, 96, 139

Catastrophe, 224, 228, 230, 235, 236, 238–240

Catechin, 8, 113, 114, 161, 162, 165, 168, 169

Cation, 99

Cavitation, 167

Cell
  cytotoxicity, 68
  death, 72, 133, 210

Cellular

changes, 69
  surfaces, 99

Cellulose, 5, 6, 8, 15, 18, 51, 95, 103, 112, 136, 199, 200, 203, 205, 207, 208, 215
  fiber, 207
  nanocrystals, 205, 207
  nanofibers, 207
  nanoparticles, 205, 207

Center and Development for Food Industries (CIDIA), 1, 159

Ceratocytis fimbriata, 111

Cereals, 193

Cervical, 68

Chelating, 24

Chemical
  cross-linking, 5, 92
  hydrogels, 94, 96
  hydrolysis, 5, 119, 203

Chicory, 49, 53, 54

Chitosan, 5, 8, 10, 11, 15, 16, 18, 21, 23, 25, 27, 44, 51, 54, 55, 91, 98, 99, 102, 103, 138, 139, 143, 199, 200, 203–205, 209–211, 213–215

Chlorella vulgaris, 77

Chlorogenate hydrolase, 110, 111, 123

Chlorogenic acid, 113, 116, 119–123

Chloroplasts, 69

Chronic, 234, 235
  arsenic toxicity, 234
  intake, 235

Cinnamaldehyde, 23, 102

Cinnamomum zeylanicum, 102

Cinnamon, 21, 23, 101, 134, 136, 139

Cinnamyl alcohol, 133

Cis-trans isomers, 66

Citral, 18

Citric acid, 18, 92, 153

Citronellol, 133

Citrus, 2, 70, 102, 134, 135, 182, 187, 188, 193

Clostridium
  botulinum, 22
  perfringens, 22

Coacervation, 50

Coal dewatering, 95

Coated guava, 12

Coating, 2–7, 9–11, 13–22, 24–28, 91, 99, 101, 102, 201, 203, 205, 208

formulation, 4, 12, 17, 19
functions, 6
Cocrystallization, 50
Co-encapsulation, 55
Coffee
  phenolic compounds, 109
  pulp (CP), 107–124
Cohesive biomaterials, 5
Collagen, 7, 10, 15, 19, 20, 91, 97
Colloidal stability, 103
Colon, 45, 48, 68
  cancer, 48
  microflora, 55
Colonic glutathione S-transferase, 48
Colonization, 47, 48
Color analysis, 11, 25
Colorants, 5, 6, 63
Colossus, 229, 230, 234
Commercialization, 13, 52, 91, 132, 153
Complexation, 201, 209, 211, 213
Compounds, 5, 7, 10, 15, 18, 22, 25, 27, 28,
  50, 51, 56, 63, 64, 70, 72, 75, 76, 78, 79,
  93, 101, 108, 109, 111, 113–118, 121,
  122, 124, 133, 136, 137, 142, 159, 160,
  162–172, 207, 231
Contamination, 11, 26, 100, 222–225, 227,
  228, 230–233, 235, 236, 239, 240
Conventional oral iron therapy, 211
Copolymer, 96, 136
Core-shell structure, 211
Corn zein, 7
Cornerstones, 227, 229, 231, 236
Coronary heart disease, 89, 90
Corrosion, 100
Corynebacterium, 22
Cosmetic
  alimentary, 63
  industry, 66, 68
Cost-effectiveness, 44
Covalent, 92, 97, 98, 120, 209
Crops homogeneity, 191
Cross-linking, 11, 92–94
Crucial juncture, 223, 228
Crude fiber, 112
Cryptococus, 65
Crystalline
  character, 97
  nuclei, 97

solid states, 5
structure, 207
Cultivars, 117, 187, 190, 191, 193
Curcumin, 212
Cyclodextrins, 55
Cymene, 133, 140
Cytotoxic potential, 212
Cytotoxicity, 214

D

Daptomycin, 210
Deboning, 90
Decontamination, 12, 227, 237
Delineates, 223, 230, 234
Denaturation, 19, 73
Dendritic cells, 48
Derivatives, 4, 90, 107, 113–116, 118, 119,
  121, 124, 133, 168
Deterioration, 14, 17, 20–22, 44, 89–91, 97,
  101, 142
Detoxification, 48, 109, 202
Diabetes, 64, 67, 68
Diarrhea, 48
Dietary supplements, 68
Diffusion, 15, 17, 49, 75, 136, 142, 168
Digestive enzymes, 51
Dipping, 18
Discoloration, 8, 22
Dispersibility, 212
Dissolution, 92, 94, 103, 204, 207, 210
Distillation, 132, 162
Domain, 93, 94, 224, 227, 231, 234
Double bond, 64–66, 123
Drug delivery systems, 95, 214
Dry jet-wet spinning, 207

E

E. fetida, 110
Ecological biodiversity, 228, 229
Edibility, 7
Edible
  casings, 103
  coating, 1, 3, 4, 10–13, 15, 17–19, 23, 24,
    26, 28, 91, 102, 104
  films, 2–15, 17, 22, 24–28, 141, 143
Eisenia andrei, 110
Elasticity, 7, 92, 140
Electric pulses, 74

Electrospinning, 56
Electrostatic
  charges, 4
  forces, 56
  interaction, 214
Elongation, 9, 10, 140
Emancipation, 221–226, 229, 231, 239
Embryogenesis, 188
Emission, 10, 69, 167, 231
Emitters, 21
Emulsion, 2, 6, 12, 14, 140, 201
Encapsulated, 44, 49–52, 56, 139
  probiotic, 55
Encapsulating agents, 44, 51
Encapsulation, 43–45, 49–51, 53, 55, 56,
  210
  processes, 49
  systems, 44
Endeavor, 221–225, 227, 228, 231, 232,
  234–238
Endocrinological effects, 47
Endogenous enzymes, 20
Endometrial, 68
Energy
  crisis, 229
  intake, 47, 48
  security, 229
  sustainability, 228–230
Enigma, 226, 232, 238
Enigmatic issue, 223, 226, 232, 236
Enrobed fish sticks, 21
Enterobacter, 22
Enterobacteria, 10, 25
Enterobacteriaceae, 8, 136
Enterococcus
  fecium, 49, 54
  SF68, 46
Enterocytes, 47
Enterohemorrhagic *E. coli* (EHEC), 22
Environmental
  crisis, 221, 224, 228, 230
  disaster, 222, 225, 238
Enzymatic
  activity, 14, 47
  lysis, 74
  modification, 108, 109
  reactions, 21, 72, 124
  treatment, 5

Enzyme, 21, 73, 90, 93, 110, 120, 161, 209,
  210, 212
Epicatechin (EC), 10, 114, 115, 164, 165,
  168, 169
  gallate (ECG), 164, 165, 168, 169
Epidemiological studies, 64, 68, 116
Epigallocatechin (EGC), 164, 165, 168, 169
  gallate (EGCG), 164, 165, 168, 169
Epithelial
  defensin production, 48
  tight junctions, 48
Escherichia coli, 22, 46, 136
Essential oils (EOs), 21, 22, 97, 131–137,
  140, 142, 144
Esterase, 110, 111, 119, 120, 123
Esterified, 116, 120, 123
Esters, 10, 116, 119, 123, 133
Ethanol, 27, 75, 79, 111, 118, 121, 163, 166,
  200, 207, 210
Ethylene, 2, 12, 16, 89, 91, 136
  vinyl alcohol (EVOH), 136
Eugenol, 18, 134
Extraction, 25, 74–77, 107–109, 111, 112,
  117–120, 124, 134, 159, 160, 162, 163,
  166–171, 206, 212
Extrusion, 24, 50, 207

F

F. oxysporum, 213
Fabrication, 102
Fatty acid esters, 5
Fermentable, 111
Fermentation, 44, 47, 72, 78, 91, 108, 109,
  111, 120, 124, 161, 162, 168, 170
Fermentative route, 70
Ferrous, 211
Ferulic acid, 11, 113, 119, 120, 123
Feruloyl, 110, 111, 113, 116, 119, 123
Fibers, 5, 56, 103, 203, 207
Film, 2–7, 9, 12–17, 19–26, 28, 90, 93, 101,
  102, 136–140, 143, 144, 207, 212, 213
  conditioning, 10
  forming, 11, 19
    materials, 4, 5
  integrity, 4
  network, 7
  solubility, 12
  thickness, 10, 12

transparency, 12
Firmness, 8, 11, 12, 18, 90, 91, 102
Flavanols, 107, 113–115, 118, 168
Flavobacterium, 22
Flavonoids, 114, 117, 118, 122, 124, 160, 165, 169
Flavonols, 107, 113–115, 162, 168
Flavor, 5, 6, 17, 142, 155
  releasers, 21
Flexibility, 5, 7, 9, 19, 93, 94
Flow injection synthesis, 203
Fluconazole, 213
Fluidization, 24
Folin-ciocalteu, 117
Food,
  additives, 21, 90, 93, 95, 131
  development, 149, 156
  Drug Administration (FDA), 26, 133, 167
  industry, 4, 26, 44, 45, 50, 52, 53, 55, 56, 66, 68, 77, 99, 116, 132, 171, 182, 185, 187, 192, 193, 199, 200, 202, 205, 206, 209, 210, 212, 215
  packaging, 6, 28, 100, 132, 142
  quality, 2, 4, 28
  sensory quality, 51
Formaldehyde, 97, 208
Formulate active packaging, 23
Fragile, 4
Freeze
  drying, 50
  thaw cycles, 92
Frequency, 160, 167, 170, 188, 189, 207
Fresh cut fruits (FFC), 90, 91
Frozen meat, 3
Fructan, 48
  prebiotics, 47
Fructooligosaccharides (FOS), 45, 49, 54
Functional
  foods, 43–45, 50, 56, 149
  ingredients, 2, 44
Functionalization, 202, 207
Fungal
  growth, 25
  infections, 15
Fungi, 19, 65, 69, 72, 97, 99, 101, 109, 111, 133, 209
Fungicides, 3, 102
Fusarium, 22, 139

G

Galactonannans, 10
Galactooligosaccharides (GOS), 45, 54
Gallic, 116, 118, 120, 168
Gallocatechin (GC), 10, 161, 168
  gallate (GCG), 168
Gamma radiation, 3
Gangrene, 235
Gas transfer rates, 10
Gastric damage, 68
Gastrointestinal
  conditions, 50
  diseases, 48
  microbiota, 45
  tract, 45, 46, 50, 53, 55
  transit, 47
  zone, 43
Gelatin, 2, 7, 8, 10, 12, 15, 19–21, 23, 25, 27, 51, 52, 91, 97, 98, 102, 103, 139, 140, 143
  coatings, 19
  pectin, 102
  types A and B, 97
Generally regarded as safe (GRAS), 6, 55, 133
Generic structures, 114
Genes, 69, 188
Genetic improvement, 191
Genotoxic compounds, 48
Genotype, 69, 189
Geology, 225, 233
Geraniol, 133
Geranyl-geranyl diphosphate, 65
Germplasm, 188, 189
Ghatti gum, 13
Glacial acetic acid, 8
Glucose, 72, 78, 79, 95, 116, 168, 207
Glucosides, 114
Glutaraldehyde, 97, 209, 210
Gluten, 7, 9, 15
Glycerol, 8, 10–12, 15, 18, 19, 25, 27, 72, 123
Golden apple, 193
Graphenes, 231
Grave, 224, 225, 236, 237, 240, 241
Grinding, 203
Groundbreaking, 232, 239, 240
Groundwater, 221–241

poisoning, 223–227, 235, 236
remediation, 222, 223, 225–227, 232, 234, 236, 239, 240
Gum, 3, 5
  locust bean gum matrix, 55
  tragacanth, 13
Gumminess, 21
Gut hormones, 47

H

Hallmarks, 226, 229, 238
Handling, 4, 5, 69, 102, 110, 163
Hazelnut meal protein, 9
Health hazards, 163, 223, 234, 235
Heavy metal, 209, 221, 222, 224–228, 230–232, 235–237, 239
Helicobacter pylori, 48
Hepatoprotective properties, 116
Herbivores, 132
Heredity, 188, 189
Heterogeneous structure, 22
Hexane, 75, 76, 118, 163
Hexyl acetate, 10
Hidroxypropyl methylcellulose, 12
High
  performance liquid chromatography (HPLC), 77, 114, 115, 122
  speed counter-current chromatography (HSCCC), 122
Hipotiocianita analysis, 25
Homeostasis, 191
Homogeneous distribution, 50
Hormones, 47, 66
Human
  civilization, 221–233, 235, 236, 239, 241
  scientific endeavor, 225
Humidity, 7, 16, 91, 110, 184
Hydrocarbon, 64, 65
  carotenes, 64
  compounds, 65
Hydrochloric acid, 208, 211
Hydrodistillation, 132
Hydrogel, 89, 90, 92–102, 104, 105, 207, 213
  structure, 92
  variants, 102
Hydrogen, 24, 65, 90, 92, 94, 97, 98, 208
Hydrogenated fats, 51

Hydrolysis, 45, 96, 97, 111, 118, 119, 124, 163, 210
Hydrolyzable, 116, 165
  tannins, 116
Hydrolyzation, 19
Hydrophilic
  character, 28
  exterior, 55
  functional groups, 94
  head, 52
  moieties, 5
  motifs, 52
Hydrophobic
  acyl chain, 52
  components, 15
  interior, 55
  materials, 14
  nature, 92
Hydrothermal models, 203
Hydroxybenzoic acid, 115, 116
Hydroxycinnamic acids, 113, 115, 116, 118, 119
Hydroxyl, 5, 65, 66, 114, 115, 122, 200, 207, 209
  groups, 5, 207, 209
Hydroxymethyl starch, 15
Hydroxymethylation, 208
Hydroxypropyl starch, 15
Hypercholesterolemia, 89, 90
Hyperpigmentation, 235

I

Illumination, 72
Immense
  lucidity, 233, 235
  scientific
    girth, 233
    prowess, 223–225, 229, 234, 238, 239
    regeneration, 230
Immune function, 66, 67
Immunological effects, 47
Immunomodulation properties, 48
Immunomodulatory capacity, 48
Implementation, 28, 155
In vitro, 68, 95, 116, 138, 188, 205
In vivo, 67, 68, 95, 116
Indo-Bangladesh Gangetic delta, 235
Inflammation, 68, 159

Ingenuity, 222, 223, 225, 229, 234, 236, 237, 240
Inhibition, 8, 12, 78, 123
Inhibitory concentration, 25
Inorganic fertilizer, 110
Integrity, 5, 21, 43, 68, 91, 93, 142, 172
Intensity, 26, 72, 77, 78, 167, 204
Interchain interactions, 101
Intermittent microwave radiation, 76
Internal network structure, 92
International Scientific Association of Probiotics and Prebiotics (ISAPP), 45
Interpenetrated network hydrogels (IPN), 92
Intestinal
    microbiota, 44
    zone, 49
Intricacies, 221, 223, 240
Intrinsic probiotic factors, 46
Introspection, 223, 224, 226, 228, 230, 231, 237, 239
Inulin, 45, 47–49, 53, 54
Inverted sugar, 25
Iodide, 8
Ionic gelation, 201, 210, 211
    method, 211
Ionized carboxylate groups, 99
Ionotropic gelation, 51
Iron encapsulation, 211
Irradiation, 4, 97, 204
Isoelectric point, 11, 213
Isomerization, 65, 66

K

Kaempferol, 168, 211
Karaya gum, 13
K-carrageenan gum, 18
Keratosis, 234, 235
Ketones, 133
Klebsiella, 22

L

Laccase, 110, 111
Lactic acid bacteria, 55
Lactobacillus, 22, 44, 46, 48, 49, 53–55
    acidophilus LA1, 46
    casei (L), 46, 49, 54, 55
    plantarum, 44, 46, 49
    reuteri, 46, 48, 55

rhamnosus GG, 46
    sporogens, 46
Lactoperoxidase system, 25
Lactulose, 45, 48
Landscape, 230
Larding, 3
Larrea tridentata, 77
Laurel, 10, 23
Lentinula, 111
Lethal effects, 69
Leucoanthocyanidins, 114
Leuconostoc spp., 22
Light harvesting, 66, 69
Lignin, 21, 112, 161, 199, 200, 203–205, 208, 215
Limonene, 18, 134, 135
Linear
    low-density polyethylene (LLDPE), 136
    polyanions, 99
Linseed, 8
Lipid, 4–6, 9, 12–14, 17, 25, 27, 51, 52, 91, 96, 121, 133, 135, 137, 140, 142, 161
    membranes, 69
    oxidation, 8, 21–23, 25, 27, 137
    solubility, 2
Lipidic moieties, 52
Lipophilic
    nature, 63
    pigment, 64
Liquid
    extraction, 75, 118, 162, 168
    fraction, 101
    nuclear wastes, 231
Listeria monocytogenes, 22, 23, 27, 136–138, 212
Low-methoxyl (LM), 8, 27, 99
Lucidly, 222, 231
Lutein, 66–70
Lycopene, 64–66, 68–70, 76–78
Lycopersicum esculentum, 70
Lyocell process, 207
Lysine, 189, 191
Lysozyme, 10, 22, 205, 210

M

Maceration, 74, 132, 162
Macromolecular chains, 93
Macromolecules, 4, 94, 116, 209

Macular degeneration (AMD), 63, 67
Magnesium, 64, 168
Magnetic resonance imaging (MRI), 202, 205
Maize, 44, 95, 136, 188–191, 193
Maleic acid, 92
Mammalian gelatins, 9
Mango, 15, 96, 189, 191, 193
Marauding domain, 224
Mastic resin, 13
Matrix, 9, 44, 50, 53, 56, 75, 76, 79, 92, 97, 137–140, 163, 166, 200, 212
Maturation, 12, 69, 90, 102
Meatballs, 10
Membranes, 101, 103, 167, 214
Mesquite gum, 13, 103
Metabolism, 46, 47, 49, 72
Metabolization, 46
Metal
  ions, 5, 24, 231
  oxides, 202, 231
Methanol, 97, 118, 121, 163, 166
Methotrexate, 212
Methoxy, 65, 134, 200
Microbial
  cells, 71, 133
  contamination, 16, 28, 137
  count, 23
  degradation, 21
  growth, 19–22, 25, 26, 73, 74, 112, 138
  membrane, 133
  pigments, 77
  production, 70, 77, 79
  spoilage, 6, 16, 18, 20, 21
Microbiological
  analysis, 27
  safety, 16
Microbiology, 8, 12, 25
Microbiota, 46, 47, 49
Microcapsules, 44, 51, 53
Microencapsulation technique, 53
Microfibrillated, 205, 207
Micrometric size, 49
Microorganisms, 8, 22, 25, 26, 44, 46, 49–51, 63, 64, 71–74, 77–79, 89, 90, 96, 100, 111, 120, 132, 136, 139, 142, 162, 170, 209
Microwave, 76, 118, 121

assisted extraction (MAE), 118, 121, 166, 169
Migration, 5, 19, 65, 135
Milling, 203, 204
Mitigation, 235
Modified atmosphere packaging (MAP), 90, 136
Moisture, 2–4, 9, 10, 13, 17, 19, 21, 24, 25, 50, 78, 100, 109, 111, 120, 138, 140
  barrier, 2, 9, 100
  contents, 10, 25
  loss, 2, 13, 25
Mold growth, 16
Molecular
  identification, 77
  mobility, 5
  structure, 4, 64
  weight, 5, 15, 19, 117, 200, 209
Molecule, 51, 52, 63, 65, 74, 93, 99, 114, 166, 201, 202, 212
  backbone, 101
Molten wax coating, 2
Monilia, 22
Monoaldehydes, 97
Monocyclic, 133
Monoterpenes, 133
Montmorillonite, 25
Moraxella, 22
Morphology, 150, 189, 204, 207
Mucilage solution, 8
Multicomponent systems, 94
Multiple sclerosis, 63
Multivalent cations, 5
Muscle foods, 17, 21–23
Mutagenic, 48
Mycobacterium spp., 22

N

Nano-aggregates, 211
Nanocarriers, 214
Nanocatalysts, 231
Nanocomposite, 25, 207, 212, 213, 231
  coating, 25
Nanocrystals, 203, 205, 207
Nanoemulsion, 12
Nano-engineering, 231
Nanofillers, 213
Nano-intermediates, 201

Nanomaterials, 56, 201, 202, 214, 230–232, 238, 241
Nanometric materials, 214
Nanoparticles (NPs), 52, 101, 201–215
Nano-range, 56
Nanoscale, 56, 201, 211
Nano-science, 231
Nanosized, 231
Nanostructured, 207
Nanostructures, 199, 201–203, 205, 213, 215
Nanosystems, 201, 203, 213, 214
Nanotechnological use, 214
Nanotechnology, 201, 202, 213, 215, 230–232, 238
Nanotube, 201
National Council of Science and Technology of Mexico (CONACyT), 193
Natural
   colorants, 63
   killer cells, 48
   materials, 6
   matrix, 66
   plasticizers, 5
   polymers, 6, 215
Neoplastic changes, 48
Network chains, 94
Neuroblastoma, 67
Neurological effects, 47
Neuronal differentiation, 67
Neutral, 5, 168
   carbohydrate, 5
Nitrogen, 79, 112, 119, 153
   source, 79
Noncarbohydrate compounds, 45
Non-encapsulated
   bacteria, 44
   probiotics, 51
Non-toxic, 51, 95, 99, 132, 199, 200, 209
Nontoxicity, 2, 44
Nopal cactus, 8
Nozzles, 26
Nucellar
   apomixis, 187
   embryos, 186, 191
   plantlets, 189
   plants, 191
   science, 226

Nutraceuticals, 5
Nutrients, 13, 14, 17, 20, 109, 110
Nutritional
   supplement, 79
   value, 7, 9, 66, 102, 112, 150

O

Obesity, 47, 89, 90
Ocular epithelia, 210
Odor, 16, 21, 25
Ohmic heating, 78
Oil diffusion, 17
Olea europaea L., 190
Oligofructose (OF), 47
Oligomeric, 114, 117
Oligosaccharides, 45, 54, 123
Olive, 13, 138, 190, 193
Opaque, 4
Optimum conditions, 77, 122
Opuntia ficus-indica cladodes, 8
Oral
   cavity, 68
   ingestion, 101
Oregano, 12, 13, 19, 23, 134, 136, 138–140, 143
   essential oil, 12, 13, 136, 140
Organic
   compounds, 202, 212, 231
   fertilizer, 91, 110
   solvents, 75, 121, 209
Organisms, 46, 56, 69, 91, 99, 182
Organoleptic, 1, 6, 23, 154, 168
Origanum virens, 25
Oryza sativa L., 190
Oscillatory movement, 233
Osmotic pressure, 51
Oxidation, 6, 17, 21, 22, 24, 25, 27, 28, 66, 69, 119, 142, 161, 162, 227, 240
Oxidative
   damage, 64, 68
   stress, 67
Oxygen, 2, 7, 12, 15, 19, 22, 50, 65–70, 73, 74, 91, 100–102
Ozone, 90

P

P. pinophilum, 77
P. purpurogenum, 77, 78

Packaging, 4–7, 9, 16, 17, 20–22, 25, 26, 28, 52, 89–93, 97, 99–103, 131–133, 135, 136, 139, 142, 144, 201, 202, 210
Paclitaxel (PTX), 211, 212
Pallbearers, 235
Pancreas, 68
Paradigm, 225, 226, 234, 238, 240
Parameter, 73, 140, 227
Partial alkaline hydrolysis, 97
Path-breaking, 226, 230
Pathogens, 12, 26, 50, 132, 133, 210
P-coumaric acid, 113, 116, 119, 120, 168
Pectin, 5, 6, 8, 10, 18, 44, 45, 51, 99, 102, 103, 166, 168
Pectinase, 110, 111, 120
Peeling, 16, 26, 89, 90
Pellets, 110
Pencillin sp., 213
Penetration, 75, 76, 166, 167
Penicillium purpurogenum, 77, 120
Perfumery, 133
Pericarp, 109, 150
Perionyx excavates, 110
Perishable, 13, 20, 24, 155
Permeability, 6, 10, 15, 17, 19, 25, 52, 133, 137, 138, 140, 167
Permeation, 5, 210
Peroxide, 10, 90, 143, 205, 208
Petroleum, 3, 4, 75, 121, 230, 231
pH, 8, 10–12, 19, 20, 25, 27, 43, 49, 51, 55, 66, 74, 78, 91, 93, 97, 99, 101, 120, 152, 153, 208, 209, 213
Pharmaceutical, 66, 95, 170
Phenol, 23, 133, 134
Phenolic
    acids, 115, 119, 120
    compound, 97, 101, 107–122, 124, 140, 142, 159
    type, 133
Phosphorus, 64, 168
Photochemiluminescence (PCL), 143
Photogenic microorganisms, 26
Photooxidations, 208
Photoprotectors, 69
Photosensitizers, 72
Photostability, 212
Photosynthetic apparatus, 69
Phototrophic bacteria, 69

Physicochemical
    characteristics, 52, 95
    parameters, 8, 120
    properties, 27, 51, 52
Physiological fluid, 92
Phytochemical, 13, 161
    compounds, 149
Phytoplasma, 191
Pigments, 64, 69, 71, 74, 75, 77, 78, 121
    production, 77, 78
    protein complexes, 69
    recovery, 74
Pigskins, 19
Pitahalla, 150
Pitajalla, 150
Pitajaya, 150
Pitalla, 150
Pitaya fruit, 156
Pivotal importance, 222
Plants, 9, 63, 64, 69, 71, 95, 108, 110, 116, 132, 150, 159, 160, 162, 163, 181, 182, 184–186, 188, 191–193, 199, 213
    extracts, 22
    matrix, 75, 166
Plantago psyllium, 49
Plasma vacuoles, 69
Plasticizer, 5, 7, 9, 14, 15, 19
Pleorotus, 111
    ostreatus, 53
Poli(vinyl alcohol) (PCA), 96
Pollutants, 167, 231
Poly(lactic-co-glycolic acid) (PLGA), 211
Polycarboxylates, 99
Polycarboxylic acids, 101
Polyelectrolyte multilayer, 8
Polyembryonic, 185, 188–191
Polyembryony, 181–193
Polyester, 4
Polyethylene glycol, 18, 19, 51, 212
Polyfunctional monomers, 92
Polyhydroxyalkanoates (PHA), 91
Poly-isoprenoid structure, 65
Polylactic acid (PLA), 91
Polymer, 1–7, 9, 14, 16, 21, 22, 44, 49, 52, 56, 92–101, 117, 119, 136, 140, 200, 202, 208, 211, 212, 215
    chains, 94, 97
    molecules, 5

Polymeric
  applications, 208
  backbone, 94
  matrix, 5, 207
  mixture, 55
  packaging, 6
  plastics, 4
  proanthocyanidins, 114
  solids, 93
  structures, 90, 93
Polymerization, 24, 92, 96
Polyol, 116, 203, 204
  methods, 203
Polyphenolic compounds, 114
Polyphenols, 23, 45, 64, 107, 120–124,
  161–163, 168, 172
Polypropylene, 136
Polypyrrole, 206, 212
Polysaccharides, 4–6, 8, 9, 14, 15, 19, 44,
  51, 52, 91, 95, 102, 120, 137, 168, 200,
  201, 212
Polyurethane foam, 78
Polyvinyl alcohol (PVA), 96–98, 101, 103,
  136
Porter's reagent method, 117
Postharvest protection analyses, 18
Post-mortem storage conditions, 20
Potassium sorbate, 27
Potential, 9, 12, 28, 55, 56, 92, 99, 100, 103,
  109, 112, 116, 123, 149, 154, 155, 160,
  163, 182, 187, 188, 192, 193, 201, 202,
  205, 208–210, 212, 223, 239
Prebiotic, 43–50, 52–56
  compounds, 44
  fermentation, 48
Pressurized hot water extraction (PHWE),
  167
Proanthocyanidins, 113, 115, 117, 118, 121,
  124
Probiotic, 43–56
Profundity, 222, 223, 227–229
Prolongation, 14, 28
Prostate, 68
Proteinaceous, 2
Proteinantioxidant interactions, 22
Proteinpectic coating, 18

Proteins, 4–7, 9–11, 14, 15, 17, 19, 20, 51,
  52, 56, 91, 96, 97, 103, 112, 116, 134,
  137, 153, 165, 201, 214
Proteus spp., 22
Protocatechuic acid, 115
Provision, 222, 227, 229, 238
Provitamin A, 65, 67
Prunus persica, 70
Pseudomonas, 22, 23, 136
Psidium guajava, 70
Pullulan, 5, 8
Pullulanase, 209, 212
Pulmonary, 214
Pulp, 107–110, 117, 121, 124, 150, 152, 207
Purge accumulation, 20
Purification, 107–109, 117, 121, 122, 209,
  211, 227, 237

Q

Qualitative alterations, 48
Quality parameters, 12
Quercus sp, 77
Queretaroensis, 149, 150, 152–154
Quinoa, 11
  protein, 9, 16

R

Radicles, 189
Rancidity, 2, 4, 20
Rayon, 207
Reactive groups, 93
Reclamation, 237
Recycling, 91, 100
Red crimson, 8
Reduction, 16, 23, 49, 67, 75, 137, 204, 237
Refrigeration, 3, 21, 90
Remediation, 222–227, 229–241
Replete, 225, 232, 234
Reproducibility, 212
Resins, 5, 6, 9, 13
Resistance, 5, 7, 16, 43, 47, 51, 55, 94, 100,
  133, 163
Respiration, 2, 12–16, 21, 90, 91
  rate, 2, 12, 14, 15, 21, 91
Respiratory processes, 133
Revamped, 226, 228–230, 232, 238
Rheological
  behavior, 207

characteristics, 2, 17
properties, 5, 205
Rheology, 24, 28
Rhizopus, 22
   stolonifer, 12
Rhodophyceae, 201
Rhodosporidium, 65
Rhodotorula, 65, 72, 73, 77, 78
Rice, 13, 72, 95, 190, 193
Rigidity, 97, 103, 200
Ripening, 18, 69, 108, 150
   effects, 18
Rootstock, 188, 191
Roseburia sp., 49
Rosemary, 19, 23, 134
Rutin, 114, 120

S

Sabinene, 133, 134
Saccharomyces boulardii, 46
Salmon, 20, 21, 23, 66
   fillets, 21
Salmonella infections, 211
Salmonellosis, 25
Sanitation, 237
Scavengers, 21
Scientific
   barriers, 221, 222, 227, 229, 238, 240
   discernment, 224, 226, 229, 234, 239
   divination, 224, 238, 240
   failures, 225
   firmament, 227, 230, 232, 239, 240
   fortitude, 222–226, 232, 237, 239
   innovation, 222, 223
   passion, 225
   profundity, 226, 231, 234, 235
   rejuvenation, 229, 232, 234, 237, 239
   vision, 221–223, 226–230, 237, 239, 240
Seedlings abnormality, 188
Self-fertilizing mulching biopolymers, 103
Semidesert environments, 150
Sensorial properties, 27, 52
Sensory
   characteristics, 19, 102, 137
   evaluation, 8, 10, 11, 25, 27
   scores, 20, 54
Shelf life, 2–4, 8–18, 20–25, 28, 90, 91,
   131–133, 136–138, 142, 154

Shellac, 12, 13
Shrimp muscle proteins, 7
Sikkim mandarin, 188
Sodium
   caseinate, 12, 51
   dodecyl sulfate, 12
Soft-solid, 5
Sol-gel, 203
Solid
   fuel, 110
   matrix, 75
   state fermentation (SSF), 78, 109, 111,
      120, 170
Soluble solids, 8, 11, 152
Sorbents, 231
Sorbitol, 8, 10, 19
Soxhlet, 75, 76, 162, 165
   extraction, 75
   process, 75
Soy protein, 7, 9, 11, 15, 18
Spheroidal nanoparticles, 210
Spinacia oleracea, 70
Spinning atomizer, 50
Spoilage, 12, 20, 22, 26
Spondias lutea, 70
Sporidiobolus, 65
Sporobolomyces, 65
Sporotichum, 22
Spray
   chilling, 50
   drying, 50
Stabilization, 43, 103, 205, 208
Stabilizers, 90
Stabilizing, 48, 201, 212
Staphylococcus aureus, 12, 22, 136
Starch, 5, 6, 11, 15, 25, 44, 49, 54, 79,
   95–97, 101, 139, 140, 206, 212
Steam explosion, 74
Stem distillation, 132
Stenocereus, 149, 150, 152–156
   queretaroensis, 149, 150, 152, 156
Sthapylococcus succinus, 49
Stimuli, 93, 101
Storange stability, 25
Subjective quality analysis, 12
Substrate, 45, 72, 73, 78, 111, 124
Sucrose, 2, 25
Sulfides, 133

Sulfur compounds, 231
Supercritical fluids, 74–76, 163, 166
Superparamagnetic, 210, 211
    alginate nanoparticles, 211
Surface
    area, 26, 56, 76, 167, 214
    charge, 207
    tension, 12
Sustainability, 152, 214, 223, 224, 227–230,
    232, 236, 238–240
Swelling, 94, 97, 101, 213
Synthetic polymers, 6, 93, 95, 102, 104, 214

T

T regulatory cells, 48
Tackling, 222, 226, 228, 232, 235
Tannase, 110, 111
Tannins, 112, 114, 116–118
Tea, 18, 23, 25, 159–162, 164, 165,
    169–172
    polyphenols, 23
Technological validation, 226
Tensile strength, 10, 101, 103, 140
Terpenes, 97, 101, 133, 168
Terpenoids, 133
Tetrahydrofuran (THF), 208
Texture, 11, 12, 25
    profile analysis, 11, 12
Therapeutic properties, 44
Thermal
    analysis, 12
    behavior, 97
    gelling, 98
    processing, 24
    stability, 103
    treatments, 26
Thermolabile substances, 76
Thiobarbituric acid value, 10, 25
Three-dimensional network, 94, 97
Thymol, 27, 133, 134, 140
Tilapia, 19, 23
Tissue, 19, 66, 76, 94, 95, 103, 185, 202,
    214
    pigmentation, 66
Titratable
    acidity, 8
    activity, 12

Torchbearers, 222, 230, 234
Torularhodin, 74
Torularodine, 72
Torulopsis, 22
Toxic
    aldehydes, 22
    metabolites, 48
    metals, 231
Toxicity, 5, 163, 233, 234
Trans-cinnamaldehyde, 18, 27
Transglutaminase, 18
Transition, 5, 69
Tryptophan, 191
Tulsi extract, 12
Tumorigenesis, 67

U

Ulcerative colitis (UC), 48, 49
Umbilical cord, 231, 232
Upregulation, 48
United States Department of Agriculture
    (USDA), 26

V

Vainilla oleoresin, 13
Vapor transmission rate, 12
Vascular function, 214
Vegetable proteins, 9
Vegetarianism, 9
Vermicompost, 108
Vermicomposting process, 110
Vinyl acetate, 96
Violacein, 211
Violaxantina, 70
Viroids, 191
Virus, 191, 201
Visionary
    coin, 226, 228–230, 232
    paradigm, 225, 232, 239
    scientific arena, 239
    timeframe, 225
Vitamin, 2, 53, 64, 168
    precursors, 66
    synthesis, 47
Volatile, 21, 92, 101, 111, 132, 162, 168,
    231
    nitrogenous bases, 21

W

Water
  activity, 25
  content, 25
  holding capacity, 10, 19, 25, 27
  purification, 237, 241
  solubility, 10
  vapor
    permeability (WVP), 10–12, 15, 16,
      19, 25, 137, 138, 140
    transmission, 10, 12
Waxes, 2, 5, 6, 9, 13–15, 51
Weight
  loss, 11, 12
  ratio, 100, 213
Wheat, 7, 15, 51, 95, 190, 193
Whey protein, 7, 10, 15, 18, 24, 25, 55, 103,
  138

X

Xanthan, 27
Xanthophyllomyces, 65, 73
Xanthophylls, 64, 65, 67, 69
Xylanase, 110, 111

Y

Yersinia enterocolitica, 22

Z

Zea mays, 70
Zeaxanthin, 66–70
Zein (corn), 7, 9, 15, 138
Zygotic
  embryos, 191
  plantlets, 189
  polyembryony, 186, 191
  rootstock, 188